ROCKET DREAMS

ALSO BY CHRISTIAN DAVENPORT

The Space Barons

As You Were

ROCKET DREAMS

MUSK, BEZOS, AND
THE INSIDE STORY OF THE NEW,
TRILLION-DOLLAR SPACE RACE

• • •

CHRISTIAN DAVENPORT

CROWN CURRENCY
NEW YORK

CROWN CURRENCY
An imprint of the Crown Publishing Group
A division of Penguin Random House LLC
crownpublishing.com | penguinrandomhouse.com

Title page and part opener art by Shutterstock.com/Paopano.

Library of Congress Cataloging-in-Publication Data
Names: Davenport, Christian author
Title: Rocket dreams : Musk, Bezos, and the inside story of the new, trillion-dollar space race / by Christian Davenport.
Description: [New York] : Crown Currency, [2025] | Includes bibliographical references and index.
Identifiers: LCCN 2025008373 | ISBN 9780593594117 hardcover | ISBN 9780593594124 ebook
Subjects: LCSH: SpaceX (Firm) | Blue Origin (Firm) | Manned space flight | Space flights | Outer space—Exploration—United States | Outer space—Civilian use— United States
Classification: LCC TL789.8.U5 D268 2025 | DDC 919.904—dc23/eng/20250701
LC record available at https://lccn.loc.gov/2025008373

ISBN 978-0-593-59411-7
Ebook ISBN 978-0-593-59412-4

Editor: Leah Trouwborst and Paul Whitlatch
Production editor: Joyce Wong
Text designer: Andrea Lau
Production: Jessica Heim
Copy editor: Lawrence Krauser
Proofreaders: Alisa Garrison, David Goehring, Alissa Fitzgerald
Indexer: J S Editorial, LLC
Publicist: Penny Simon
Marketer: Tara Gilbride

Manufactured in the United States of America

1st Printing

First Edition

The authorized representative in the EU for product safety and compliance is Penguin Random House Ireland, Morrison Chambers, 32 Nassau Street, Dublin D02 YH68, Ireland, https://eu-contact.penguin.ie.

For Annie, Harrison, and Piper

Far better it is to dare mighty things, to win glorious triumphs, even though checkered by failure, than to take rank with those poor spirits who neither enjoy much nor suffer much, because they live in the gray twilight that knows neither victory nor defeat.

—THEODORE ROOSEVELT, *THE STRENUOUS LIFE*

CONTENTS

TIMELINE OF EVENTS

SEPTEMBER 2000: Jeff Bezos founds Blue Operations LLC, the precursor to Blue Origin.

MARCH 2002: Elon Musk incorporates Space Exploration Technologies, aka SpaceX.

DECEMBER 2013: China lands its robotic Chang'e-3 spacecraft on the moon—the first soft landing on the lunar surface since the Soviet Union's Luna 24 in 1976.

SEPTEMBER 2014: NASA awards the Commercial Crew contracts to Boeing and SpaceX to develop spacecraft capable of transporting astronauts to and from the International Space Station.

SEPTEMBER 2016: Jeff Bezos announces Blue Origin is working on a rocket capable of reaching orbit, called New Glenn.

NOVEMBER 2016: Donald Trump defeats Hillary Clinton in the U.S. presidential election.

MARCH 2017: Bezos pitches NASA to allow Blue Origin to build it a moon lander.

SEPTEMBER 2017: Elon Musk gives a presentation on SpaceX's new BFR (Big Falcon Rocket) and calls for the creation of "Moon Base Alpha."

SEPTEMBER 2017: Bezos hires Bob Smith as CEO of Blue Origin.

OCTOBER 2017: Vice President Mike Pence hosts the first meeting of the reconstituted National Space Council.

DECEMBER 2017: President Trump signs Space Policy Directive-1, instructing NASA to return astronauts to the moon.

FEBRUARY 2018: SpaceX launches its Falcon Heavy rocket for the first time.

APRIL 2018: Jim Bridenstine is confirmed as NASA administrator.

JANUARY 2019: China lands its Chang'e-4 spacecraft on the far side of the moon.

MARCH 2019: Pence calls for NASA to accelerate its return of astronauts to the moon by four years, to 2024, and to include a woman in the first crew.

APRIL 2019: SpaceX's Crew Dragon capsule explodes during an engine test.

MAY 2019: Bezos gives his "vision speech" in Washington, D.C., and says Blue Origin can help astronauts land on the moon by 2024.

MAY 2019: SpaceX launches the first batch of sixty Starlink satellites.

MAY 2019: Bridenstine names NASA's moon program Artemis, after the twin sister of Apollo in Greek mythology.

APRIL 2020: Blue Origin wins the lion's share of the first round of a NASA contract to build a moon lander.

MAY 2020: SpaceX launches NASA astronauts Bob Behnken and Doug Hurley in a test flight to the International Space Station, restoring America's human spaceflight program.

NOVEMBER 2020: Joe Biden defeats Trump in the presidential race.

FEBRUARY 2021: The White House announces that Artemis will continue under President Biden.

APRIL 2021: SpaceX wins the moon lander contract over Blue Origin.

MAY 2021: SpaceX lands its Starship SN15 prototype.

MAY 2021: China lands a spacecraft on Mars.

JUNE 2021: China launches the first crew to inhabit its new Tiangong space station.

JUNE 2021: China announces a partnership with Russia to build the International Lunar Research Station.

JULY 2021: Richard Branson reaches space on a suborbital trip flown by his space tourism company, Virgin Galactic. Bezos follows suit nine days later.

NOVEMBER 2022: NASA's Space Launch System rocket launches the uncrewed Orion spacecraft around the moon in the Artemis I mission.

APRIL 2023: SpaceX launches the Starship Super Heavy booster and spacecraft for the first time.

MAY 2023: Blue Origin wins a NASA contract to build a lunar lander.

SEPTEMBER 2023: Dave Limp replaces Bob Smith as CEO of Blue Origin.

JUNE 2024: China's Chang'e-6 spacecraft returns first-ever samples from the far side of the moon.

NOVEMBER 2024: Trump defeats Kamala Harris in the presidential election; Musk is slated to become head of the White House's Department of Government Efficiency.

JANUARY 2025: Blue Origin launches its New Glenn rocket to orbit for the first time.

Starship

The rocket booster first appeared as a dot in the distance. It grew larger as it descended, leaving contrails like skid marks in the sky. Within minutes, it fell from an altitude of about sixty miles to six. As the booster became visible to the thousands of people who had packed the beaches of South Padre Island, they began cheering wildly, thrilled to witness not only a launch but a landing.

SpaceX always pushed boundaries. But flying a rocket the size of a twenty-three-story building back from the edge of space for an unprecedented landing attempt perilously close to a densely populated city— that was risky even by the standards of Elon Musk. The company's workhorse Falcon 9 rocket had made hundreds of successful landings, touching down on a ship at sea or designated landing pads with such frequency that it was considered routine. This was different. On this day, October 13, 2024, SpaceX's next-generation Starship rocket would instead attempt to fly back to its launch pad, where it would be caught with a pair of mechanical chopstick-like arms. The daring feat had never been accomplished—or even attempted.

Six-tenths of a mile above the ground, the Starship booster lit its engines once again, slowing down as it crept closer to land. The tower stood with its arms wide, like a parent beckoning a toddler. The booster edged in, flames still shooting from its engines, and nestled into the mechanical arms, which closed around it. It looked almost gentle.

An hour or so later, the Starship spacecraft made a similar descent, firing its engines to land softly in the Indian Ocean, right on target. If the booster landing was a test for rapid rocket reusability, the spacecraft landing was a rehearsal for landing on the moon or even Mars. Both feats, once improbable, now seemed one step closer.

IN THE HISTORY of spaceflight, there has been nothing quite like Starship. On the launch mount, it is a colossal pillar of stainless steel, resembling not so much a rocket—because how could such a behemoth escape Earth's gravity?—but rather a curious sculpture that could have been born from the steampunk world of Jules Verne.

Implausible though it may seem, Starship is not science fiction but very real—the largest and most powerful rocket ever assembled. NASA has placed it at the center of its deep space exploration plans, awarding SpaceX $4 billion in contracts to return astronauts to the surface of the moon. China has paid SpaceX the ultimate compliment by trying to copy Starship's design, and intends to use their version of the rocket in their quest to establish a lunar base.

Cloning Starship, however, won't be easy. It's an extremely complex vehicle, comprised of two stages—the Super Heavy booster and the Starship spacecraft—that together stand some four hundred feet tall, taller than the Statue of Liberty. The payload bay—which would house astronauts on a crewed mission, satellites or cargo otherwise—is comparable to the volume of the International Space Station. Fully fueled, it holds more than ten million gallons of liquid methane and liquid oxygen. Its booster is powered by a staggering thirty-three engines known as Raptors, and the spacecraft has six of its own. A single Raptor produces twice as much thrust as a Boeing 747. All told, Starship has more than double the thrust of the Saturn V rocket that propelled Neil Armstrong, Buzz Aldrin, and Michael Collins to the moon during NASA's Apollo program. In every respect—power, capability, cost, design, technology—Starship surpasses every other rocket before it, representing a light-years' leap forward—the difference between horse-drawn covered wagons and locomotives.

For perspective: Assembling the International Space Station required forty-two rocket launches; Starship could do it in five.

Just as important is the fact that Starship is designed to be fully reusable, a feature that will drastically reduce the cost of access to space. Instead of being ditched into the ocean, as has been the practice for most of the history of space exploration—something Musk has compared to throwing away an airplane after each flight—both the booster and the spacecraft are designed to land and be reflown. SpaceX mastered the art of rocket reusability years ago with its Falcon 9 booster, built to hoist satellites and astronauts to low Earth orbit. But there was one big inefficiency: Because it lands on a ship at sea or on a separate landing pad, the Falcon 9 has to be transported back to its launch site, a cumbersome process that can last days. In Musk's world, where he wants rockets departing multiple times a day, that's far too slow. "You need to be looking at your watch, not your calendar," he said.

Starship, by contrast, is designed to land at its original launch site and get prepped for another ascent in a matter of hours. Precise landings, Musk knows, would also be critical for landing on the moon, or Mars.

IN EARLY 2018, Starship was still just one of Musk's obsessions, which also included the Boring Company, which was digging underground tunnels to alleviate traffic, and Neuralink, the venture that would plant computer chips into people's brains. Then called "Big Falcon Rocket," or BFR (though everyone knew the *F* didn't actually stand for *Falcon*), it existed more in Musk's imagination than on SpaceX's launch pads. But Musk wasn't the only one dreaming about a new generation of rockets, capable of taking humans farther into space than ever before. As it sought to compete with SpaceX, Amazon founder Jeff Bezos's Blue Origin wanted to know whether Starship could indeed become a next-generation vehicle that might lead to SpaceX further dominating the space industry, or was a delusion that only Musk could see.

Blue Origin's leadership dispatched Greg Allen, a Harvard Business School graduate working as Blue's senior manager of market strategy, to investigate. His conclusion came in an eight-page memo delivered to Bezos and Bob Smith, Blue Origin's CEO at the time, in April 2018. With each page marked "Blue Origin Proprietary & Confidential," the report served as a warning that BFR could indeed disrupt the space industry and render every other rocket almost instantly irrelevant. Once operational, it could "be the best choice for nearly every customer requirement," Allen wrote, adding that it could "plausibly achieve a monopoly" in the U.S. launch market.

Allen also noted that there "are strong reasons to be skeptical of nearly all of SpaceX's BFR claims" on cost and performance, and that "SpaceX has a long history of publicly overestimating these for new programs. However, a BFR vehicle that underperforms on all of SpaceX's initial claims for performance, pricing, and schedule could still have a major impact on the future competitive landscape."

SpaceX was moving fast and was "farther along than would normally be expected," Allen wrote. He concluded with a call to action, urging the company to "consider SpaceX's BFR as a potentially serious medium-term competitive threat and undertake efforts to meet the challenge by a BFR-class competitor." In a section marked "Potential Blue Origin Response Strategies," Allen suggested a series of options, including that Blue could play hardball: "Seek to deny, delay and diversify government development funding for BFR," he wrote. Blue could also accelerate its own technological advancements, including building a fully reusable rocket that could be refueled in space.

Allen presented his memo at a meeting that began with Blue Origin executives and program managers reading it in silence. As the team discussed BFR and debated how best to counter SpaceX, Bezos, unafraid to offer unsettling observations to employees, asked a provocative question that made it clear he didn't believe his team was pioneering and pushing forward enough: "Are they just smarter than us?"

In an email to employees a few months later, Smith wrote that the development of the massive rocket—as well as Musk's focus on Starlink, an ambitious program to flood Earth's orbit with thousands of satellites to beam Internet signal to ground stations—meant SpaceX was stretching itself thin and that there were "opportunities for us to catch up and surpass them as they get distracted."

That was a miscalculation. SpaceX did not get distracted, as Smith had hoped. Rather, in the years that followed, SpaceX pulled even further ahead. And Bezos would eventually concede that Blue Origin hadn't been moving fast enough to meet the challenge posed by SpaceX, which has become one of the most improbable success stories in the history of American entrepreneurship.

In early 2025, however, Blue Origin proved it could finally compete when, on the first try, it successfully launched its New Glenn rocket to orbit. Starship it was not. Still, it was a very large and powerful vehicle— by no means a starter rocket—and its launch even drew admiration from Musk, who had long goaded Bezos and lamented Blue Origin's lack of progress. Now the situation had clearly changed.

The primary reason Bezos stepped down as the CEO of Amazon in 2021 was "so I could come focus full-time on Blue Origin," he told me in early 2025 during an interview in the company's sprawling factory in Cape Canaveral, Florida. The company was driving "extraordinary urgency," he said, and becoming "a more decisive company" that moved fast. That renewed commitment to the company, which Bezos has called "the most important work I'm doing," was evident by the noise coming from the factory floor below us, as engineers worked on multiple rockets in various stages of production.

The new ethos was embodied in a private memo crafted in 2021 to chart a way toward a future where the company would build the rockets, spacecraft, and infrastructure that would allow for a settlement on the moon. Dated in the year 2032 and written in the form of a fictional press release from the future, the memo began:

Blue Origin announced today the successful commissioning of humanity's first permanently off-world base on the Moon. Armstrong Station has begun 24/7 operations for various exploration, science, research, manufacturing and tourism business. Located on Shackleton Ridge at the Moon's South Pole, the base hosts government customers and now commercial customer visits for the first time. The newly christened base is "one giant leap" toward Blue Origin's vision of enabling millions of people living and working in space to benefit Earth.

Just a few years earlier, the idea of humans building an "off-world" base seemed ludicrous. The Space Shuttle retired in 2011, leaving NASA without the means to fly people anywhere. Astronauts, once feted in ticker tape, were now largely anonymous. Interest in space was at a nadir since its peak in the late 1960s. Instead of reaching out to building a permanent presence on the moon and eventually Mars, NASA was confined to low Earth orbit, a place where, as astrophysicist Neil deGrasse Tyson has said, "we boldly go where hundreds have gone before." Today, however, SpaceX and Blue Origin together represent a paradigm shift that has remade space exploration. They have touched off an emerging space economy that some analysts think could be worth $1 trillion by the middle of the 2030s—a new wave of economic activity that could be as transformative as the advent of the Internet. The new space age unfolding before us started almost two decades ago, the beginning of which I chronicled in my last book, *The Space Barons,* which posed a question: Is the commercial space industry for real? The answer comes in this one: a resounding yes.

This book tells the story of the emergence of a growing and dynamic private sector, as well as a competition between superpowers—namely the United States and China—to build the infrastructure, mining, transportation, power generation, and in-space manufacturing that would allow a permanent presence in orbit, on the moon, and eventually on Mars. It's the story of the fight for billions of dollars in government contracts and national prestige that has pitted Musk against Bezos once again; the maneuverings of one of the savviest and most productive NASA administrators in decades; and the birth of an armada of new rockets and spacecraft more capable and powerful than any built before.

It's also the story of failure and disappointment, the danger of space travel, and the perils of entrusting missions of national significance to billionaires susceptible to, say, the distractions of running a social media company or a midlife crisis. It's the story of space itself—the vast commons that has inspired generations of stargazers and is now a theater of conflict; a vital part of national security and the economy; and home to a proliferating number of satellites that beam signals to our phones and warn of hurricanes as well as adversaries' missile launches.

Musk has long been lured by the prospect of building a city on Mars, allowing humanity to become what he calls a "multiplanet species." Bezos wants to help "build the road to space," to foster a "dynamic, entrepreneurial explosion in space just as I've witnessed over the last 20 years on the Internet." Will either live long enough to see their vision become reality?

MUSK AND BEZOS are not the only ones building big rockets and spacecraft, with big ambitions in space. Between 2019 and 2024, six nations—China, India, Israel, Japan, Russia, and the United States—attempted to land a total of eleven robotic spacecraft on the lunar surface. China intends to land its first crew before 2030, a timeline that has not wavered. It also has pursued a steady campaign toward the lunar South Pole that's included going four-for-four in moon landings since 2013, as well as becoming only the second country after the United States to land and operate a rover on Mars. Given its clandestine nature, it's possible China would launch a crew to the moon for a landing in secret, revealing carefully curated images of its taikonauts on the surface and surprising the world.

Still, NASA has assembled an international coalition of more than forty-five nations supporting it. The newly established U.S. Space Force is giving space resources and attention it previously lacked as it works to thwart threats from China, Russia, and others, reflecting how space has moved from a peaceful sanctuary to the front of the next war. SpaceX has shown how a commercial space company can be profitable, setting the standard for a resurgent Blue Origin and the rest of the commercial space industry. All of which means the long-delayed return to the moon, and the beginning of the space race after that—to Mars—is finally morphing from long-dormant rocket dreams toward inevitability.

Here's how it happened.

PART I

· · ·

Earth

2016–2018

"I Get This Angry Twice in a Year"

Jeff Bezos was fuming—so angry his eyes bulged. The multibillionaire was aboard his private jet, a Gulfstream G650 with a flight attendant and a bar stocked with Tomatin thirty-year-old, single malt Scotch. The other passengers were executives from Blue Origin, the space exploration company he'd founded. They were en route to Washington, D.C., to pitch the leadership of NASA on a plan to build a spacecraft that could bring cargo and supplies to the surface of the moon. As Bezos reviewed the proposal his team had worked up, he found it so deficient as to be enraging. The wording was sloppy, the ideas timid, not fully formed. At Amazon, such a weak effort would never be tolerated, he said. "I get this angry twice in a year and it's always because of the decisions Blue Origin makes," Bezos said.

It was March 1, 2017, and it was hard to believe that just three days earlier Bezos had been basking in the televised limelight of the Oscars. Several Amazon-produced movies had been nominated for awards, including *Manchester by the Sea*, the dark but compelling feature film starring Casey Affleck. "It's a huge honor, and we're seriously fortunate," Bezos said on the red carpet, wearing a custom, slim-fitting tuxedo. Nothing conferred acceptance to Hollywood society like an appearance in the Oscar host's monologue, and cameras caught Bezos in the audience laughing after Jimmy Kimmel made a joke at his expense: "If you win tonight, you can expect your Oscar to arrive in two to five business days, possibly stolen by a Grubhub delivery man." Amazon won three Academy Awards, among the first ever awarded to a streaming service, and Bezos later showed up in triumph at the ultimate insider event: Madonna's after-party, where he marveled at the singer's stamina on the dance floor.

Now, on his private jet, his Oscar glow had vanished. The document in front of him—Blue Origin's proposal to sell its lunar lander to NASA—was all wrong.

Bezos was big on memos. At Amazon, meetings often began with "study hall," in which a team leader passed around a six-page document outlining an idea, and the group sat quietly reading for as long as thirty minutes. Bezos valued clear writing as a means to deep thinking, and six pages was long enough to showcase both. Anything shorter and the author could get away with faking it. Six pages allowed Bezos to see ideas in full—which ones were good and which were bad. As he wrote in a 2004 email to Amazon employees, "The narrative structure of a good memo forces better thought and better understanding of what's more important than what, and how things are related. PowerPoint-style presentations somehow give permission to gloss over ideas, flatten out any sense of relative importance and ignore the interconnectedness of ideas." Bezos himself spent weeks, even months, crafting a letter to Amazon shareholders each year. The title of the 2017 memo was "Building a Culture of High Standards," and in it he discussed the value of writing well. "The great memos are written and re-written, shared with colleagues who are asked to improve the work, set aside for a couple of days, and then edited again with a fresh mind."

On his plane, Bezos berated three of the Blue Origin executives behind the NASA memo—Rob Meyerson, A. C. Charania, and Brett Alexander—who were falling short, he said, of the level of leadership that he had built at Amazon. "Amazon operates on a world-class level in terms of its decision-making. I want Blue Origin to be on that level."

It wasn't the first time the Blue Origin executives had found themselves in their boss's crosshairs. More than one recalled him saying that meeting with them "is like shoving bamboo shoots under my fingernails." At Amazon, as they knew, he was famous for the outbursts that could quickly turn cruel.

"I'm sorry, did I take my stupid pills today?"

"Are you lazy or just incompetent?"

"We need to apply some human intelligence to the problem."

Along with another zinger that could have applied here on the plane: "This document was clearly written by the B team. Can someone get me the A-team document? I don't want to waste my time with the B-team document."

Comparing Blue Origin to Amazon, however, wasn't entirely fair. One of the executives, Charania, had only been at the company for a couple of months. And while Bezos had built Amazon from the ground up, shaping every aspect of its culture and ethos as it moved from books to retail to movies, he spent just one day a week—Wednesday—working at Blue Origin, a start-up that did not even have a chief executive officer. Its employees had gotten used to Bezos's pointed criticisms. Their lunar lander was called Blue Moon, and Bezos hadn't been happy with that, either. *Couldn't they come up with something better?* Bezos had wondered at an earlier meeting. There was the Elvis Presley song with the same name, and a beer as well. The moniker felt like a cliché, not the sort of thing that would roll smoothly out of the mouth of a modern-day Walter Cronkite narrating the next lunar landing to a live audience. Ultimately, Bezos had swallowed his pride and embraced one of the mantras he had created at Amazon: *Disagree and commit*, allowing the project to move forward.

This time, Bezos refused to compromise. The NASA memo, particularly the first few pages, didn't convey nearly enough vision or technical detail, he thought. Finally, he ended his diatribe, took one of his employees' laptops, and spent the rest of the plane ride editing it himself.

THE REASON BEZOS and his team were flying to NASA headquarters could be traced directly to the fact that Donald J. Trump had won the

2016 presidential election a few months earlier. Under his administration, NASA would be tasked with returning astronauts to the moon. If many of Trump's initial policies were unpredictable and chaotic, the product of infighting among his hastily assembled team of advisers, the small cadre involved in space policy had rallied around the moon as NASA's next destination.

Of course, the United States had already been to the moon—on six different Apollo missions almost a half century earlier. But the retro-sounding idea to return to the lunar surface was actually not so retro. The moon the Trump administration sought to return to was not the cold, dead rock that America chose to stop visiting after the Apollo era. Instead, it was a new moon, one that scientists now knew had been guarding precious resources for eons—a potential oasis with water hidden in the permanently shadowed craters at the poles. Since the moon's axial tilt is only about 1.5 degrees—compared to 23.5 degrees for Earth—the sun barely rises over the lunar horizon, giving the region its most defining feature: an ethereal, curious light that casts long, jagged shadows, creating a checkered panorama of light and dark—with peaks illuminated by continuous sunlight and caverns in perpetual Stygian blackness.

Water is vital to sustain human life, but its component parts, hydrogen and oxygen, could also be used as rocket fuel, potentially making the moon a gas station on the highway to the rest of the solar system, namely Mars. Not only that, the water, which has existed in the form of ice for billions of years, is like a Rosetta Stone time capsule that scientists say could tell the story of the formation of our closest neighbor in the solar system, as well as answer vital questions about Earth itself. How it was formed. How we got here. Where we came from.

A group of Republicans, including former House Speaker Newt Gingrich, had been crafting a moon plan that would involve the commercial space sector and make NASA a part of Trump's Make America Great Again movement. The goal, according to a slide Gingrich helped prepare for the incoming administration, was that "America becomes the dominant power on the space frontier; and uses space to become the dominant power on Earth."

In an op-ed that had just been published in *SpaceNews,* two of the Trump campaign's advisers—Robert Walker, the former chair of the House Committee on Science, Space, and Technology; and Peter Navarro, an economist—echoed some of Gingrich's points and said that under Trump, the White House would embrace the private space sector at a time when NASA had grown sclerotic and slow. The United States "must recognize that space is no longer the province of governments alone," they wrote. "Public-private partnerships should be the foundation of our space efforts."

Walker and Navarro went on to lament that NASA "has been largely reduced to a logistics agency" confined to low Earth orbit, and lacked ambitious efforts in deep space, which the United States was quickly ceding to its adversaries. The need for an "ambitious space program is existential," they wrote. "While the American government's space program has suffered from under-investment, both China and Russia continue to move briskly forward with military-focused initiatives. Each continues to develop weapons explicitly designed, as the Pentagon has noted, to 'deny, degrade, deceive, disrupt or destroy' America's eyes and ears in space. To maintain our strategic advantage in space and defend our troops and homeland, we must reinvigorate our space program."

Walker and Navarro had a point. NASA had once symbolized American innovation and exploration. But by now, many felt the space agency had in middle age lost its swagger. To help revive the nation's efforts in space, the Trump administration would resurrect the National Space Council, disbanded in 1993, which Vice President Mike Pence would lead. Trump may not have cared about space as much as the corporate tax rate or immigration, and he likely didn't have a clue as to what a moon expedition would require. But Pence was a true space enthusiast, eager to return astronauts to the moon and make space a priority at a time when geopolitical tensions were transforming into astropolitical ones.

THE TRUMP ADMINISTRATION was encountering a space race far different from when the United States competed against the Soviet

Union at the dawn of the Space Age. That sprint had a clearly defined starting point—the Soviet launch of Sputnik in 1957—and finish line: the Apollo 11 moon landing twelve years later. This time, a growing number of countries had decided that space was a vital domain and were vying to compete there, including India, Brazil, Australia, Japan, Argentina, Korea, and the United Arab Emirates. Between 2000 and the start of the Trump administration, some twenty-seven countries created national space agencies, all looking to bolster their economies while harnessing innovation, at a time when their national security and prestige increasingly relied on the expanse outside the atmosphere. The list has since grown considerably longer, as has the number of countries with at least one satellite in orbit. Dramatic improvements in technology have lowered the barriers to entry, and the potential to make money and wield power in space has only become clearer.

As a result, the new space race wasn't so much of a race but rather more like an Olympics, with multiple nations competing in multiple events. As noted by the aerospace analyst Todd Harrison, there were three essential competitions, the first of which involved commercial space. Private satellites allowed for unprecedented kinds of communication and Internet access, as well as the ability to closely monitor virtually every nook and cranny on Earth with high-resolution imagery. At the same time, a new fleet of commercially developed rockets were starting to hoist more mass to space more frequently and at lower costs.

The second event was militarization, based on the realization that space is vital for modern warfare. Rockets and missiles had always shared technology; it was only what was on the tip of the rocket—a nuclear warhead or a capsule carrying an astronaut—that was different. Satellites had become the Pentagon's primary missile warning system. They also provided essential reconnaissance, monitoring the movement of troops and ships at sea. GPS satellites also guided missiles and bombs so that they precisely hit their targets. A robust military space force was, in short, "the difference between being a regional power and a global power," as Harrison has written, allowing militaries "to see farther and act over longer distances with greater precision."

The third event in this space Olympics was lunar exploration. Whoever got there first, under this new paradigm, would not just have a head start in accessing its resources, but would set the rules of engagement for lunar exploration—and, more broadly, the deep space economy—for generations to come. Would it be free nations pushing transparency and open markets—or ones more autocratic?

NO COUNTRY WAS moving faster to catch up to the United States in space than China.

The country had gotten off to a slow start and didn't launch its first satellite until 1970. Its first astronaut, or taikonaut, as they are called, didn't reach orbit until 2003, forty-two years after Russian cosmonaut Yuri Gagarin became the first person to fly to space. It was still a major coup. State media broadcast footage of Yang Liwei orbiting Earth fourteen times in his Shenzhou 5, or "Divine" capsule. Officials described the flight as flawless, and after landing, cameras captured Yang emerging from the craft, looking healthy as he smiled and waved. The voyage revealed China's growing capabilities—as well as its penchant for secrecy and deception. It later emerged that the neatly crafted storyline had gone badly off script. When Chinese engineers had first opened the Shenzhou 5's doors, they found a disturbing sight. During the return to Earth, a problem with the design of the capsule had apparently exposed Yang to excessive G-forces that split his lip open and splattered blood all over his face. The engineers quickly cleaned him up, strapped him back into his seat, and opened the doors again so that the cameras could record a clean-faced Yang this time, smiling and waving, unblemished by so much as a smudge.

China's space agency later published a paper making plain its intention to rival the United States as the world's dominant space power. China would expand its program to "meet the demands of economic development . . . national security and social progress," as well as "protect China's national rights and interests and build up its national comprehensive strength." That included plans for a human lunar landing. In 2007, China's moon program, known as Chang'e for the Chinese moon goddess, had taken a major step forward with the launch of

Chang'e-1, a probe that mapped the entire lunar surface. The satellite also used sensors to characterize the depth of the moon's soil and what resources—including helium-3, the stable isotope that could be used as fuel in nuclear fusion—might be stored there. In 2010, a second probe, Chang'e-2, provided high-resolution images of the moon's surface that scientists used to determine where China should land its first spacecraft.

China's military was growing active in space as well. In 2007, it fired a rocket that destroyed a defunct weather satellite, littering space with more than three thousand pieces of debris that would stay in orbit for years, even decades. Traveling at 17,500 miles per hour, or five miles a second, even the smallest pieces were like bullets, threatening other satellites as well as the International Space Station. Another launch, in 2013, shook the Pentagon. In May of that year, China sent a rocket into orbit at an altitude of roughly 22,000 miles, where the United States parks some of its most sensitive national security satellites. Known as "geosynchronous orbit," the satellites' paths match the Earth's rotation. That allows the spacecraft to stay over a fixed point, continuously monitoring a particular location—say, Beijing, Moscow, or Pyongyang. The Chinese rocket didn't hit any satellites, but the message to the United States was clear: *Your satellites are not safe.*

"We have considered space a sanctuary for quite some time," Robert O. Work, the deputy U.S. defense secretary, said after the 2013 launch. "And therefore, a lot of our systems are big, expensive, enormously capable, but enormously vulnerable."

A few months after this orbital warning shot, China launched another rocket, this one carrying a spacecraft that landed on the moon and deployed a six-wheeled, solar-powered rover. The mission, known as Chang'e-3, made China the third country, after the United States and Soviet Union, to land a spacecraft softly on the lunar surface. And it was the first time the feat had been achieved since 1976, when the Soviet Union completed a lunar sample return mission.

The moon landing was celebrated as a triumph in China. "The dream of the Chinese people across thousands of years of landing on the moon has finally been realized," the state-run China News Service reported. And there were hints of much larger ambitions. The Chang'e-3 spacecraft was capable of carrying payloads more than twelve times the size of the 300-pound rover. China also announced plans to visit the moon's far side, which no nation had explored, and bring samples back. "By successfully joining the international deep-space exploration club, we finally have the right to share the resources on the moon with developed countries," the news service reported.

It didn't mention which resources it intended to harvest on the moon. But certainly, China was thinking about the water. China had also been eyeing the moon's vast reserves of helium-3 for its ability to help meet the energy demands of its growing population. One official in the department of lunar and deep space exploration thought the substance could power the country for some ten thousand years.

Some NASA scientists agreed that helium-3 could prove astonishingly valuable. A 2011 study for the agency that was authored by Harrison Schmitt, the Apollo astronaut and geologist, found that helium-3 could be worth $1,400 per gram, or almost $40,000 per ounce. "That compares with about $28 per gram ($800 per ounce) for gold at the beginning of 2009," according to Schmitt's report. Just 100 kilograms of helium-3 would be enough to power a 1,000-megawatt electric plant, serving nearly one million homes, for a year.

Accessing helium-3 and bringing it home would require enormous resources and new technologies. The scale of the endeavor would rival the European conquest of the Americas centuries before, with a lunar base that would serve as an outpost not only for science but for mining—and the financial returns that have long driven exploration. Under the Outer Space Treaty of 1967, no nation could claim sovereignty over the moon. But they could lay claim to its resources—at least according to the United States. In 2015, Congress passed a law giving private companies the right to the materials they mine on celestial bodies.

If any nation were able to harness such a valuable resource as helium-3, it could affect the balance of power on Earth, opening up a new source of energy and touching off an economic dynamism—all while bringing the moon into its sphere of influence. Unlike the United

States, which carefully separated its civil and military space efforts between NASA and the Pentagon, China's space program had no such distinction. And its actions and statements—the anti-satellite tests, the moon landing, and ambitions to compete for lunar resources—set off alarms at the highest levels of government.

THE TRUMP ADMINISTRATION believed that NASA alone could not match China, and that it would need to further rely on America's growing commercial space industry. In a private memo to the incoming Trump administration, Gingrich proposed taking that a step forward, with the government funding a space race that would pit companies like Bezos's Blue Origin against the industry's biggest star: Elon Musk's SpaceX. "The entire world will watch in awe as American billionaires race back to the Moon in partnership with NASA, in a race created by President Trump," Gingrich wrote. He added that Trump should give a speech on his first day in office and "challenge American industry (in partnership with NASA) to put humans in lunar orbit before the end of his first term, and on the Moon before the end of his second term. Declare that competition is critical, so he wants at least two companies, and that each company needs to accomplish this for a total price of $5 billion or less."

The White House did not run with Gingrich's idea to stage a made-for-television moon race. But if it had, SpaceX would have been a front-runner. The company had become a once-in-a-generation success that completely disrupted the rocket launch market, upended the military-industrial complex that had held a vice grip on it for decades, and ended government's long-held monopoly on space exploration, all while innovating new technologies, like rockets that flew back to Earth so that they could be refueled and reused over and over. Most significantly, SpaceX had gone on an incredible winning streak, raking in billions of dollars in government contracts from both NASA and the Pentagon, and it was poised to win more. The company was pressing at such a ruthless and unrelenting pace that it seemed Musk's only real competition was time and gravity. And he had attracted the attention of the new president.

On December 14, 2016, just before he took office, Trump invited America's most prominent technology executives to meet with him in New York. In the garish, Vegas casino–like lobby of Trump Tower, the press was cordoned off behind a red velvet rope, within view of the building's golden-hued elevators. From there they could film and photograph those CEOs whom Trump deemed worthy of an invitation. In all, about a dozen made the cut, including Tim Cook of Apple, Sheryl Sandberg of Facebook, and Larry Page and Eric Schmidt of Alphabet, all looking somewhat uncomfortable as reporters shouted questions and recorded every gesture until the elevator doors mercifully closed. C-SPAN, the dedicated chronicler of Washington minutiae, installed a Trump lobby cam, offering live coverage as the cortege of CEO suitors arrived.

Musk, of course, was among the chosen. That morning, he had met with the editorial boards of *The New York Times* and *The Wall Street Journal,* and afterward his aides suggested that given the horrendous New York City traffic, he should take the subway to make Trump's Midtown Manhattan meeting on time. Musk made his way through the lobby and up to the twenty-fifth floor, where there was a conference room with views of the Plaza Hotel and Central Park. Musk sat at one end of the long table, Bezos at the other. Both were in close enough proximity to Trump, but had to lean in slightly to make eye contact with the president-elect.

Both had a distinct set of interests—and challenges. Musk's SpaceX was increasingly dependent on the federal government for hundreds of billions in contracts, and was seeking to deepen its ties to NASA and the Pentagon. Bezos, who in 2013 had purchased my employer, *The Washington Post,* for $250 million, had come under attack for the newspaper's relentless coverage of the thin-skinned Trump during his 2016 campaign. "The @washingtonpost, which loses a fortune, is owned by @JeffBezos for purposes of keeping taxes down at his no profit company, @amazon," Trump wrote on Twitter early in the campaign. Uncharacteristically, Bezos fired back, writing on Twitter: "Finally trashed by @realDonaldTrump. Will still reserve him a seat on the Blue Origin rocket." He added a hashtag, #sendDonaldtospace, and a link to a video of the first time the company's New Shepard rocket successfully landed back at its West Texas launch site after flying to space. A couple of weeks before the election, Bezos said at a conference that Trump's rhetoric—threatening retribution against the media, refusing to concede if he lost—"erodes our democracy around the edges. . . . These are not acceptable behaviors."

After Trump won the election, however, Bezos quickly took a conciliatory tone, writing on Twitter: "Congratulations to @realDonald Trump. I for one give him my most open mind and wish him great success in his service to the country." Now, in Trump Tower, any hint of acrimony was gone. The meeting lasted some ninety minutes, as the participants discussed a range of topics, including cybersecurity, improving the country's aging infrastructure, taxes, technology in schools, trade with China, immigration, and more.

Afterward, Bezos issued a statement that again seemed designed to appease the man he had threatened to shoot into space just a year earlier.

"I found today's meeting with the President-elect, his transition team, and tech leaders to be very productive," it read. "I shared the view that the administration should make innovation one of its key pillars, which would create a huge number of jobs across the whole country, in all sectors, not just tech—agriculture, infrastructure, manufacturing—everywhere."

But Trump invited Musk to another meeting, this time in private with several of his closest advisers, including Gingrich, Jared Kushner, and Steve Bannon, followed by a session with Pence and members of the vice president's team. The conversations were freewheeling and wide-ranging, and Trump told Musk twice that he wanted "to get NASA going again." Musk responded by saying his administration should go big—and send people to Mars. He extolled the importance of building reusable rockets, and explained how they were vital to bringing down the cost of space travel.

Trump had just sent shockwaves through the aerospace industry, when he inserted himself in the middle of a major Pentagon procurement program—a $4 billion contract with Boeing to build the next-generation presidential airplane. "Cancel the order!" he wrote on Twitter, and he told reporters that Boeing's costs on the new Air Force One were "totally out of control." Musk was inclined to agree. Just as egregious, he felt, was Boeing's performance on the huge rocket it was building for NASA.

Called the Space Launch System (or SLS), it relied on 1970s technology, was years behind schedule and hundreds of millions of dollars over budget. Musk pointed out all those flaws to Trump, plus the fact that the Orion spacecraft it would hoist to the moon was not capable of actually landing on the moon. NASA, Musk said, would need a system, such as the one SpaceX was developing for Mars, that would also be able to get astronauts to the lunar surface.

To some in Trump's team, SLS was a symptom of a government agency that had become too risk averse and too bureaucratic to do great things anymore. If the space agency was increasingly outsourcing such high-priority missions to the private sector, they wondered, why not turn over the whole enterprise to companies like SpaceX? The privatization of NASA was a radical idea but one that would pique Trump's interest. The new president would soon marvel in public at SpaceX's reusable rocket boosters, and he had an affinity for the wealth and power that Musk was accumulating.

Indeed, SpaceX was moving much faster than NASA. The California start-up was developing innovations that many had assumed would only come out of a federally funded agency. Musk had been frustrated with the government's pace of progress; one of the reasons he decided to start a space company, after all, was because he had gone to NASA's website to see when it planned to send astronauts to Mars and realized that there was no plan. But he was also one of NASA's biggest fans. The fact was, SpaceX would not exist without NASA's help and large infusions of taxpayer money. In 2006, when it was scrambling to send its first rocket to orbit, NASA had saved the company with a $278 million contract that prompted Musk to change his login password to *ilovenasa*.

Two years later, SpaceX won a $1.6 billion government contract to fly cargo and supplies to the International Space Station.

At the time, the thinking was that NASA would never entrust flying its astronauts to the private sector. But in a radical shift, it awarded contracts to SpaceX and Boeing to do just that in 2014 under President Obama. True, NASA had always relied on contractors. Companies like Grumman built the lunar module that landed Neil Armstrong and Buzz Aldrin on the moon. Rocketdyne and Morton Thiokol built the engines and solid rocket motors for the Space Shuttle. Boeing and Rockwell Collins helped build the orbiters.

The difference now was that NASA didn't own or operate the rockets or spacecraft; the companies did. NASA supervised the development of their rockets and spacecraft and had influence upon the design. But the companies were free to build and operate them as they saw fit, so long as they met NASA's requirements. As a result, some of the nation's premier exploration programs were falling increasingly into the hands of private industry, beholden not only to the national interest but to the bottom line. NASA's long, slow decline since the height of Apollo may have become an example of government's lackluster performance, but NASA was still the world's preeminent space agency. It would have to coexist with the commercial sector in whatever Trump's space program turned out to be. As Pence later wrote in his memoir, "If their companies' technology could get Americans back to the moon, NASA shouldn't hesitate to work with them."

FLYING TO WASHINGTON on his G650 jet in March 2017, Bezos was furiously trying to position Blue Origin to be one of those companies. The proposal he was now editing himself showed on its cover a spacecraft standing on the lunar surface bearing NASA's meatball logo, Blue Origin's feather emblem, and an American flag, with Earth looming bright and blue in the background. The image, and the moment—one of the richest and most powerful men in the world working on a pitch to sell NASA on building a moon base supporting a mining operation at the lunar South Pole—might have seemed just a bit too preposterous for even the most daring of the Hollywood studio executives Bezos had been hobnobbing with just a couple of days earlier at the Oscars.

Then again, Bezos's imagination was fueled by all the science fiction he had devoured as a child, and he had long believed in the power of big, audacious ideas. "I believe the dreamers come first, and the builders come second," he once said. "A lot of the dreamers are science fiction authors, they're artists. They invent these ideas, and they get catalogued as impossible. And we find out later, well, maybe it's not impossible. Things that seem impossible if we work them the right way for long enough, sometimes for multiple generations, they become possible."

Far-fetched though his proposal may have seemed, Bezos now believed this was the multigenerational work that would define his legacy. And he wanted NASA to take him seriously. "It is time for America to return to the Moon—this time to stay," he wrote in the executive summary of the proposal. "A permanently inhabited lunar settlement is a difficult and worthy objective. If such an endeavor is to be successful and practical, we must be able to soft land large amounts of mass onto the lunar surface, and we must be able to do so cost effectively. Any credible first lunar settlement will require this capability." As his jet neared the nation's capital, Bezos said they would reconvene the next morning, at his suite at the Four Seasons Hotel in the posh neighborhood of Georgetown, to rehearse the presentation.

CHAPTER 2

Elon's Real Superpower

There is perhaps no one Elon Musk likes taunting more than Jeff Bezos. Over the years, he's unleashed a string of insults that have run the gamut from juvenile to vulgar to nerdy. Much of Musk's catalogue of derision was born from a frustration that Blue Origin seemed lazy and entitled—and simply wasn't moving fast enough.

That was the case when in 2013 the pair sparred over who should be granted the right to lease Launch Complex 39A, the famed launch site at NASA's Kennedy Space Center that hoisted Neil Armstrong, Buzz Aldrin, and Michael Collins to the moon in 1969. The agency had originally awarded it to SpaceX, but Blue protested. An infuriated Musk pointed out that Blue Origin had yet to fly a rocket to orbit. In an email to *SpaceNews,* he made it clear that such a feat lay beyond the company's engineering capabilities: "Frankly, I think we are more likely to discover unicorns dancing in the flame duct."

Blue ended up losing that particular squabble, as well as a fight over a patent that it had filed for a plan to land rockets and reuse them. Musk called it "ridiculous" and filed a lawsuit, leading Bezos to back down. The schoolyard jabs continued over the landings themselves, which devolved into an unseemly spat about the size and velocity of rockets and culminated with Bezos giving it back to Musk. "Welcome to the club," Bezos tweeted after SpaceX landed its Falcon 9 booster for the first time—a reminder that Blue Origin had done it first, even if it was only with a rocket on a suborbital trajectory.

For all the public feuding, however, Bezos told his team not to get distracted by SpaceX. At Amazon, he preached an almost religious mantra of focusing on the customer, not rivals. Meetings always had at least one empty seat, reserved for an imaginary shopper, as a reminder of whom they were working for. "The number one thing that has made us successful, by far, is obsessive, compulsive focus on the customer as opposed to obsession over the competitor," he once said.

Growth was more important than profit. Patient, long-term vision was more important than quarterly results. Reacting to competitors was not only foolish, it was heresy. That ethos served Bezos exceptionally well, as his repeated reinventions of Amazon—from online bookseller into a retail and Web services behemoth—made him the wealthiest person in the world. If it worked for Amazon, it would work with Blue Origin, which given the complexities of rocket science would require an even more farsighted approach. The company's motto—*Gradatim Ferociter,* Latin for "step by step, ferociously"—embodied that thinking. As did the company's mascot: a tortoise.

In 2004, Bezos drafted a memo outlining the core principles he hoped to enshrine in Blue Origin's culture and how they would guide the company through an uncertain future. "We believe local hill climbing is our best way forward," he wrote. "We have been dropped off on an unexplored mountain, without maps and the visibility is poor. Every once in a while, the weather clears up enough for us to glimpse the peak, but the intervening terrain remains largely obscured."

The only the problem was that Musk had already planted a flag on that mountaintop, and all the other ones on the horizon. Blue Origin wasn't forging a new path, but rather following the one bushwhacked by SpaceX. While Blue Origin was focused on the long term, Musk capitalized on the present. Ironically, he'd done so by coopting another Amazon motto: Get big fast. With the lucrative NASA and Pentagon contracts it won, SpaceX had beefed up to 5,500 employees by 2016. Blue Origin had 700.

The race between the companies had become so one-sided, it was difficult to even describe it as a race. Early on, a lot of journalists, including me, had used the tale of the tortoise and the hare as a metaphor to describe the rivalry, casting Musk as the hare and Bezos as the tortoise. Two wildly successfully billionaires duking it out as their space companies reprised the roles of the United States and the Soviet Union during the 1960s Cold War space race to the moon, one sprinting ahead, the other plodding along, slow and steady. But we had failed to account for something: Elon Musk doesn't get caught napping. He doesn't nap. He barely sleeps. He runs hard, all the time, and drives his employees to do the same.

Musk was not like any competitor Bezos had ever faced, and in the spring of 2016 he decided he could no longer be ignored. During a Saturday meeting that May at Blue Origin's headquarters outside of Seattle in Kent, Washington, Bezos broke his rule and focused on his competitor. SpaceX had recently won a $33 million contract from the Air Force to develop its next-generation Raptor engine, which would ultimately power a vehicle that Musk had dubbed the "Big Fucking Rocket."

"Why didn't we bid on that?" Bezos wanted to know.

The answer was that it had, but as a subcontractor to bigger, more established aerospace giants. Blue Origin was still something of a start-up, but Bezos wanted to change that. Ever since he had founded the company in 2000, two years before Musk started SpaceX, it had existed essentially as a nonprofit, funded almost entirely by his vast fortune, which now stood at about $140 billion. He had ceded enough territory to SpaceX. Now it was time to compete.

"Elon's real superpower is getting government money," Bezos told his executives. "From now on, we go after everything that SpaceX bids on."

TRUMP AND HIS team admired Musk and SpaceX, and if they wouldn't give them the keys to NASA, they would at least be happy to include the company in their plans for a return to the moon. Trump, however, had an entirely different opinion of Bezos, whose ownership of *The Washington Post* put him in Trump's crosshairs.

After Trump's election, Blue Origin's executives were desperate to make inroads with the new administration. On December 28, 2016, two of Blue Origin's top officials in the Washington, D.C., office, Brett Alexander and Clay Mowry, met with the transition team for NASA, which included Sandy Magnus, a former NASA astronaut; Greg Autry, then a business professor at the University of Southern California; and Charles Miller, a former NASA official and longtime commercial space advocate.

By then, Blue Origin had successfully landed and reused its New Shepard booster several times, demonstrating that rockets did not have to be thrown away after each flight, which would dramatically lower the cost of space travel. But that was not what the transition team wanted to talk about.

"We're going back to the moon," Miller said bluntly. "What can you do relative to the moon by 2020?"

The honest answer would have been *Nothing* or *Very little,* especially by 2020, the end of Trump's first term. Instead, Alexander responded, "We just landed New Shepard in the desert five times in the last twelve months. We can land something like it on the moon."

Something like it? Such a spacecraft didn't exist—not even on paper. Alexander was completely off-script, eager to somehow align Blue Origin with whatever NASA was doing. The farthest Blue Origin had ever gone was just over 60 miles; the moon was 240,000 miles away. It had no atmosphere, nothing to slow down an incoming spacecraft. Its surface was littered with mountains and craters, making a touchdown perilous—as Neil Armstrong discovered in 1969 as he scouted a suitable landing site with just seconds of fuel left. Landing on the moon was far more difficult than landing on Earth.

Everyone at Blue Origin knew Bezos was fascinated with the moon, though. They knew his childhood heroes were the Apollo astronauts. They knew that Bezos had even commissioned a pair of multimillion-dollar expeditions to the bottom of the Atlantic Ocean to find and re-cover the F-1 engines that powered the Saturn V rockets. Only a true obsessive would do that.

But that was Bezos's *personal* passion. At Blue Origin, engineers rarely discussed the moon. It just didn't come up. They were focused on getting the New Shepard rocket ready to fly again and eventually certify it as safe for human spaceflight. While New Shepard only scratched the edge of space, flying suborbital flights that went straight up and down, there were plans for a far larger rocket, code name Very Big Brother, which would be able to fly to Earth orbit. For a company like Blue, which was content to move slowly and deliberately, the moon was an almost inconceivable leap forward.

Still, Bezos's decree from a few months earlier—"From now on, we go after everything that SpaceX bids on"—resonated. Blue Origin had passed on several of NASA's most high-profile public-private partner-ship programs focused on developing spacecraft to deliver cargo and astronauts to the International Space Station. SpaceX had won those, hauling in contracts that were not only worth billions of dollars but also allowed the company to benefit from NASA's expertise as they de-signed and built its Falcon 9 rocket and Dragon spacecraft. Perhaps most importantly, after years of working side by side on high-stakes missions, SpaceX had earned NASA's trust. Some at Blue had thought it was a big mistake to sit out those programs. And here, suddenly, was a chance to get in with NASA before there was even a formal contract on which to bid.

The Trump transition team loved the idea of a New Shepard–like lander on the moon. "Can you get us a white paper?" Miller asked.

As soon as the meeting was over, Alexander called Rob Meyerson, Blue Origin's president. "You're not going to believe what just hap-pened," he said.

Until that point, Bezos had wanted to stay true to the company's step-by-step approach. But when Meyerson told him about the NASA meeting that evening, Bezos got excited. He spent a part of the New Year's holiday working with his team on the white paper. He wanted them to pursue landing a small spacecraft on the rim of Shackleton

Crater, one of the largest at the moon's South Pole. It was a logical destination given that NASA had discovered water in the form of ice in its permanent shadows. On the ridge outside the crater there was continuous sunlight, an ideal spot to set up solar arrays. A later mission might even be able to bring samples of the water home.

This would mark a significant step in the space race that was developing between the United States and China—and put Bezos at the forefront. While other technology companies were shunning the Pentagon for fear of their technology being used in conflict, Bezos felt it was Amazon's patriotic duty to help serve the nation. It was lucrative as well. In 2013, Amazon Web Services won a $600 million cloud computing contract from the Central Intelligence Agency. And Bezos would defend his decision while going after another cloud computing contract, this one worth $10 billion. "We are going to continue to support the DoD, and I think we should," he said in 2018. "One of the jobs of a senior leadership team is to make the right decision, even when it's unpopular. If big tech companies are going to turn their back on the U.S. Department of Defense, this country is going to be in trouble."

Getting to the lunar South Pole ahead of China would be a herculean effort. The key, Bezos decided, was to do it inexpensively, and he wanted the team to adopt Amazon's tactic of thinking about the project backward—starting with the press release they would write after a successful mission. The team worked straight through the weekend and by January 4, 2017, had a seven-page white paper ready to deliver to NASA, outlining a robotic lunar expedition plan that just days before had never been considered.

Bezos wrote the introductory line himself, editing it on a whiteboard in one of the company's conference rooms on January 3 until he felt like he had it just right. The first draft began: "First landing on rim of Shackleton Crater followed by first return of lunar ice to Earth, enabling human settlement of the moon. . . ." Then he crossed out "First" and replaced it with "Precision." He added "on a peak of eternal light in 2020," as well a couple other small tweaks. By the time he was done, the final version read: "A precision landing at Shackleton Crater on a peak of eternal sunlight in mid-2020, followed by first sample return of lunar ice to Earth, enabling human settlement of the Moon." Underneath was an image of what looked like a modified New Shepard rocket, casting a long shadow on the moon with the expanse of Shackleton Crater behind it.

The white paper called for a series of increasingly ambitious follow-up missions, including the development of a human colony, enabled by Blue Moon's ability to deliver supplies, energy, and "habitation modules" to the lunar surface. While the paper was big on ambition, it was light on technical details. "Within a few months, a second Blue Moon can land and perform a groundbreaking sample return of lunar ice from Shackleton Crater," it stated, without any explanation of how the ice would be located or extracted from one of the most forbidding regions of the moon. The paper made no mention of the tumultuous topography of the South Pole—its high ridges and deep valleys, its low-angled sunlight casting long, cold shadows that would make any expedition more perilous. The whole plan was a fictional sales job, but it read like something Blue had been crafting quietly for years. More importantly, Bezos loved it. He even reached out to Pence, eager to smooth things over with the new administration as well as put Blue Origin in a prime position. "During the transition I took a call from Bezos," Pence later recalled. "When I congratulated him on his rocket innovation, he told me, 'Mike, I started Amazon just to feed my space habit.'"

Blue Origin now had some real momentum. The transition team was intrigued, and so was the leadership of NASA, which scheduled a meeting for March 2, 2017. Bezos gathered his team on his private jet and headed to Washington, eager to pitch the Blue Moon plan himself.

ABOUT A WEEK before that meeting, I obtained a copy of the white paper. It was a juicy scoop for my employer, *The Washington Post,* made sweeter by the fact that Blue Origin was a notoriously secretive company that was difficult to penetrate. Then there was the sticky fact that Bezos owned the *Post,* but we covered him and his companies without fear or favor, as we would anyone else. Here was further evidence of that: Blue Origin desperately did not want this news to get out.

When I first approached Blue and told them I had the paper, they responded by saying that they would offer up Bezos for an interview—but only when his schedule permitted it. An interview with Bezos was a rare get, well worth holding the story, as long as Blue promised to let us know if any other news outlets were pursuing it. A few days passed without word, until eventually Blue said that Bezos would only answer written questions. That was more than a lame cop-out, I thought; it was reneging on the deal. Grudgingly, we decided to hold the story once again, since quotes directly from Bezos in any format were so rare that they would still be valuable for readers. I submitted a series of questions and waited.

Bezos was stalling, intent on sitting down with NASA before the story got out. After he arrived in Washington on March 1, two of his staffers stayed up all night putting the finishing touches on his presentation. At an office supply store in Northern Virginia, they also printed out poster boards for Bezos to use, showing images of the lander, a diagram of the course it would take, and a New Shepard booster landing in Texas. The proposal itself was twenty-four pages, each marked "Blue Origin Proprietary & Confidential / Subject to Export Control." On the cover was another image depicting the Blue Moon lander on the lunar surface. But this time it looked more like the lander used during the Apollo era, not the modified New Shepard that was on the cover of the white paper—a last-minute tweak that revealed how Bezos and his crew were making this up as they went along.

The proposal claimed that Blue Origin would be able to land on the surface by July 2020 and eventually set up an Amazon-like delivery service for the moon. "The United States will build momentum for its space program with the *Blue Moon* mission architecture, resulting in a cadence of missions to multiple locations on the lunar surface," it read. Given the spacecraft's size, it would be able to transport rovers even larger than the one China had landed on the moon, which could be used for prospecting. Later missions could help NASA test what's known as "in situ resource utilization"—essentially, the technologies needed to build things and generate power on the moon, rather than transporting them from Earth. That would be essential for creating the first lunar settlement.

The place to do this, Bezos felt, was at Shackleton Crater at the lunar South Pole—"an ideal location for a permanent settlement offering mineral compounds to develop structures, lunar ice for propellant production and life support, and near-continuous periods of lunar sunlight for power generation and thermal stability."

The BE-7 engine that would power Blue Moon would use hydrogen as its fuel, a decision made in large part so that it could be refueled on the moon using the water extracted from the craters and broken down into its component parts. If the moon was going to become a gas station in space, Blue Moon would be its first customer.

The proposal went on to describe, in far greater detail than the white paper, how the landing would be accomplished. It would start with an Atlas V rocket, the reliable workhorse operated by the United Launch Alliance (ULA), a joint venture of Lockheed Martin and Boeing. A "transfer stage," powered by engines burning liquid oxygen and liquid methane, would perform the "Trans Lunar Injection" burn that would propel the spacecraft out of Earth's orbit to the moon, as well as the "Lunar Orbit Insertion" burn. The lander itself would perform "the braking and final descent maneuvers and landing on the Moon."

On the morning of March 2, the Blue Origin team met in Bezos's suite at the Four Seasons in Georgetown. The room was tidy, the bed was made, and Bezos was finally happy with the proposal. He was in a cheery mood, as if the diatribe on the plane the day before had never happened. They rehearsed again and again, and kept going as they rode to the meeting in a black SUV, which looked as if they were escorting a head of state and his entourage. Bezos had it down. They were confident.

THE A-SUITE CONFERENCE room on the ninth floor at NASA headquarters was packed with more than a dozen top NASA officials, nearly every seat taken. Robert Lightfoot, a longtime agency official who was serving as the acting chief while the Trump administration chose its nominee, sat at the head of the table. He was flanked by Thomas Zurbuchen, the head of NASA's science mission directorate, and Ken Bowersox, a former astronaut then serving as the head of human exploration. Several members of the transition team were there as well. As they went around the room and introduced themselves, Bezos said, "I'm Jeff Bezos, and I send lots of brown boxes to your house."

The Blue Origin team stood up the poster boards, lining them up in order from start to finish, easy for everyone to follow. But from the start, Bezos went off course. He was rushing, and took the first two visual aids out of order. It wasn't until Lightfoot interrupted with a question that Bezos found his stride, and as more queries came, the more confident and relaxed he got. Bezos was a self-taught rocket scientist who could keep up with even the best of his engineers. He was most comfortable delving into technical details, and smart, insightful questions were just what he needed to allow his deep expertise to come through. He explained why their choice of propellent was the best and how the spacecraft could land itself vertically. He talked about the "inverted pendulum problem": The larger the spacecraft, the easier the landing, for the same reason it's easier to balance a broomstick on the palm of your hand than a pencil. He extolled the scientific virtues of landing on the edge of Shackleton Crater and why building a sustainable presence on the moon was the logical next step for NASA.

Free-flowing and loose, he even took a shot at Musk. Every time Musk tweeted yet again about Blue Origin's slow progress, Bezos joked, he'd sell $1 billion of Amazon stock and invest it in the start-up. He punctuated the comment with his full-bodied cackle of a laugh.

Bezos was winning the NASA officials over. The meeting, scheduled for forty-five minutes, went for nearly double that. Zurbuchen said that having a commercial company land on the lunar surface would allow his team to conduct more science experiments, particularly ones that probed for water. "This will give us shots on goal," he said excitedly, in a thick Swiss accent. What really got the attention of NASA leadership was when Bezos explained his personal commitment to the effort, pledging to invest a large portion of his own money, starting at $200 million to get it going.

In a private moment after the meeting, Lightfoot joked that if Bezos was worth $78 billion, as he was at the time, that would be like a parent with $78 in their pocket. If the kids wanted one dollar, it would be no big deal to "just give them two dollars." Bezos laughed but made no promises.

ONLY AFTER THE meeting ended did I get the responses to my questions. The *Post* broke the news late that afternoon, under the headline "An Exclusive Look at Jeff Bezos's Plan to Set Up Amazon-like Delivery for Future Settlement of the Moon." It quoted Bezos's written responses to my questions, which in many cases were taken directly from the proposal he submitted that day, including a verbatim repetition of the opening lines: "It is time for America to return to the Moon—this time to stay. A permanently inhabited lunar settlement is a difficult and worthy objective." And he stressed that the mission could "only be done in partnership with NASA. Our liquid hydrogen expertise and experience with precision vertical landing offer the fastest path to a lunar lander mission. I'm excited about this and am ready to invest my own money alongside NASA to make it happen."

That evening, Bezos and his team were at the National Building Museum in downtown Washington to receive an award from *Aviation Week & Space Technology,* the venerable aerospace publication. As part of the black-tie ceremony, Bezos agreed to a fireside chat with Joe Anselmo, the editor of *Aviation Week,* who right off the bat asked him about my story: "Just a couple of hours ago, it came out that you're planning a moon mission with a moon lander as well."

Bezos, now well practiced in the pitch, hit all the talking points but added that the moon is a key step in expanding even farther into the solar system. "If you go to the moon first, and make the moon your home, then you can get to Mars more easily," he said.

Afterward, Bezos posed for photos with several young scholarship winners honored at the event, joking at one point that they were so young they probably had never seen anything like the pay phone still installed at the back of the museum. It had been a long couple of days. The painful cross-country flight. The edits on the proposal. The pitch to NASA. Keeping my *Washington Post* story at bay. Now, after the *Aviation Week* ceremony, Bezos and the Blue Origin team were ready for a late-night party. They had reserved a room across the street at Denson Liquor Bar, an old-school, speakeasy-type basement joint with bottles stacked to the ceiling behind the bar. With Bezos's security personnel keeping watch, they drank well into the night, as Bezos regaled them with stories from the Academy Awards and the after-party hosted by Madonna, where the other guests included Leonardo DiCaprio, Johnny Depp, Katy Perry, Ed Norton, and other A-listers. But no photos allowed. Madonna wanted her guests to feel free to let loose on the dance floor without fear of unflattering images being spread on social media. The Material Girl set the tone for the evening with a salsa dance, then stayed out there for hours, Bezos recalled, amazed she didn't pause once for a bathroom break.

Partying with celebrities seemed out of character for a man who was a genuine space nerd, whose childhood had been spent devouring science fiction novels, watching *Star Trek* and cheering his hero, Captain Kirk. His previous brush with Hollywood had been a cameo in the movie *Star Trek Beyond,* in which his costume and heavy makeup transformed him into a freakishly bulbous-headed alien.

"There's no relative direction in the vastness of space," a Starfleet officer tells Captain Kirk in the film. "There's only yourself, your ship, your crew. It's easier than you think to get lost." The counsel came during a moment of introspection for Kirk, when his future as captain of the *Starship Enterprise* appeared uncertain. Now, as Blue embarked on a quest to the moon, it was advice Bezos could have heeded as well.

CHAPTER 3

Converting the Impossible to Late

I f Jeff Bezos was obsessed with the moon, he viewed Mars as a hell-hole. "My friends who want to move to Mars, I say, do me a favor and go live on the top of Mount Everest for a year first, and see if you like it, because it's a garden paradise compared to Mars," he once said.

"Sometimes my friends say, 'Would you move to Mars?'" he said at another point. "Not in the near term. Think about it: no whiskey, no bacon, no swimming pools, no oceans, no hiking, no urban centers. Eventually Mars might be amazing. But that's a long way in the future."

Mars was Musk's muse, not Bezos's, and the comments served as something of a rallying cry for the employees at Blue Origin, who were eager to differentiate themselves from their counterparts at SpaceX. One employee even printed up T-shirts that read: NO WHISKEY? NO BACON? NO THANKS, with an image of the Red Planet replacing the o in the last NO.

Musk, the South African–born entreprenuer, had founded SpaceX in 2002, after earning about $180 million from the sale of his company PayPal to eBay. Since reading science fiction in his childhood, notably Isaac Asimov's Foundation series, he had been fascinated with space and certain that Mars was key to humanity's future. With SpaceX, Musk's goal was reducing the cost of access to space, an endeavor that would help bring about the most significant achievement that humanity could accomplish in his lifetime: becoming a multiplanetary species.

In June 2017, I traveled to interview Musk at SpaceX headquarters in Hawthorne, California, just outside of Los Angeles. If there was an aesthetic to the decor there, it could have been called Mars chic. There were images of the Red Planet all over the place. The doormats depicted boot prints in Martian clay, and as I made my way past security into the building, I was greeted by a stark, white wall with a pair of images. One showed Mars as it exists today, beautiful and barren, a forbidding other world the color of rust. The second showed Mars as Musk imagined it could be transformed: an Earth-like planet of blues and greens, even clouds, budding with life.

Musk's workspace was nearby—a cubicle like any other in the office space. He preferred to see and be seen without the barrier of an office. The design of the place, like the design of the rockets and spacecraft and space suits, had to have style as well as functionality. The two went together. That was one of Musk's mantras and one of the things Gwynne Shotwell, one of his initial hires, who would go on to become SpaceX's president and chief operating officer, learned from him early on.

"Beauty in the office—really important," she once said. "This is one of the areas where I feel really bad for government employees. Many government offices are not beautiful. They're not inspirational. I think that should be re-thought. It doesn't have to cost a lot of money for there to be beauty in the office. You don't want to work in a depressing place. How do you do extraordinary work if you're working in a dungeon?"

I had a lot of time to observe the SpaceX offices because Musk was, as usual, horrendously late. The list of the adjectives often used to describe Musk—*mercurial, brilliant, eccentric, intense, impulsive*—has never included *punctual*. As a child growing up in South Africa, his older brother, Kimbal, would tell him the bus was coming several minutes ahead of schedule to ensure he caught it and got to school on time. In his adulthood, a Tesla board member coined the term Elon Time to describe what was becoming a chronic problem, a condition Musk did not deny. "I do have, like, an issue with time," he once said. Like their creator, his products, too, were often late, in some cases years behind schedule. Musk justified it with a pithy turn of phrase: "At SpaceX, we specialize in converting the impossible to late."

His personal assistants quickly learned that rather than keeping the boss on schedule, their job was to make everyone else—no matter how important—adapt to Musk's internal clock, however inconvenient that might be. For example: A Hollywood producer from Ron Howard's Imagine Entertainment once came to SpaceX to pitch Musk on a National Geographic show about Mars. Though the meeting had been scheduled for weeks, Musk, of course, was tardy, his assistant contritely explained, by now well versed in the art of apology. An hour passed. Then two. Finally, the assistant returned. Musk was too busy to see the producer after all. Could he please come back tomorrow? (He did.) Another time, Musk made William Shatner, Captain Kirk himself, wait four hours.

When Musk and I finally sat down, I was worried that the meeting would be cut short given his hectic schedule. I'd heard horror stories of him walking out of interviews he felt were a waste of time. Or that he would spend the entire session with his back to the interviewer as he banged out emails on his computer. And so I was surprised when Musk came across as thoughtful, even gracious and vulnerable. He was introspective, going quiet for long periods and then emerging with ruminations that belonged to someone with a depth born not just from brilliance but, it seemed, from reservoirs of pain as well.

"The rocket business is definitely one that leaves an enormous amount of emotional scar tissue," he said at one point. "You know, when one of these things explodes, and you have your heart and soul in the thing, it's crushing. Crushing."

"But then when it succeeds?"

"It's great," he said. "A lot of highs and lows."

IT WAS A version of Musk many didn't see—a kinder and gentler Musk, not high or low but, it seemed to me, on balance. (I'd remember this interview later, when he'd turn his X feed into a torrent of rapid-fire, manic belligerence that targeted all manner of perceived enemies, and became a divisive figure embedded in Trump's "Dark MAGA" world during the 2024 presidential campaign.) Our interview, initially scheduled for forty-five minutes, stretched to nearly an hour and forty-five minutes, covering everything from the early days of SpaceX to the broader commercial space industry, to the recent successes the company had flying missions to the International Space Station for NASA. But there was one topic he kept coming back to over and over, often unprompted: the BFR.

The Big Fucking Rocket, also known as the Mars Colonial Transporter or the Interplanetary Transport System, was a preposterous vehicle that would stand some four hundred feet high, taller than a cruise liner stood on end, taller than two Space Shuttles stacked on top of each other. At the time, the plan was for it to have a staggering forty-two engines and be capable of transporting as many as a hundred people at a time to Mars. But as Musk had laid out his design for the rocket at an international space conference eight months earlier, he admitted he had no idea how to pay for such a gargantuan piece of machinery—or a Mars mission—joking at one point that he'd start a GoFundMe campaign.

The rocket was a fantasy bordering on delusion, or at least that's how it appeared to me. Even Musk admitted the whole thing was "extremely improbable." But to his legions of fans, it was a vision of the future, one he stoked by creating sleek animated videos showing the BFR landing on Mars.

At the conference, held in Guadalajara, Mexico, Musk had reserved time at the end of the presentation for what he'd hoped would be "hardcore technical questions," as he told me now. Instead, the questions were ridiculous, including one from an audience member who asked if she could come on stage and kiss him, and another who wanted to give him a comic book about the first person on Mars.

"I was really hoping to get critical feedback like, 'Hey, I see what you did there. But I think you're wrong, and you should consider this other thing.' That's the kind of feedback I was actually looking for," he told me. "There were definitely people in the audience who could have done that, but unfortunately, the hardcore science engineers were not going to beat the lunatics to the mic."

Sitting in the conference room at SpaceX headquarters, it was becoming clear that Musk had, in the months since, asked himself those hard technical questions. And the answers had led to some significant revisions. Initially, he was reluctant to share them with me. "I wouldn't want this to get out prematurely," he said. But I was curious to get a glimpse of whatever it was that was making him so excited, so we went off the record with the agreement that I wouldn't share them until he revealed them publicly himself.

The size of the rocket would still be huge, ridiculously huge, the most powerful ever built. But the diameter of the booster would shrink from twelve meters to nine meters, or about thirty feet. That meant the number of engines that could fit would be reduced from forty-two to thirty-one. That meant that the rocket could be useful not just for sending lots of people to deep space but also for serving the customers—commercial satellite operators and the government—who were providing SpaceX the revenue it was using to grow so dramatically.

Those missions to low Earth orbit were being flown on SpaceX's Falcon 9 rockets. But with a built-in revenue stream, the BFR could become economically viable. That's why Musk was so jazzed. "I'm really excited about the new architecture for BFR," he said. "The thing that I hadn't figured out was, like, how do we pay for this?"

The previous design was simply too big for anything but missions to deep space. A smaller rocket could do everything—take people to Mars, fly satellites to Earth orbit, even supply the International Space Station. That, in turn, would allow SpaceX to replace its existing line of rockets and spacecraft—the Falcon 9, Falcon Heavy, and Dragon spacecraft. "The key is, we can obsolete our entire existing production line," he said.

I was dumbfounded. SpaceX had worked painstakingly for years to develop those vehicles. The Falcon 9—so named for the number of engines mounted to the base of its first stage—had become a trusted workhorse, used by NASA and the Pentagon. Dragon, the spacecraft that was flying cargo to the space station, would in a few years start transporting NASA's astronauts with upgrades designed for human spaceflight. The Falcon Heavy, essentially three Falcon 9 boosters mated together for a total of twenty-seven engines, was to be the massive rocket of SpaceX's future and was preparing for its first launch a few months later.

No longer content to simply disrupt the space industry, Musk aimed now to disrupt himself.

Of course, this new rocket would need a new launch facility, a home where SpaceX could fly and test the BFR, or whatever it would be called, without interference. Musk was planning to shoot rockets into the sky at an astonishing rate—perhaps as frequently as once a week, if not more, when the standard for launches by U.S. rocket companies was closer to once a month, if that. SpaceX had been growing frustrated with the constraints of government facilities like Cape Canaveral, and Musk wanted his own site—a private spaceport and the freedom that would come with it.

It was something he had been thinking about since soon after the company had launched its first rocket, the single-engine Falcon 1. In 2005, SpaceX had been working with the Air Force to prepare a launch pad at Vandenberg Air Force Base, investing $7 million. But Lockheed Martin and other big military contractors complained, saying they were worried the Falcon 1 would blow up and destroy the pads they used for missions for the Pentagon. That irritated Musk, who moved his Falcon 1 operations to a government site in the most remote corner of the world he could find, Kwajalein Atoll in the middle of the Pacific Ocean. In 2011, he told an audience that he still intended to create his own base one day. "It only makes sense," he said. Commercial launch activity should happen at a "commercial launch site, just as occurs with aviation."

He might have found the spot at an end-of-the road, forgotten fantasyland at the southernmost tip of Texas, outside one of the most impoverished cities in America. It was a marshy, sparsely populated spit of land that ran straight into the Gulf of Mexico and out to the horizon, a mirage as implausible as the rocket that had become Musk's obsession.

. . .

THERE ARE NOT many places in the United States suitable for launching large rockets to orbit. The Federal Aviation Administration regulations effectively require launch pads to be located by water, away from populated areas. That way, if the rocket explodes, the debris rains down on fish, not people.

Being as close as possible to the equator is also a plus. The surface of the Earth rotates faster there, and rockets can take advantage of the spin and get a running start to space. As Musk and his team started studying the map to see where such property might exist, they quickly realized there were few choices. Just north of Cape Canaveral in Florida was a possibility, and so was the coast of Georgia. There was also a site in Puerto Rico that might work. And then there was the mysterious corner at the very tip of Texas. There was a city there, Brownsville, but also a fair amount of largely uninhabited land just beyond it.

In early 2011, Musk directed one of his most trusted engineers, Tim Buzza, to fly to Brownsville, get in a car, and start driving south. "See what's out there," Musk said. If it was at all possible to build a launch site, he wanted to know.

NASA could have told him. The space agency had studied an area not far away at the dawn of the Space Age. A report from 1961 described it as "ranch country with little population (only 884 in 1960); it remains much as it must have been when the first Spanish explorers saw it." It was only "four to six feet above sea level, with scattered drifting dunes. Along the coastline proper, mostly sand, dunes and stream deposits are found. None of these formations contain any solid rock. In fact, the bed rock lies as far down as 25,000 to 30,000 feet." That was a major problem. Launching a rocket powerful enough to get to the moon required a solid foundation.

Still, there were some positives. "Land in the area is relatively inexpensive and sparsely settled," the report noted. There was also "a large pool of unskilled labor" in nearby Brownsville. NASA ultimately passed on the site, opting for the Florida coast instead, and Cape Kennedy was born.

On his trip to Texas, Buzza brought along Hans Koenigsmann, a rocket safety engineer; Steve Davis, who ran the Dragon spacecraft program; and Caryn Schenewerk, an attorney who served as SpaceX's director of legislative affairs in the company's Washington, D.C., office. From the airport, they drove along a two-lane road out of Brownsville that ran past a Border Patrol checkpoint, past a gun shop and firing range, through a wildlife protection area and a state park, and finally arrived at the Gulf of Mexico. There they parked and surveyed what was one of the most bizarre places they had ever been. There was no electricity, no Internet. Water surrounded them, and the land was low. One good Gulf-fueled hurricane would wash everything away. The ground was porous and soft, the consistency, in some places, of Jell-O. How would they ever launch a rocket from here? Buzz wondered.

I guess we'll have to launch at low tide, Buzza was thinking, just as Border Patrol agents pulled up in a truck, demanding to know what they were doing. Buzza wasn't sure how to respond. "If we said we were looking for a super-heavy launch site to get to Mars, they would have detained us," he later recalled. Instead, they said something about looking for land for commercial purposes.

The agents said they should be on their way. But before the SpaceX team left, they turned into a one-street residential village called Boca Chica, a tiny hamlet that didn't show up on most maps, whose few residents made it clear they cherished their lonely quiet and end-of-the-road anonymity.

Half of the homes looked abandoned. There seemed to be more barking dogs than people. Since the residents had no running water, they relied on the county to truck it in once a month in big blue barrels that sat on their lawn. Buzza stepped forward to get a better look at one house, then quickly retreated after seeing a sign that read MEET MY FRIEND SMITH AND MY OTHER FRIEND WESSON.

Boca Chica was on the water, but the Gulf of Mexico was not the vast, empty Atlantic Ocean. Any rocket launching from Boca Chica could in a matter of minutes find itself over Florida or Cuba. To protect the people who lived there from the possibility of falling shrapnel, the rockets would have to bank right, like a golfer hitting a slice on the drive of a dogleg par 4.

The site's other neighbors posed a problem as well. The land was situated between a Texas State Park and a U.S. Fish and Wildlife refuge, home to the piping plover, a diminutive, at-risk shorebird vigorously defended by environmentalists who would, no doubt, object to a rocket business moving into the neighborhood.

Even if SpaceX were able to somehow solidify the spongy soil, if it could acquire the property and buy out the nearby residents, if it could find a way to launch without flying over Florida, it would still have to build the infrastructure—rocket pads, manufacturing sites— from scratch. The launch facilities it used on Cape Canaveral and Vandenburg Air Force Base might have been fixer-uppers, but at least the basic necessities were there. At Boca Chica there was nothing; it was an empty canvas of waterlogged sand.

When Buzza returned to SpaceX and gave Musk his report, he didn't sugarcoat the challenges. He laid out every single one—the floodplain, the lack of services, the environmental concerns, the soft, marshy land. The degree of difficulty was a 10. But then again, so was everything SpaceX had ever done. The degree of difficulty in starting a space company from scratch was a 10. So was building a rocket capable of getting to orbit. So was building a rocket that could fly back from orbit and land on Earth so that it could be reused.

Buzza knew a challenge would not deter Musk. He'd only want to know: Was it possible? Could SpaceX build a launch site there?

The answer, Buzza believed, was yes.

Musk nodded and listened quietly as Buzza reeled off all the problems, and said finally, "Okay, let's tackle them one by one."

THE FIRST HURDLE was to acquire the land—quietly. SpaceX did not want people to know it was scouting real estate in South Texas and drive up the price. A small team, led by Steve Davis, one of Musk's most trusted confidants, looked at the real estate possibilities. At first, they studied the playbook Disney used in 1964 to acquire the property outside Orlando that became Disney World. The company used an array of real estate agents who did not know the identity of their client. Disney also set up dummy corporations—Latin American Development and Management Corp., and Reedy Creek Ranch Corp.—to scoop up vast parcels, and for cheap, in some cases less than $100 an acre.

Jeff Bezos did something similar when he was buying land in West Texas for Blue Origin in 2003, though his dummy corporations used the name of explorers: Coronado Ventures, the James Cook and William Clark Limited Partnerships. These were linked to a holding company that used a Seattle post office as its address. It was called Zefram LLC, after Zefram Cochrane, the *Star Trek* character who built the first spacecraft that could travel faster than the speed of light.

Hiring lawyers, setting up fake companies, and enlisting real estate agents might have worked well for Disney and Bezos. But that was not the SpaceX way. It was too pricey. Too inefficient. Too corporate. The team Musk assigned to acquire the land enjoyed the fact that they would do it differently—that is, they'd do it themselves. They started poring over county property records and decided against hiring the real estate agent they had interviewed, concluding they were more adept at locating opportunities than she appeared to be.

Working from county rolls, they made cold calls and wrote letters asking people to sell—never mentioning SpaceX, of course. Even Davis, one of SpaceX's most hardcore engineers, was intimately involved. He had impressed Musk in his initial interview by solving what he later called "math-based technical brain teasers" in his head, and then spent years immersed in developing the Dragon capsule's propulsion systems, avionics, and heat shield. He also helped write the software that would allow the self-driving capsule to autonomously dock with the International Space Station. Later, he'd be dispatched to revamp Twitter after Musk purchased it, and then to slash the size of the federal government as part of Musk's Department of Government Efficiency. But now the Boston-born, Penn- and Stanford-educated wonder kid had tapped a skill set no one, including himself, knew he possessed. The aerospace engineer transformed himself into a real estate prospector, dialing random property owners and charming them into selling him their land, as if he were amassing a portfolio of penny stocks.

One advantage SpaceX had was that many of the properties were delinquent on their taxes, or the owners were dead, meaning many parcels were just sitting there vacant, contributing nothing to the county's already depleted coffers and undesirable to everyone but the apparently delusional CEO of an emerging space company. The first purchase was recorded in the Cameron County property database on June 6, 2012—just over half an acre for $2,500 from the local school district. The buyer used the address of SpaceX headquarters, at 1 Rocket Road in Hawthorne, California, but not the company's name. Instead, it was Dogleg Park LLC, a curious moniker chosen by Schenewerk in honor of the turn the rockets would have to make after launching from Boca Chica to avoid flying over land.

In September of that year, SpaceX bought a couple more properties, this time at auction, one for $6,400 and the other for $15,000. This time a name was recorded alongside Dogleg Park: Lauren Dreyer, SpaceX's thirty-one-year-old director of business affairs and compliance. By then, Emma Perez-Trevino, a reporter for *The Brownsville Herald,* was on to SpaceX. On September 23, she broke the news that the company was snatching up land. "Public Records Show Local Purchases for Musk's Company," read the headline. She followed up with another story in November: "Land Grab: SpaceX Buys More Land Near Boca Chica."

A few months later, on March 8, 2013, Musk came to Austin to appear before a committee in the state legislature as it was working on an incentive package to help lure SpaceX to the state. He cautioned that his company was still looking at other sites, possibly in Florida and Georgia. "But I would say Texas is probably our leading candidate right now," he told the panel. "We're talking about something that's really the big leagues here. This would be a commercial version of Cape Canaveral. It would be a historical first in the world." Musk said he was "optimistic about making this work in Texas, in the Boca Chica area. It's looking quite good. So, any support that Texas can offer would obviously be helpful."

Local politicians, including the economic development authority, were doing their best to woo the company to an area that had one of the highest levels of poverty in the country. City officials drooled at the possibility of some six hundred new jobs with a starting salary of $55,000, well above the average for the area. "This is money from the heavens," Gilberto Salinas, the executive vice president of the Brownsville Economic Development Council, said to city officials during a town council meeting.

Musk's visit to the state capital generated headlines across Texas, which only meant buying property would be more expensive now that the word was out. To avoid a land rush, SpaceX had to move fast. When Cameron County held another auction for tax-delinquent properties a few months later, SpaceX snatched up nine more. An employee dispatched to the county courthourse became a regular fixture at the auctions, one hand on the paddle ready to bid, the other holding a cell phone, with Davis on the other end at his desk in the D.C. office, instructing her on how high to go. SpaceX created another shell company—The Flats at Mars Crossing—but it didn't take long for the area land investors to figure out who the employee was working for. "Hey, rocket girl," they taunted. "We're going to beat you, and you're going to have to buy it from us for three or four times the amount."

On at least one occasion, Dreyer came down to help out at the auctions. But, worried that she'd be recognizable as the SpaceX employee helping to lead the development effort in South Texas, she arrived at the auction in disguise. She wore a wig to cover her blond hair; large Jackie O. sunglasses to shield her blue eyes; and a colorful poncho.

THE TEXAS LEGISLATURE came through, offering SpaceX $15.3 million in incentives, luring a company that Governor Rick Perry said in August 2014 would "pump $85 million in capital into the local economy." And Musk announced that he would indeed build his private spaceport in Boca Chica.

The following month, the governor and dignitaries from across the state traveled to South Texas to celebrate with a ceremonial ground-breaking with Musk. The few employees on-site scrambled to get it ready and even hired a local resident who, to their surprise, cut the waist-high grass while wearing only a pair of boxer shorts. To keep everyone out of the sun, SpaceX erected a tent and, for the VIPs, an RV with air-conditioning. Musk, as usual, was running late. His private jet had landed at the Brownsville airport about a half hour away, and a police escort had been arranged. But he was just sitting on the plane, apparently immersed in a phone call. SpaceX employees started growing nervous their guests would be offended—or melt. It was ridiculously hot. Everyone was sweaty and uncomfortable, and some supersized species of Texas mosquito was out in force. One stung a young SpaceX government liaison staffer on the forehead, drawing blood that dripped onto his pressed white dress shirt.

Finally Musk arrived, and despite his tardiness he was greeted with enthusiasm, a messiah from Silicon Valley who would spark an economic revival in a region that desperately needed it. If the area once was on the western frontier, "it is the frontier again," Perry said in his Texas accent during the ceremony. The effort symbolized the state's "pioneer heritage" and "our tradition of thinking bigger, dreaming bolder, and daring to do the impossible."

Wearing a white shirt open at the collar and a dark suit, Musk was in a great mood, basking in the attention. The long-term goal of SpaceX, he said, "is to create the technology necessary to take humanity beyond Earth, to take humanity to Mars and establish a base on Mars. It could very well be that the first person that departs for another planet could depart from this location." But he warned that rockets wouldn't be shooting from the site anytime soon. "It is going to take several years to build out a spaceport because this is going to be quite a significant building endeavor," he said. Standing side-by-side, Musk and Perry hoisted the first ceremonial shovelfuls of dirt.

With the ceremony concluded, SpaceX's engineers got to work transforming the swampy marsh into a launch site. They drilled down hundreds of feet, then more than a thousand, but couldn't find bedrock. "All we got were alternating layers of silt and sand, silt and sand," one former employee recalled. At one point an excavator got swallowed up by a sinkhole. Employees spent days trying to get it out, but it only sank more. Finally, they rigged the heavy piece of machinery to two bulldozers and a thirty-ton dump truck and got it out right before a rainstorm that would have made it impossible to save.

What SpaceX needed was more dirt. A lot more. In 2015, crews started hauling it in with a constant stream of heavy trucks that rumbled down Route 4 and left the lonely road pockmarked with potholes. The effort took months. Eventually, 310,000 cubic yards, or enough to cover the area of a football field fourteen stories high, stood on the site as a man-made mesa.

In a process known as "soil surcharging," the giant pile would slowly compress the soil beneath it, eventually to a density that could serve as a foundation solid enough for rocketry. But that would take years. Mars, or any rocket launch from Boca Chica, would have to wait.

THE DIRT WAS still settling by the time Musk took to the stage at yet another space conference, in September 2017, four months after I had interviewed him at SpaceX headquarters. Speaking in Adelaide, Australia, he was ready to reveal publicly what he'd told me in private. He had figured out a way to pay for the BFR—reduce the size of the rocket so that it could be used for commercial and government launches in low Earth orbit, as well as going all the way to Mars.

He repeated the idea that had left me dumbfounded: "Essentially, we want to make our current vehicles redundant. We want to have one system, one booster and ship, that replaces Falcon 9, Falcon Heavy, and Dragon. So if we can do that, then all the resources that are used for Falcon 9, Heavy, and Dragon can be applied to this system."

Then he said something that took me by surprise again. The new platform could also be used to go to the moon and "enable the creation of Moon Base Alpha or some sort of lunar base," he said, as an image of BFR on the moon flashed on the screen behind him. *Moon Base Alpha?*

This was bordering on heresy; Mars had been so embedded in SpaceX culture since its founding that it carried the weight of dogma. But now, just like that, the moon was in SpaceX's sights.

"It's 2017," he said. "I mean, we should have a lunar base by now. What the hell is going on?"

As Musk envisioned it, the BFR wasn't just a rocket and a spacecraft, but a novel transportation system, capable of lifting enormous numbers of people and cargo into space. In order to travel throughout the solar system, the rockets would need to have their propellant refilled in orbit by a fleet of tanker ships—something that then existed only in science fiction. At the time, the total number of rockets leaving Earth was a few dozen a year. To build a base on the moon or Mars, Musk believed, "you need thousands of ships and tens of thousands of retanking or refilling operations. Which means you need many launches *per day*. In terms of how many landings are occurring you need to be looking at your watch, not your calendar."

At the time, SpaceX had landed its Falcon 9 booster sixteen times in a row, on barges at sea or separate landing pads, proving that rocket reusability was indeed real. No more did boosters need to be discarded into the ocean, as they had been for decades. But in order to meet the unheard-of launch cadence Musk was now detailing, the rockets would need to land with such precision and reliability that they would be "on par with the safest commercial airliners," he said. The rockets would be so precise that they would no longer need the landing legs the company had installed on the Falcon 9; instead they would land on their launch mounts, where they could turn around and take off again.

Musk tantalized his audience by asking: What if that capability was applied not to deep space, but to Earth? "We looked at that and the results are quite interesting," he said. Flying some 18,000 miles per hour, the BFR could launch to space, descend back into the atmosphere, and land at a new spot on Earth, allowing people to travel vast distances across the globe in a blink. Los Angeles to New York in twenty-five minutes. Tokyo to Singapore in twenty-eight minutes. New York to Paris in thirty minutes. London to Hong Kong in thirty-four minutes.

Thousands of ships, bases on the moon and Mars, even the rocket itself—all of it was still a Muskian dream that lived only in his head and in fictional animations. But during the presentation in Australia, Musk also showed an image of an enormous propellant tank able to hold 1,200 tons of liquid oxygen kept extremely cold, at minus 400 degrees Fahrenheit. The tank was made of a specially designed carbon fiber matrix that SpaceX had designed in-house. It was not only suited for extreme temperatures, but it was lightweight as well, a key ingredient in any spaceship design. And it was tough.

"We tested it up to its design pressure. And then—" Musk paused as a video showed the tank exploding "—a little bit further. We wanted to see where it would break. And we found out. It shot about three hundred feet up into the air." He added that SpaceX was also well into the testing campaign of the high-performance, methane-fueled Raptor engines that would be used to power BFR. Another video showed a purplish flame shooting out of a nozzle.

Blue Origin may have had a well-written proposal on its Blue Moon lunar lander that carried Bezos's personal touches. But SpaceX, it was now clear, was already *building*.

CHAPTER 4

On a Jihad

Jeff Bezos owns a compound on Indian Creek Island, the exclusive Miami enclave, where he snatched up one, then two, then three properties for a combined $230 million. He owns an $80 million penthouse in New York City that has a stunning view of the Empire State Building from the bathtub. His mansion in Beverly Hills, purchased for $165 million, spans twelve acres, while his $78 million waterfront estate in Maui is slightly bigger, at fourteen. His home in Washington, a former textile museum in D.C.'s tony Kalorama neighborhood, is a few blocks from the Obamas and holds much of his art collection, which is so extensive that visitors can take a self-guided tour using a printed pamphlet detailing the works. Then, of course, there's his floating home—the $500 million yacht *Koru*, with a mermaid bust designed to look like Lauren Sánchez, his paramour, and an accompanying support boat with a helicopter pad.

And yet of all his mansions and estates, it's perhaps his relatively unassuming ranch house in West Texas where Bezos feels most at home. It has proximity not to celebrities or the ocean but to the bull sheep roaming the mesquite brush and, most importantly, Blue Origin's suborbital rocket launch site. Not that the place isn't impressive. There's a helipad and an entirely separate house for the caretakers. The most stunning feature is a rooftop observatory, where Bezos installed a powerful telescope to study in more detail the stars he had gazed at with his naked eyes as a kid. The property is meant for entertaining, since, well, where else would guests stay? The nearest town, Van Horn, population 1,900, has limited accommodations. So Bezos built a hotel-like guesthouse, where each room has its own en suite bathroom. There's also a courtyard with a pool and a firepit that at first blush looks normal but is made from the titanium nose cone of a nuclear submarine—the sort of thing only a billionaire could acquire. Nearby is a bar stocked with bottles of tequila and scotch, where Blue Origin employees would gather after successful launches of the company's New Shepard rocket and spacecraft.

The capsule was designed to fly to an altitude of more than sixty miles, past the so-called Kármán line where many believe space begins. The first flight, in April 2015, had no humans on board, but Bezos stashed in the capsule a bottle of Johnnie Walker Blue, which he broke out by the firepit during the post-launch party. Once it was empty, Bezos signed and dated the label and put it back on the shelf, a notch on his belt, in a tradition that would follow after future successful New Shepard flights.

Toasting his team, Bezos said, "I had tears in my eyes. It was one of the most magnificent things I had ever seen. This is a huge milestone, but it isn't the end, it is the beginning. This is the start of something amazing. This is truly a great day, not just for Blue Origin but for all of civilization. I think that what we have done today is going to be remembered for thousands of years, and you should be so proud of yourselves."

In a blog post, Bezos wrote that "any astronauts on board would have had a very nice journey into space and a smooth return." The company's attempt to land the booster had been unsuccessful—the booster crashed into the ground after losing pressure in its hydraulic system, but he promised Blue Origin would fly again soon. And he hinted at a much larger rocket to come, the one code-named Very Big Brother.

Building a rocket and a spacecraft capable of reaching the edge of space was a triumph, even if it was just a suborbital, up-and-down flight that lasted about ten minutes. Blue followed that up with another successful mission in November 2015, in which the booster flew back to its landing site successfully so that it could be reused and flown again. In a tweet, Bezos called the machine "the rarest of beasts," and he and his team celebrated it with Mr. Walker and another label-signing.

It would have been considered even more momentous, though, if not for the fact that by then SpaceX had already been flying its Falcon 9 to orbit on a regular basis and was charging ahead at breakneck speed. A month later, SpaceX landed its Falcon 9 booster after a flight to orbit—a far more difficult challenge. Musk had tweeted at Bezos after New Shepard's landing: "Not quite rarest." He also pointed out that SpaceX's Grasshopper prototypes had previously completed several suborbital vertical takeoff-and-landings. There was a growing chasm between the companies' performance, a reflection of how intensely Musk drove his staff. SpaceX was notorious for churning through employees, who were expected to keep up with Musk, who worked insane hours, didn't sleep much, and often spent nights at the office.

Blue, by contrast, had a homey, comfortable vibe. Employees hung out in a communal kitchen and lounge. They brought their dogs to work. They enjoyed weekends and vacation time. Parking was first-come, first-served; even the president didn't have a reserved spot. Bezos invested a limited amount of his own time and money into Blue Origin, as it moved with a slow but a steady pace. In a way it was the opposite of Amazon, which had followed Bezos's mantra to *get big fast* and was constantly expanding into new markets.

The New Shepard flights, however, marked a new phase for Blue Origin and gave Bezos confidence in its future. He started telling people that Blue Origin was "the most important work I'm doing." He began thinking that Blue, even more than Amazon, would define his legacy, and that the opportunities in space could rival those produced by the Internet.

"If I'm eighty years old, and I'm looking back on my life, and I can say that I put in place, with the help of my teammates at Blue Origin, the heavy-lifting infrastructure that had made access to space inexpensive so that the next generation could have the entrepreneurial explosion like I saw on the Internet, I'll be a very happy eighty-year-old," he said during an event in Washington.

IF BLUE WERE going to compete with SpaceX and open space to the masses, it would need a rocket that, unlike New Shepard, would be able to make it to orbit. That would allow it to launch satellites and transport passengers, gaining customers in not only the commercial sector but also the government. And, like SpaceX's Falcon 9, Blue's Very Big Brother would be reusable.

In September 2015, Bezos announced that Blue was taking over a defunct launch site on Florida's Cape Canaveral, just a few miles from the pads SpaceX occupied—a big upgrade from the company's West Texas site, which was a bit of a backwater. The new orbital rocket, Bezos vowed, would fly from here by the end of the decade. "The thing I'm most excited about is humans in space," he told me in an interview at the time, "and the vision for me is millions of people living and working in space."

To complete the company's transition from a passion project to a genuine rival to SpaceX, however, Blue Origin would need not only a new rocket but a new leader. Rob Meyerson had been at the company since 2003, but his title was president. Blue Origin had never had a chief executive officer. Bezos wanted to find one, a visionary who could start bringing in the precious resource that had so far eluded the company: revenue. In November 2015, two weeks before New Shepard flew to space and landed for the first time, Bezos turned to Meyerson and told him to hire himself a boss.

Bezos wanted a CEO that the entire company could learn from; someone who felt passionately about space and who had a technical background (but didn't necessarily have to be an engineer); someone who knew how to shepherd the company during a period of growth and put big projects into production; someone, he said, "who doesn't let reputation overcome instinct, who doesn't let their head overrule their heart."

The news came as a gut punch to Meyerson. The New Shepard launches were Meyerson's triumphs as much as anyone's at the company, and the culmination of fifteen years of work. Bezos tried to soften the blow, assuring Meyerson he did not want him to leave. "I want you to be a part of this," he said. What he was looking for, he said, was a "leader of leaders." Meyerson wanted the right to be able to at least press his case to Bezos, to show that he could be that leader. He asked if they could discuss it at their next one-on-one meeting. Bezos said they could. But in the weeks that followed, those meetings kept getting canceled, and Bezos kept emailing Meyerson to ask about his progress on the search and when he was going to inform his team that soon they would have a new chief.

MEYERSON'S PATH TO Blue Origin had begun with a tantalizing phone call. He'd worked for ten years on the Space Shuttle and other programs at NASA's Johnson Space Center in Houston, and then at a start-up known as Kistler Aerospace. In early 2003, a friend from NASA called and mentioned that he'd been consulting for a secretive space company. "I can't tell you the name because I signed a nondisclosure agreement, but you'll love what we're doing," the friend said. They hung up, and Meyerson called back a few minutes later to say the start-up sounded intriguing. If Meyerson would send him his résumé, the friend said, he'd get it into the right hands. He did.

A few days later, Meyerson received another call, this time from a man named Jim French, a longtime space engineer, who let him in on the secret. The company was called Blue Origin; it was being funded by Jeff Bezos, and they wanted him to come in for an interview to lead the team. On April 17, 2003, Meyerson went to Amazon's offices in the Beacon Hill neighborhood of Seattle. The conference room was unlike any Meyerson had ever seen. The tables were actually doors taken off their hinges and sitting on sawhorses, and behind the clutter of boxes and paper there was a cutout of Yoda standing in the corner and a framed quote from Dr. Seuss's *Oh, the Places You'll Go!* on the wall.

Meyerson found it "immediately disarming," he later told me.

Bezos made a strong first impression—a curious man with an intense, bug-eyed stare and an odd cackle of a laugh. His passion about space was evident and infectious, and he had big ideas for Blue Origin. Bezos appreciated Meyerson's deep background working with space systems, and he also had a calm, thoughtful manner that Bezos liked almost immediately.

They spoke for about forty minutes. Bezos asked Meyerson about "judgment"—how Meyerson defined the word, and examples of judgment calls he had made. He asked about hiring and where the best aerospace talent came from. Bezos and others who interviewed Meyerson also asked about firing. Had he ever gotten rid of anyone? Apparently, Bezos had acquired a few hangers-on he was looking to shed as he entered this next, more serious phase for his space company. Whoever got the job would not only have to share Bezos's passion for space and his optimism for building a better future, but also his ability to cut a few throats. An empire like Amazon was not built without a ruthless adherence to what Bezos liked to call "operational excellence."

Still, Meyerson was sold. Bezos's passion about space was evident and infectious, and he had big ideas for Blue Origin. "It was very clear that what Jeff wanted was to have someone come in and build the company," Meyerson later recalled. Under his tenure, Blue would never get the staffing or resources that SpaceX received. Amazon was still Bezos's main focus while Blue remained something of a hobby. Meyerson was the ideal person to lead this version of the company. He was an engineer's engineer, who remembered the names of his employees' spouses and children, who parked where they parked.

But he had never run a company before. And Bezos was eager to move to the next phase and made it clear—again—that he wanted someone to come in and build the company. He said he hoped Meyerson would stay. But any new CEO would build their own team and likely possess a measure of the corporate mercilessness that had been a hallmark at Amazon but had not yet taken hold at Blue. The new CEO would have to not only be able to answer the question—*Have you ever gotten rid of anyone?*—that Bezos had asked Meyerson when they first met all those years ago, but be willing to act on it with the unemotional detachment Bezos was now displaying toward the person he had first hired to lead his space venture.

AS THE SEARCH for the CEO continued, Blue Origin sent an email to the thousands of people signed up to receive updates from the company. These notes arrived from time to time, usually under Bezos's name, offering glimpses into his interest in engineering and rocketry as much as what was actually going on at the otherwise secretive venture. Often the updates detailed technical progress on various projects—reading more like an aerospace engineering textbook than a press release—such as the benefits of a "regeneratively cooled" engine chamber and the performance of its "preburner."

But on this day, September 12, 2016, the email had big news. Under the title "A Next Step. . . ." Bezos wrote: "Our mascot is the tortoise. We paint one on our vehicles after each successful flight. Our motto is Gradatim Ferociter—step by step, ferociously. We believe slow is smooth and smooth is fast. In the long run, deliberate and methodical wins the day, and you do things quickest by never skipping steps. This step-by-step approach is a powerful enabler of boldness and a critical ingredient in achieving the audacious. We're excited to give you a preview of our next step. One we've been working on for four years. Meet New Glenn."

What followed was an artist's rendering of the Very Big Brother rocket Bezos had been quietly developing. Named for John Glenn, the first American to reach orbit, it was pictured not on a launch pad or lifting off under a pillow of fire. Rather it was standing alongside a series of other rockets, sorted by size. On the left stood the pipsqueak of the bunch, the Antares rocket, then operated by a company called Orbital ATK. Then came the Russian Soyuz and the Atlas V, operated by the United Launch Alliance, the joint venture of Lockheed Martin and Boeing. Then came SpaceX's rockets, the workhorse Falcon 9 and the even more powerful Falcon Heavy.

At 270 feet, Bezos's New Glenn was nearly as tall as the Statue of Liberty, bigger than all the other rockets in the lineup except for the biggest and most powerful ever to fly, the NASA Saturn V rocket that launched the Apollo astronauts to the Moon. New Shepard, named for Alan Shepard, who became the first American in space when he flew on a suborbital trip that lasted just fifteen minutes in 1961, peaked at about Mach 3, or three times the speed of sound. New Glenn would be able to push to orbital velocity, reaching Mach 22. It was, Bezos said in his email, a natural evolution for the company. "Building, flying, landing and re-flying New Shepard has taught us so much about how to design for practical, operable reusability," Bezos wrote. "And New Glenn incorporates all of those learnings."

THE ADVENT OF new rocket technologies was a needed bright spot in an industry that had seen little progress in the previous couple of decades. And Bezos's announcement was coming as a new international space race was heating up, a contest that would require new and bigger rockets.

The United States had begun to cede its dominance in space, many felt, and NASA was starting to look like a tired and bloated bureaucracy that did not at all resemble the agency that had put men on the moon within a decade of receiving the charge. The Space Shuttle had retired in 2011 without a replacement, forcing NASA into the costly and embarrassing position of relying on Russia to fly its astronauts to the International Space Station. NASA's follow-on program, called Constellation, was canceled because of cost overruns and schedule delays. And the rocket NASA was directed to build instead, the Space Launch System, had yet to fly despite billions of dollars of taxpayer investment; it appeared to be more of a way to keep engineers employed in key congressional districts than to launch astronauts. Instead of being reusable, it would be thrown away after each flight and, worse, relied on the engines that powered the Space Shuttle. In other words, NASA's new rocket relied on technology developed in the 1970s. Adding to the ignominy was the fact that the Atlas V rocket, operated by Lockheed and Boeing's ULA, and used by the Pentagon to launch sensitive national security satellites for missile defense and reconnaissance, was powered by a Russian-made engine, the RD-180. In 2014, Dmitry Rogozin, the bombastic Russian deputy prime minister and a target of U.S. sanctions after Russia's recent annexation of Ukraine's Crimea region, threatened to cut off supply of the RD-180. "I suggest to the USA," he said, "to bring their astronauts to the International Space Station using a trampoline."

As a result, China was catching up. Russia was raising the price to fly American astronauts. Both were threatening U.S. satellites in space, alarming Pentagon officials. NASA, meanwhile, had started to feel irrelevant. If the space agency couldn't fly astronauts, why did it even exist?

At the same time, U.S. officials had been warning of increased Russian military activity in space, notably a series of mysterious satellites that had been acting strangely. Instead of flying in a fixed circular orbit, as virtually all satellites do, the Russian spacecraft started moving, flying close to the spent rocket stage that had deployed them. The practice, known as "rendezvous proximity operations," was seen as a potentially aggressive maneuver, allowing spacecraft to sidle up next to their prey, like a pirate ship getting uncomfortably close to a merchant vessel.

In 2015, a Russian military satellite did just that, parking itself for several months between two communications satellites operated in geosynchronous orbit by Intelsat, an American company. While there was no disruption to the Intelsat satellites, the incident concerned officials at the Pentagon, which operated some its most sensitive satellites in the same orbit. Soon there were a series of high-level classified briefings and an extraordinary response: the Defense Department had been secretly developing surveillance satellites of its own in a classified program that would serve as a "neighborhood watch program for everything that goes on in that high-value orbit," as Air Force general John Hyten put it. The decision to make it public, he said, was intended to put China and Russia on notice, and to "send a message to the world that says: Anything you do in the geosynchronous orbit we will know about. Anything."

. . .

IN ANNOUNCING NEW Glenn, Bezos made no mention of whether the company would offer it to the Pentagon, or what role it would play as the government looked to stay ahead of Russia and China, which was also threatening U.S. satellites. But if it were as big and powerful as Bezos claimed, it could very well become a direct competitor with SpaceX and further showcase the power of American industry.

New Glenn would be powered by seven mighty engines, known as the BE-4, that had been under development for years. It was the most difficult project the company had ever attempted. To reach orbit, a spacecraft or satellite must be traveling almost incomprehensibly fast, some 17,500 miles per hour, or about five miles every second. Word of the development of a new American engine was exactly what ULA needed to hear at a time when it was desperately seeking an alternative to its Russian-made RD-180.

After Russia's annexation of Crimea, the RD-180 went from an obscure fixture of admiration within the rocket industry to a symbol of the growing tensions between the United States and Russia. The fact that Russian engines were being used to loft satellites the Pentagon used to spy on other nations, including Russia, was becoming untenable. In particular, the arrangement attracted the ire of the irascible chairman of the Senate Armed Services Committee, John McCain, who often used his perch to criticize Russian president Vladimir Putin and his "corrupt cronies." The only target bigger than Russia, in McCain's eyes, was the military-industrial complex that made billions off the government, and the Pentagon brass who did business with them. "It seems to me that we should be encouraging a capability to manufacture rocket motors here in the United States of America, rather than being dependent on Vladimir Putin," McCain fumed from the dais during a hearing in 2014.

Musk was more than happy to point out that the rockets SpaceX built used engines made in the United States. SpaceX had filed a lawsuit against the Air Force for the right to compete against ULA for the lucrative contracts—and it won, opening up a new line of business, potentially worth billions of dollars. As teams from SpaceX and the senator's staff worked to end ULA's access to the Russian-made RD-180 engine, Musk and McCain became a pair of unlikely allies.

Eventually they prevailed, and Congress cut off ULA's supply of RD-180s, after allowing it to retain enough to complete its missions for the Pentagon. Having lost its monopoly and soon its engines, the company was suddenly fighting to stay alive.

IN 2014, ULA announced it would buy Blue Origin's BE-4 engine. Privately, the contract called for ULA to invest $100 million into Blue Origin's development of the BE-4. But ULA was also retaining Aerojet Rocketdyne, one of the country's premier rocket engine manufacturers, as a potential backup in case Blue Origin couldn't deliver. In April 2016, Blue Origin threw its annual party at the Space Symposium, the conference held each year at the Broadmoor Hotel and Spa in Colorado Springs. As usual, Blue Origin had rented one of the resort's cottages, and the popular event was especially well attended because Bezos himself was there. On the terrace overlooking the golf course, space royalty were drinking cocktails. There was Al Worden, the astronaut who had flown to the moon during Apollo 15. General Hyten, the commander of the Air Force Space Command, attended, as did Deborah Lee James, the Air Force Secretary. And, of course, there was Tory Bruno, the chief executive of ULA, and his team.

Bezos welcomed everyone with a short speech. Toward the end, he acknowledged his friends from ULA but couldn't help pointing out they had still not made the engine deal official and were keeping Aerojet Rocketdyne on as a possible supplier as well. Staring straight at Bruno, he said, "We. Will. Not. Fail. You." His deliberate tone and lingering glare made it clear this was more a prophecy than a plea; to some in the room, it carried the weight of admonition.

If ULA executives left the party feeling chastised by the world's second-richest man, they were heartened to hear that he would not compete against them for Pentagon launch contracts, ULA's main source of revenue. During a fireside chat at the symposium that week, Alan Boyle, the respected space and science journalist, asked Bezos if the company would be entering the national security launch market.

"Our contribution to national security is going to be making a great BE-4 engine, and then the United Launch Alliance is going to use that engine in their vehicle for national security payloads," Bezos responded. "I'm very excited about that." This seemed pretty definitive, especially to the ULA executives sitting in the audience: Blue would not pursue national security contracts.

But Bezos went on to add: "The C.I.A. is a big user of Amazon Web Services. It has been very gratifying to be part of that national security mission. Not that I really need any additional passion or motivation for space, but I have to tell you that all of us at Blue Origin find the fact that we are going to get to help with the national security missions incredibly motivating." This was a hint of Blue's future ambitions. Later, Bezos would say he never explicitly ruled out one day competing for Pentagon launch contracts—a claim that was tenuous at best.

When ULA executives learned at the next Space Symposium, in 2017, that Bezos had reversed course and ordered Blue Origin to compete for those contracts, they were angry and disappointed, but they were not surprised. The move fit Blue Origin's new, more aggressive ethos—as it pursued SpaceX and Musk. If it was going to do that, it could not afford to ignore the Pentagon and its nearly $600 billion budget. For all the fanciful talk about opening space to the masses, Bezos was, of course, a ruthless businessman who had built his Amazon empire by battling Barnes and Noble, Wal-Mart, eBay, and other giants— and won. "We were all waiting for the Amazon Jeff to show up," one former Blue Origin employee told me. Now he had.

The Pentagon had the possibility to be a far better customer than ULA would ever be. From a business standpoint, going after the lucrative national security launch contracts was a no-brainer, even if it enraged some at ULA. Blue again held its annual party at the cottage overlooking the Broadmoor golf course in 2017. Once again, it attracted the luminaries of the space industry. This time the ULA team did not attend.

. . .

IF THE "AMAZON JEFF" had been absent for much of Blue Origin's existence, he showed up at a company meeting on Saturday, June 24, 2017. The weekend gatherings had become a tradition in which executives and program directors provided updates on different aspects of the company. Outsiders would sometimes visit as well, to talk about topics such as additive manufacturing, software development, or advancements in composite materials. About once a year, Bezos, standing in a common room by a fireplace designed to look like flames shooting from a Jules Verne–style rocket, would take questions in an "ask me anything" format. On this Saturday, the questions on employees' minds centered around the new direction the company seemed to be heading in, the company's culture, and its reputation of plodding along slowly.

"Do we need a sense of urgency to go with 'step by step'?'" one worker asked. Bezos's answer made it clear he was unhappy with Blue's pace of progress and that he intended to do something about it.

"The motto was not just 'step by step,'" he said. He had added the *Ferociter* on purpose, but it didn't appear to be resonating. "In my opinion, Blue is not yet good at making high-quality decisions with high velocity," he said. "I think we make high-quality decisions, but slowly. I do believe that what we're trying to do is so ambitious that it cannot be achieved if we make high-quality decisions slowly. We need to get better at that. I'm on a jihad to get better at that. When you're good at that, it's way more fun."

Blue, he said, was not bad at "decision-making speed by aerospace standards. I just think that aerospace standards suck, and I want us to be way better than that."

THE CEO SEARCH moved slowly. Bezos wanted to maintain a high bar. This was a big decision, and big decisions, as Bezos liked to say, were one-way doors—once you made them, they were difficult to reverse. Small decisions, with less consequential outcomes, were two-way doors; make a mistake and you could go back through and try again, no problem.

Blue Origin had hired an executive search firm, Heidrick & Struggles, and formed a committee that included Bezos, Meyerson, and Jeff Ashby, a former NASA astronaut who had flown three Space Shuttle missions and had been with Blue since 2010. The fourth member, Susan Harker, wasn't an employee of Blue Origin but rather Amazon's vice president of global talent acquisition.

Early in the process, Harker took the initiative to reach out to one of the biggest stars in the space world, Gwynne Shotwell, the president and chief operating officer of SpaceX. If Musk was the visionary behind the company, the one plotting its course into the future, Shotwell was its soul, the one who made SpaceX work on a day-to-day basis. She met with customers, maintaining relations with leaders from NASA and the Pentagon, a sometimes thankless task that included mopping up when Musk made a mess with an outburst or an ill-advised tweet in the middle of the night.

A smart, highly educated engineer who could hold court in public, Shotwell represented the company at conferences while inspiring the employees at SpaceX with straight talk, such as the implementation of a strict "no assholes policy." She was employee number seven at SpaceX and held a special relationship with Musk, who famously ripped through even the most loyal lieutenants if he felt they were not meeting his expectations. While others tiptoed around the mercurial CEO, or tried to avoid him completely, she dealt with him firmly, speaking to Musk in a way only a few others could. Most of all, she understood him, could read his moods and find ways to transform his impossible directives into achievable tasks.

"First of all, when Elon says something, you have to pause and not immediately blurt out, 'Well, that's impossible' or 'There's no way we're going to do that. I don't know how.' So you zip it. And you think about it. And you find ways to get that done," she once said.

Every time she helped SpaceX implement some impossible goal, and the company was recovering from the effort, "Elon would throw something out there, and all of a sudden, we're not comfortable and we're climbing that steep slope again. But then once I realized that that's his job, and my job is to get the company close to comfortable so he can push again, and put us back on that slope, then I started liking my job a lot more, instead of always being frustrated."

It was a special relationship—their cubicles were even situated next to each other. There was no way she was going to leave SpaceX. She turned Harker down in an instant.

Blue Origin's search then largely focused on executives from major aerospace companies: Lockheed Martin, Boeing, Northrop Grumman. In the end, Bezos chose Bob Smith, an executive at Honeywell, the aerospace and defense giant. Smith shared Bezos's passion for space and earned his trust not just as a true believer in the mission but also because he had business acumen. For someone who routinely asked job candidates at Amazon their SAT scores, Bezos was impressed by Smith's academic credentials: an undergraduate degree in aerospace, aeronautical, and astronautical engineering from Texas A&M, a master's from Brown in applied math and engineering, a Ph.D. in aerospace engineering from the University of Texas (his thesis was titled "The Onset of Chaotic Motion in the Restricted Problems of Three Bodies"), and a master's in business from MIT.

Like Bezos, he had a profound intellect. And, like Bezos, he had spent the bulk of his career in the corporate world—At the United Space Alliance, which served as a prime contractor on the Space Shuttle fleet for NASA, he worked on everything from the spacecraft's wheels, brakes, and cockpit avionics, making it safer and better performing. At Honeywell, he rose from vice president of advanced technology to chief technology officer, to the head of the company's mechanical systems and components. He had experience with NASA as well, overseeing all the space agency program at the Aerospace Corporation, a federally funded research center.

When he started at Blue Origin in September 2017, almost two years after Bezos told Meyerson he was going to get a boss, the company made no public notice that it finally had a CEO. There was no press release.

Bezos had given Smith the task of turning Blue Origin into a real company, establishing formal processes and procedures, winning government contracts, and attracting customers. Bezos had handed Smith an extensive mandate that included flying people on New Shepard, getting the BE-4 engine ready, finishing development of the orbital New Glenn rocket, competing for Pentagon launch contracts, and developing the Blue Moon lunar lander.

Soon, Smith, empowered by Bezos, would begin a reorganization designed to accomplish these complex tasks, putting his imprint on every program, expanding the company's reach and ambitions while forcing it to grow up and embrace the more formal operations of a venture seeking to become profitable. Under Meyerson, Bezos had communicated freely with a host of Blue Origin employees, who enjoyed his engagement and his ability to problem-solve, as well as access to the boss and the status it conferred. That would soon end. All communications would run through Smith—at least that's how many of his employees felt. It was one small part of a tumultuous culture change in the years to come, as the company grew from a few hundred to several thousand, split into four divisions, each with its own executive in charge. Meyerson would be put in charge of a division called Advanced Development Programs, that would oversee projects such as the Blue Moon lander. The arrangement would not last, however, and Meyerson, one of Blue Origin's longest-serving employees, would leave the company within about a year.

Smith's reorganization would start with something much simpler, however. The new CEO wanted a reserved parking spot. Meyerson had been content to park wherever, but Smith wanted a VIP spot, close to the building, one befitting his stature as CEO. He started parking his Audi SUV in one of the few choice spots designated for visitors. Employees at Blue noticed. This was not the culture that had guided Blue for so long and was a sign, they thought, that the new boss meant business. The facilities manager noticed as well. Not wanting to get crosswise with the new CEO, he finally just painted the word RESERVED on the spot in bright white paint, signaling the start of the Smith era.

CHAPTER 5

The Dark Side of Space

Vice President Mike Pence was giddy. It was October 5, 2017, and he was in a prep room at the National Air and Space Museum's Udvar-Hazy Center in Chantilly, Virginia, where he would soon officially announce the Trump administration's space policy goal: a return to the moon.

The event was to be the first meeting in twenty-five years of the newly resurrected National Space Council. Pence asked that the room be cleared except for the principals—much of the president's cabinet, including the Secretaries of State, Commerce, Energy, Transportation, and Homeland Security, as well as the Director of National Intelligence and the Deputy Secretary of Defense. Major members of the private space industry were also in attendance, from the CEOs of stalwarts like Boeing and Lockheed Martin, to those of the entrepreneurial SpaceX and Blue Origin, all of whom sought to align their companies with the goals of the new White House and its enthusiastic space cheerleader, the vice president.

"He pulled us together almost like a football huddle," one person recalled. "He was just so excited." Pence told them that even though such whole-of-government events are highly scripted, he might not be able to contain himself and warned that people might need to think on their feet. "I might ask you questions," he said, "so make sure you pay attention."

The event was designed to be the ultimate space flex, a way to demonstrate to allies and adversaries alike—from the United Kingdom, France, Canada, and Japan to China—that America was going to take space seriously again, both as a boundless expanse meant to be explored and as an emerging theater of conflict. Like most presidents, Trump understood little about space except for its political value. In his case, he'd make it part of his MAGA agenda, vowing to "make space great again."

The White House advance team measured the stage so that Pence would deliver his remarks perfectly positioned in front of the Space Shuttle Discovery—its nose pointed as if to the future. Discovery, which first flew in 1984 and completed thirty missions, was the crown jewel of the museum's collection. A reliable workhorse of a spaceship, a symbol of resolve and perseverance, it had been tasked by NASA to fly the first missions after Space Shuttle Challenger exploded in 1986 and Columbia came apart in 2003.

Space Council staffers liked the idea of enshrining their first policy directive in front of NASA's last great human space vehicle. There was also a certain symbolism to holding the meeting among the relics of America's spacefaring past—which also included one of the rockets flown by Robert H. Goddard, the father of rocketry; and the Gemini capsule flown by NASA astronauts Jim Lovell and Frank Borman—as if they would whisper their secrets and guide the way forward from behind their glass-encased exhibits. John Dailey, a retired Marine Corps general and the museum's director, introduced Pence and kicked off the event by addressing the handpicked audience of space industry and government officials. "By choosing this location to begin their work, the council is sending a strong message about its intentions for our future in space," he said. "Rather than an executive boardroom or an agency conference room, this first chapter in the new space age will be written here, among the icons of our nation's highest, proudest achievements."

He quoted Eugene Cernan, the last man to walk on the moon, as saying that the Apollo program was "a decade out of time—a twenty-first-century event born into the middle of the twentieth century." Fifty years on, he said, "the time is at hand for the heirs of Apollo to take up the mantle."

Taking the podium, Pence at first seemed to channel John F. Kennedy's "because it is hard" speech that propelled the country toward the moon and gave birth to the Apollo program. "We will turn our attention back toward our celestial neighbor," Pence said. "We will return American astronauts to the moon, not only to leave behind footprints and flags, but to build the foundation we need to send Americans to Mars and beyond. The moon will be a stepping stone, a training ground, a venue to strengthen our commercial and international partnerships as we refocus America's space program toward human space exploration."

But soon the tone became less lofty and more prosecutorial, an indictment of decades of failed leadership by Republicans and Democrats alike. Since the glory days of Apollo, America's leaders were lost, Pence said, and "the question became 'What should we do? Where should we go next?'

"In the debate that followed," he continued, "sending Americans to the moon was treated as a triumph to be remembered, but not repeated. Every passing year that the moon remained squarely in the rearview mirror further eroded our ability to return to the lunar domain and made it more likely that we would forget why we ever wanted to go in the first place. And now we find ourselves in a position where the United States has not sent an American astronaut beyond low Earth orbit in forty-five years. Across the board, our space program has suffered from apathy and neglect."

As a result, he said, other nations had caught up to the United States in space in an unprecedented erosion of power and prestige that threatened the country in ways few people realized. "Our adversaries are aggressively developing jamming, hacking, and other technologies intended to cripple military surveillance, navigation, and communication systems," he said. "In the face of these actions, America must be as dominant in space as we are here on Earth."

. . .

IN ADDITION TO establishing NASA's next exploration goal, the meeting showcased the national security space enterprise, which was starting to become more public with its concerns about emerging threats. The history of space exploration had largely been told through civil programs, from Mercury, Gemini, and Apollo to the Space Shuttle era and now this new age involving commercial start-ups. But space has had an equally important hidden history: the militarization of space, including satellites for spying and guiding bombs, as well as new weapons to counter those satellites.

In at times blunt language, speaker after speaker at the Space Council laid out in detail the stakes of what the country was facing. More than fifty years earlier, the U.S. population had been paranoid over the Soviet Union's launch of Sputnik, a small satellite that did little more than emit radio signals. Here were far more dire scenarios.

Mike Griffin, a former NASA administrator and a top Pentagon official, warned Pence about a hypothetical attack on the thirty-one Global Positioning Satellites the United States operates in orbit 12,000 miles above Earth. They were used, of course, by everyone with a smartphone—that little blue dot denoting your location and allowing you to find your way using Google Maps or Waze comes courtesy of the U.S. military, free of charge. GPS was also used by the military to steer bombs and munitions in the kind of precise modern warfare we've become used to, where an autonomous drone could hit a pickup truck full of combatants and not the nearby school.

But the positioning part of GPS was just one aspect. In addition to longitude, latitude, and altitude, GPS provided a critical, but little understood, fourth dimension: time. Using carefully calibrated atomic clocks synced with the U.S. Naval Observatory, GPS was an invisible, space-based timing system that provided time so precise it became the standard used in an array of fields. The electrical grid relied on GPS timing, using it to distribute power to homes at intervals during peak usage so that the system doesn't get overwhelmed. Cell towers used it to send data to phones at the right time so that it doesn't pile up. Banks used the

GPS network to record exact transaction times, stock markets used it to record trades, air traffic controllers used it to guide jets. In short, GPS became a utility, like electricity or water, vital in modern society.

And yet, as Griffin noted, "GPS is vulnerable in a number of ways." Interrupting GPS was shockingly easy and widespread. A truck driver in 2013 was able to block the GPS signal used by his employer to monitor his whereabouts. But when he drove by busy Newark Airport in New Jersey on I-95, his jammer also interfered with the airport's GPS system. Criminals also regularly use the jammers, which can be purchased for as little as fifty dollars, to hijack trucks hauling goods. Nation-states could do far worse, taking out the entire GPS system and sowing chaos—disrupting the banking industry, shutting down stock markets, blacking out the electrical grid.

"I don't think the general public realizes the extent to which the Global Positioning System's timing signal is critical for these ATM transactions and every other point-of-sale transaction conducted in the United States and throughout most of the world," Griffin said. "I have to ask the question: To what extent do we believe that we have defended ourselves if an adversary can bring our economic system near collapse?" The United States needed to "be extremely clear," he said, that an attack on the GPS system and other critical space satellites "is an attack on the United States."

The military had put GPS and many other critical satellites in orbit at a time when space was regarded as a peaceful sanctuary. Now that space was a warfighting domain, the Pentagon realized it had made an epic strategic blunder, like leaving gold on the front steps of Fort Knox. And everyone—the Russians, Chinese, North Koreans, terrorists, hackers, and criminals—knew it.

Other speakers elaborated on the point. "The ability to threaten our national security space assets is not solely resident in nation-states; it is also increasingly available to non-state actors and individuals," retired admiral James Ellis, the former commander of U.S. Strategic Command, told the vice president. "As with other of our national security challenges, a few dragons have been replaced by a hundred snakes." Not that the dragon-level threats were entirely gone. "They include the electromagnetic pulse impact of the detonation of a nuclear weapon in space, but also the potential for cyber or terrorist attacks on the satellite ground control systems."

Ellis's warning was prescient. In 2024, reports that Russia was developing a nuclear weapon designed to destroy satellites in orbit would alarm White House and Pentagon officials. Congressman Michael Turner, the chairman of the House Intelligence Committee, compared the threat to the Cuban Missile Crisis, saying it could be a "day zero" catastrophe that could bring "the end of the Space Age." He urged the Biden administration to declassify information relating to the weapon so that "our allies can openly discuss the actions necessary to respond to this threat." A nuclear bomb in space would be devastating. The Pentagon knew this because it had actually detonated one there. In 1962, the United States exploded a 1.4-megaton nuclear weapon in orbit in a test called Starfish Prime. At the time, there were only twenty-four satellites circling the globe, compared to thousands today, but the blast knocked out one-third of them, left a belt of radiation that lingered in orbit for months, disabled streetlights in Hawaii, and damaged the electrical grid.

This—nuclear weapons in orbit, the vulnerability of GPS, the potential for economic collapse—was not the version of space that inspired hope and giant leaps for mankind. But the public needed to be aware of the risks of catastrophe, as Dan Coates, the former director of National Intelligence, said to the audience at the Air and Space Museum.

"As wonderful as it is to hear about the vision in the future for space and what it can provide commercially, what it can provide for human exploration, we end up here sobered up by the fact that like other innovations that have occurred, the Internet and so forth, and all the blessings that it brings and all the optimism it brings about the future, it's a double-edged sword," Coates said. "There's a dark side. There's a dark side to this and that forces us to be prepared. And, so I think one of the important things this council can do is to ensure that we achieve the dominance in space necessary for us to protect our people."

. . .

PUBLICLY, THAT SPACE dominance seemed to be provided in part by NASA, which beat the Soviets to the moon and led the world in exploration, from operating the International Space Station to rovers on Mars and probes throughout the solar system. But NASA had long had a "secret sibling," as the military historian Aaron Bateman put it, an agency that had for years been scrubbed from the official record but that was perhaps one of the most important entities in the defense and intelligence establishment: the National Reconnaissance Office. Born at the height of the Cold War, the NRO was so stealthy that its very existence was classified from 1960, when it was founded, to 1992, when it came out of the "black." Unlike the Central Intelligence Agency, the NRO relied less on human spies than on a fleet of classified satellites that used cameras to peer behind enemy lines, radar to see through clouds, and sensors that could even eavesdrop on conversations.

Created by President John F. Kennedy's defense secretary, Robert McNamara, the NRO was designed to ensure the United States would never be surprised by another Pearl Harbor–type sneak attack. By the early 1980s, as it served to spy on the Soviet Union's nuclear arsenal, its budget was twice that of the CIA, according to a *New York Times* report from the time. The *Times* also reported in 1996 that the agency had lost track of $2 billion because its secretive nature meant "the money had been hidden in several rainy-day accounts that secretly solidified into a 'slush fund.' The reconnaissance agency is really a set of secret offices—so secret that they have been shielded from each other, like safes locked within safes. Each office, and each program, had separate management and accounting systems, all 'black.'" In 1987, U.S. Rep. George Brown Jr., a Democrat from California, was forced to resign from the House Intelligence Committee after referring to the NRO by name in a speech.

The NRO's classified satellites have long been some of the most capable in the world. In the weeks leading up to the 2011 raid on Osama bin Laden's compound in Pakistan, for example, the NRO's satellites were one of the primary ways the United States military kept tabs on what was happening there. The satellites performed more than 387 "collects" of high-resolution and infrared images of the compound, according to *The Washington Post,* a critical trove of valuable intelligence.

BY THE TIME of the Space Council meeting in October 2017, the military had made a radical shift and had begun to speak more openly about the threats in space and the need to combat them. General Hyten was particularly vocal, giving interviews and speeches in which he let it be known, "I'm not NASA." Without space, the United States would be forced to revert to "industrial age warfare," Hyten said at one point. "It's Vietnam, Korea, and World War II," he said. No more precision missiles and smart bombs. "Which means casualties are higher. Collateral damage is higher," he said. "We don't want to fight that way because that's not the American way of war today."

The Pentagon had designated the Air Force secretary a "principal space adviser," with authority to coordinate actions in space across the Defense Department. Agencies had begun participating in war-game scenarios involving space combat at the recently activated Joint Interagency Combined Space Operations Center. Most of all, there had been a culture change, Hyten said. Where Pentagon officials who focused on space once operated in what was a peaceful environment, they have had to think of themselves—and space—differently.

"They are warriors," Hyten said. "And they need to recognize that they are warfighters."

Still, some in Congress felt like the U.S. military needed to do more. Two influential members, Mike Rogers, a Republican from Alabama, and Jim Cooper, a Democrat from Tennessee, were alarmed at what they were hearing about how China and Russia were behaving, and in 2017 called for the creation of a special branch of the military dedicated to space. They called it the Space Corps, and their proposal placed it under the Air Force the way the Marine Corps was part of the Navy. Their measure was not successful. But it did shine a light on an issue that would eventually get the attention of a White House eager to make its mark in space.

"It is disturbing, the rate at which China and Russia are pursuing these capabilities," Rogers said at the time. "We have lost a dramatic lead in space that we should have never let get away from us. So that's what gave us the sense of urgency to get after this."

Russia and China "want to take our eyes and ears out," he said. "That's what's up there and that's why they are spending an inordinate amount on space-based capabilities."

To get up to speed on China's space efforts, Pence asked for information about China's space program from the intelligence community and from Scott Pace, the newly appointed executive secretary of the National Space Council. This was not a request for the usual two-page memo with the highlights bulleted. Instead, Pence asked Pace to assemble some lengthy and detailed reading material so that he could fully understand the issue.

Pace got a three-ring binder and filled it up with an exhaustive review of China's space program, detailing its history in space, its current projects, and its future plans. China was aiming to become the first nation to land a robotic spacecraft on the far side of the moon. Soon it would send a rover to Mars, and it was working to develop a space station in low Earth orbit to rival the International Space Station. The vice president didn't read it overnight. It took a few weeks, but he read it all.

PACE FELT STRONGLY that in the age of globalization, NASA was no longer just an exploration agency; it was a diplomatic one. If during the Cold War space race against the Soviet Union, NASA was an arm of the Pentagon, now it needed to be an arm of the State Department, a tool of soft power. The quest to return to the moon was not a military endeavor by any means. There would be no soldiers, no weapons. But it was very much a national security program, a way to shape the space environment in the national interest, to build the technologies that would foster American growth and set the norms of behavior in space that would last for generations.

During the Cold War, the race to the moon was "about showing leadership by doing things no other country could do," Pace said in

November 2017. "But now the measure of leadership is how many people want to work with you, how many people want to be part of your team. If we want to be a global leader in space exploration today, we need to have projects that are both challenging and realistic—but which also allow for meaningful international partnership and private sector participation."

Under President Obama, NASA developed a plan to send a crew of astronauts to an asteroid, grapple it, drag it into moon's orbit, retrieve a sample, and bring it back to Earth for study. Why an asteroid? When would this happen? And why send astronauts, as opposed to a robotic mission? The White House never said. But many felt the main problem with the program was that it lacked significant international participation. Still, the White House announced that the asteroid mission would somehow be followed by a Mars mission, though again the details were scant.

Virtually no one bought what NASA called the Asteroid Redirect Mission, or ARM. Not the scientific community. Not even many within NASA's own leadership. "It's a one and done stunt," Richard Binzel, a professor of planetary sciences at the Massachusetts Institute of Technology, groused in a *Popular Science* article. The headline: "Everyone Hates NASA's Asteroid Capture Program."

NASA had plenty of asteroid samples, many found right here on Earth. There was one—or at least part of one, the 15.5-ton Willamette Meteorite—on display at the American Museum of Natural History in New York City. "We could send an astronaut there for the cost of the subway fare," as one space industry official once groused to me. "Unless you're involving Bruce Willis, it's nonsensical."

Lee Billings, the science journalist, summed it up this way in *Scientific American:* "NASA's prioritization of ARM could become a very expensive mistake. As has happened several times before, inertia and internecine conflict again seem set to send the agency's latest plans for human spaceflight tumbling into the void, boldly going nowhere."

That's precisely how the incoming Trump administration felt about NASA's plans. Its officials considered NASA not just directionless but isolated. Space was now vital to warfighting, the economy, and even everyday life. But the United States didn't control it. It couldn't put a fence around space or plant a flag in orbit. Pace and others believed that the way to protect national interests was to convince other sovereign nations to join them in a sort of international space alliance that would counter the Chinese, Russians, and any other potential adversaries. There was also a political benefit. The more countries that joined the U.S. moon effort, the harder the program would be to kill by the next presidential administration—just as the Trump administration had put the asteroid mission out of its misery.

AT PENCE'S EVENT at the Air and Space Museum, speakers from the booming commercial sector kept the mood from being entirely bleak. The vice president had invited the heads of Lockheed Martin and Boeing to the Space Council meeting, but also those representing what some called the "new space" sector, finally advancing innovation in a field that had long relied on the same technology that sent men to the moon in the 1960s and '70s. "New and exciting companies," Pence said, "are pushing the bounds of what we thought was possible; even shaking things up a little bit in aerospace markets globally, to say the least."

The leaps in technology that transformed personal computing, put smart speakers in homes, and gave rise to artificial intelligence and machine learning were also revolutionizing space. While rockets and human exploration received most of the attention, a quiet and often overlooked shift had taken place in the way satellites were manufactured and operated. The result was an explosion of data and imagery from orbit.

Just as computers shrank from room-size behemoths to the iPhone, satellites had also gotten smaller—from the size of a garbage truck, costing as much as $400 million, to no larger than a microwave or even a loaf of bread. They were starting to cost a fraction of their predecessors, as little as $1 million or less, and could be mass-produced in factories, or in some cases a garage or college classroom. At the same time, companies like SpaceX were dramatically lowering the cost of putting them in orbit.

While the vice president had painted a gloomy picture of America's recent past in space, Gwynne Shotwell, the president and chief operating officer of SpaceX, had a rosy view of its future. While the United States had once all but ceded the commercial launch market to Europe and Russia, now SpaceX had the edge. So far in 2017, it had sent up thirteen rockets, "more than any other nation," she said. The company was preparing to fly NASA astronauts to the International Space Station from American soil for the first time since the retirement of the Space Shuttle, and it was also launching sensitive satellites for the Pentagon, NRO, and other national security agencies—including the X-37B, the Pentagon's classified autonomous space drone that would spend months at a time in orbit doing who knows what. The company was also moving quickly with its next-generation rocket, the BFR.

"In short, there is a renaissance underway right now in space," Shotwell said, one that could help make history. "A permanent presence on the moon and American boots on the surface of Mars are not impossible, and they are not long-term goals."

Bob Smith, the newly appointed CEO of Blue Origin, came next in what was his first public appearance. Bezos had wanted the company to mount a campaign to compete with SpaceX and get on equal footing, and with Smith sitting on the dais next to Shotwell, it appeared they had. In reality, of course, Blue was nowhere near SpaceX. It had not yet launched New Glenn to orbit. It didn't have meaningful contracts with NASA or the government. Its BE-4 engine was still very much in development, a time- and resource-consuming process that was facing serious setbacks.

Still, Smith did his best to position Blue as a key partner for NASA and the Pentagon. "Our New Glenn launch vehicle will be more capable than existing launch vehicles flying today and can be used not only for human spaceflight and other commercial missions but also for civil and national security payloads," he said.

Smith went on to argue that the company should be central to NASA's new flagship mission. "We also believe strongly that it is time for America to return to the moon. The moon is a key step on the path to long-term exploration of the solar system, and we have proposed the

Blue Moon lunar lander concept as a low-cost cargo delivery system to enable NASA and commercial activities on the moon," he said. He claimed that the lander would be able to reach the lunar surface "within the next five years, and we are willing to invest alongside NASA to make this happen."

THE ENTIRE MEETING was something of a show—and not just because of Smith making promises the company could not keep. No policies were signed, no directives ordered, no decisions made. The only action item was a request for the council's staff to put together a list of recommendations that could be pursued at the next meeting. But it did shine a light on the national space enterprise. Space was now a priority in a way it hadn't been in the fifty years since Apollo. Yet for all the lofty rhetoric, it was unclear whether Pence's enthusiasm would translate to success and action. In the meantime, the threats from Russia and others were not going away. China was making real progress toward the moon. The Trump administration might only have four years to put America on a path to the moon, and the first year was nearly over.

The man who would oversee NASA—its return to the moon as well as the restoration of human spaceflight from U.S. soil—was not even on the stage. Jim Bridenstine, a little-known congressman from Oklahoma who had somewhat improbably become the Trump administration's pick for NASA administrator, had spent the meeting sitting in the front row. He was still waiting to be confirmed by the Senate so that he could start his new job.

CHAPTER 6

"I'll Say You're Fired in Two Minutes"

The tornado touched down at 2:56 P.M. on May 29, 2013—just before school was to let out. It stretched more than a mile wide, with winds peaking at 210 miles per hour as it tore through the city of Moore, Oklahoma, a suburb of Oklahoma City. The twister leveled two elementary schools and a hospital and wiped out entire neighborhoods. In all, twenty-four people would die, ten of them children, including Karrina Vargyas, who was four, and her seven-month-old sister Sydnee, who were ripped from their mother's arms as they took shelter in their bathtub as the storm rolled over their house. Seven more died at Plaza Towers Elementary School, one from blunt force trauma, the others suffocating under the crushing weight of debris. The Moore tornado was an EF-5, the highest rating on the scale used to measure their power. It was brutal even by the standards of Tornado Alley, where the humid air rising from the Gulf of Mexico collides with the jet stream dipping down from Canada. Later, meteorologists would say that the force of the tornado was greater than the atomic bomb dropped on Hiroshima.

As he toured the wreckage the next day, Jim Bridenstine, then a newly elected congressman from Tulsa, had a hard time processing the damage. There were cars on the roofs of buildings, a mattress impaled on a tree, children's drawings mixed among the litter, and housing foundations cleaned bare. The vortex had dissected entire blocks of houses. It vacuumed grass from fields and bark from trees, leaving them naked and eerie. Water geysered up from broken pipes. Scores of cows and horses were dead, some of them strewn far from their fields, with eyes open and bulging. The stench of death and gas leaking from severed lines was overwhelming. Looking at the rubble, Bridenstine later recalled, he had a sickening realization: "There very well could be people inside there."

An Eagle Scout, Bridenstine wanted nothing more than to start pulling aside the bricks and debris. But the professional rescue crews who had descended on the town were already doing that, and had warned Bridenstine and the congressional delegation he had flown down with to stay away, both for their own safety and that of the people potentially trapped inside. "We couldn't do anything about it," he said. "It was heartbreaking and gut-wrenching. Devastating." He was jealous when one of his fellow congressmen, Markwayne Mullin, who owned a plumbing company, climbed through a rubble pile to shut off one of the sources of gushing water. At least he could make one small contribution.

Moore had been prepared for a tornado—as much as any community could be. Its residents knew that storms would come every spring; they had sirens and built shelters stockpiled with provisions. Just to the south, in Norman, the University of Oklahoma had one of the top meteorology schools in the country, which attracted storm chasers from all over the world, eager to study the worst of Tornado Alley. Still, the people of Moore had only received fifteen minutes of warning. Surveying the devastation, Bridenstine wondered whether science and technology could be pushed to do better—and help the next town prepare for the inevitable.

THE ANSWER, HE came to believe, was in space. There were new technologies being developed for small satellites, as well as radar that could allow for better detection of tornadoes. As chairman of the House Science Committee's subcommittee on the environment, Bridenstine had jurisdiction over the National Oceanic and Atmospheric Administration, the federal agency operating a fleet of weather satellites that provided forecasters with the data they used to make their predictions. Within a month of the Moore tornado, he introduced the Weather Forecasting Improvement Act of 2013, a bill aimed at improving tornado warning times from minutes to an hour or more.

The legislation would force NOAA to work more closely with the private sector, which was developing technology much faster than the government. At the University of Oklahoma's National Severe Storms Laboratory, for example, researchers were using advanced radar to detect particles in clouds. And small satellite companies, such as Spire and PlanetiQ, were using what was called GPS Radio Occultation to beam signals between satellites that could measure the density of the atmosphere—and therefore temperature, pressure, humidity, and electron activity, giving forecasters a much better picture of brewing storms.

Bridenstine wanted to "get these new technologies off the drawing board and into the field," as he said, framing it as a moral obligation, with lives at stake. For the next several years he held hearing after hearing on the issue.

Satellite technology and space are not issues that many members of Congress care about. Unless you represent the Johnson Space Center in Houston, or the Kennedy Space Center on Florida's Cape Canaveral, space doesn't get many votes. Citizens tend to focus on taxes, health care, immigration—not what's happening dozens of miles above their heads.

Though Bridenstine had served as the head of the Tulsa Air and Space Museum, space had been "not at all" on his agenda when he was elected to Congress in 2012. He had been a triple major at Rice University, studying economics, business, and psychology, and earned his MBA from Cornell. A veteran of the wars in Iraq and Afghanistan, he flew forty-one combat missions for the Navy in the E-2 Hawkeye, with 333 landings on aircraft carriers. His call sign was "Stem," as in *brain stem*, a way for his fellow pilots to poke fun at his bookish tendency to go deep on particularly challenging technical subjects. Bridenstine remained in the Navy Reserve; he never drank and he kept his hair high and tight, according to regulations. When he ran for Congress in 2012, he presented himself as a conservative's conservative, interested in lowering taxes, easing regulations on business, and promoting traditional values such as faith and family.

As he continued to push his forecasting bill, he began to learn more about other space issues—the problem of orbital debris, how wars would be fought in space, what China and Russia were doing to thwart the United States, how small-satellite technology could create new ways to communicate and monitor the Earth. His next major piece of legislation was a sprawling bill called the American Space Renaissance Act, with provisions as broad as giving the head of NASA a ten-year term; assigning which federal agency would be responsible for collision warnings in space; and pushing the Pentagon to fund launches for small satellites. The bill was so expansive that he admitted it would not pass as a whole, but rather be broken up into chunks. But it came as the country was beginning to awaken to what was happening in space, in large part because of companies like SpaceX and Blue Origin that were making space cool again. Soon Bridenstine was one of the leading space experts in Congress. Then again, it wasn't a particularly high bar to clear.

And so, when the Trump administration went looking for one of its allies to lead NASA, Bridenstine, then forty-two years old, was at the top of the list. When he met with Pence at the vice president's office at the Capitol, the pair hit it off immediately. They were both devout Christians and shared ties to military aviation—Pence's son was a young Marine Corps officer making his way through flight school. For forty-five minutes, they talked about the retirement of the Space Shuttle program and how to restore human spaceflight from the United States. They talked about NASA's efforts to return to the moon and the importance of doing so before China. They talked about NASA being a tool of diplomacy and a force for good, rhapsodizing nostalgically about how the dawn of the Space Age was a projection of American power and greatness.

At the end of the meeting, Pence shook Bridenstine's hand and said he thought NASA could be the right agency for him to lead. But not everyone agreed.

BRIDENSTINE MIGHT HAVE become a leading advocate for space in Congress, but he was far from the most powerful. For years, that had been Bill Nelson, the Democratic senator from Florida whose personal history was tied to the growth of the national space program. He was a fifth-generation Floridian, whose ancestors arrived to the state in 1829. His grandparents' 160-acre homestead was a few miles from what would become the Kennedy Space Center. And in the mid-1980s, when NASA decided to open up flights on the Space Shuttle to civilians, Nelson, then a member of the House, wrote the agency expressing his interest in taking a ride to orbit.

There was a precedent. Nelson was following in the footsteps of Senator Jake Garn, the head of the appropriations committee that funded NASA, who had sought a seat on the shuttle by claiming he needed to "kick the tires" of the program. The astronaut corps scoffed at having to give up a precious seat to a politician, and so did much of the public. In his influential comic strip, *Doonesbury,* Garry Trudeau called it "the most extraordinary junket in the history of Congress," while columnists derided the flight as a "phenomenal waste of taxpayer money" and said Congress was trying to "turn the space program into their own private Disneyland."

When he faced his own blowback for securing a ride to space, Nelson disputed the idea that NASA was currying favor with him because he represented the Space Coast in Congress. "Anybody who knows me, knows I'm going to make up my own mind," he said. Nelson wanted to be seen as a serious participant contributing to the mission and the experiments the astronauts would be conducting. His flight, on Space Shuttle Columbia in January 1986, orbited Earth ninety-eight times over six days and landed ten days before the next shuttle flight: Challenger, which came apart seventy-three seconds after liftoff.

The disaster shook NASA to its core and left a huge imprint on Nelson. Now, in 2017, he resolved to throw a roadblock in front of Trump's nominee. He did not want a politician leading the space agency. In theory, NASA existed above the partisan fray, and he thought the fact that Trump's nominee came from Congress was, on its own, grounds for dismissal. Still, before Bridenstine's confirmation hearing that November, Nelson was cordial, greeting the nominee, his wife, and their three children in his office. Nelson had been in politics since before Bridenstine was born, and he had a senator's grace and charm, as well as a sugary Southern accent. But both men knew where they stood.

Leading up to the hearing, Bridenstine visited as many senators as he could, stressing over and over that despite his background he would be an apolitical leader. Given the hyperpartisan divide in Congress and the ways it was exacerbated by Trump's election, Bridenstine's team knew he would face sharp questioning. He spent days running through "murder boards"—mock hearings designed to mimic the worst of what could come.

Still, Bridenstine was completely taken aback once Nelson started in on him as soon as the hearing began. For twelve minutes, the senator eviscerated Bridenstine, focusing all his attention, and ire, on him even though the committee was considering three other people nominated to federal posts that day. Nelson started by talking about the NASA family as if he were its patriarch, and invoked the Challenger disaster. "This deep respect for NASA comes from having witnessed, very directly, the tragic consequences when NASA leadership has failed us," he said. Now that the agency was about to get back into the risky business of human spaceflight, he wanted an ultracompetent administrator in charge.

"We have three new human spaceflight vehicles right now that are at the most critical phase of their development," he said. "Now, as much as ever, NASA needs and deserves an administrator who is up to the challenge of leading the agency through this critical juncture. We are about to embark on putting Americans back into space on American rockets. . . . And so failure at this particular juncture could jeopardize the lives of brave astronauts and set back the search for life beyond Earth for decades."

Bridenstine was stunned. Keenly aware that the live cameras would capture his every expression, he tried to keep his composure. He couldn't believe what he was hearing. Was Nelson really saying that if he was confirmed, astronauts would die?

Nelson wasn't done yet. He painted Bridenstine as a Trumpian conservative—a climate change denier who was against gay rights, a partisan warrior whose behavior exemplified "why Washington is broken." From the dais, Nelson's message to Bridenstine was clear: *Your presence at the witness table is an affront—to the Senate and the space agency.* Nelson would protect his cherished institution from what he considered an unfit neophyte.

"The NASA administrator should be a consummate space professional who is technically and scientifically competent and a skilled executive," he said. "More importantly, the administrator must be a leader who has the ability to unite scientists, engineers, commercial space interests, policymakers and the public on a shared vision for future space exploration." He left open the possibility that his mind could be changed. But in the written testimony that would be submitted into the congressional record, it was clear Nelson had already arrived at his decision. "Frankly, Congressman Bridenstine, I cannot see how you meet these criteria," he had written.

For two hours, Bridenstine calmly defended his record. He did believe in climate change, he said. He did not persecute any particular group of people and pledged to leave politics aside if confirmed. But it probably didn't matter what he said, or how he said it. He wasn't going to convince anyone of anything. The senators had made up their minds about him even before the hearing, and they voted along party lines to send his nomination to the full Senate.

REPUBLICANS DIDN'T BRING Bridenstine to a vote on the Senate floor until almost six months later, in April 2018. With a two-vote majority, they couldn't afford to lose a single vote, especially with Senator John McCain out while battling brain cancer, and Senator Marco Rubio threatening to vote against. Like Nelson, Rubio said it was because he didn't think a politician should run NASA; that Bridenstine would be "devastating for the space program." The Bridenstine camp suspected that it was political retribution. During the 2016 presidential campaign, Bridenstine had supported Senator Ted Cruz in the GOP primary and was critical of Rubio, attacking his stance on immigration.

Bridenstine's supporters feared his nomination was losing steam. For more than a year, NASA had been run in an acting capacity by Robert Lightfoot, a career executive at the agency, and he was eager to turn the office over to the rightful appointee. If Congress couldn't make up its mind, then Lightfoot would force the issue by announcing his resignation. In March 2018, he sent an email to NASA employees announcing that he would retire on April 30. It would leave the agency leaderless just as the Trump administration was reviving the National Space Council and attempting to make space a priority.

The news galvanized the effort to confirm Bridenstine. Senate leaders successfully leaned on Rubio to flip his vote to yes and brought the nomination to the floor for a procedural vote on April 18, ahead of the final confirmation. It would be close, but they had the numbers. Bridenstine watched on C-SPAN from his suite at the Broadmoor in Colorado Springs during the annual Space Symposium conference. Everything was proceeding on script—until Jeff Flake of Arizona suddenly changed his vote to no. That meant it was 49 for Bridenstine and 49 against. Bridenstine and his advisers were shocked. They didn't know Flake and had had no idea that he might turn on them. Bridenstine immediately called one of his allies, Senator Mike Lee of Utah, saying, "I just got defeated and need your help." He asked Lee to go to the Senate floor to see what was going on.

Lee called back a few minutes later from the cloakroom. "This has nothing to do with you," he told Bridenstine. Instead it had to do with one of Flake's pet issues—easing travel restrictions to Cuba. "He's using you as a bargaining chip," Lee said. Pence, who was at Trump's Mar-a-Lago resort in Florida with the president, was aware of the situation and talking to Flake. If needed, the vice president would return to Washington to cast the deciding vote. They'd keep the floor vote open, Lee assured him, until Pence got there.

In the end, there was no need. Flake was placated and voted yes. That allowed the full vote to move forward the following day, when Bridenstine was confirmed by another single-vote margin: 50–49.

AFTER SUCH A painful confirmation process, Bridenstine's swearing-in was sweet. Pence did it personally, at NASA headquarters. Usually, lower-ranking appointees travel up Pennsylvania Avenue to meet with the president, vice president, or another official on their turf. Here the vice president had come to him. Pence wanted to send a message—to NASA, to Capitol Hill, to allies and potential adversaries—that the White House had Bridenstine's back. "We need the whole world to see we are all-in on space," Pence told Bridenstine.

During the ceremony, Pence made a show of it, saying on the stage of the NASA auditorium that "today marks a new and exciting chapter for this storied agency." After the ceremony, Pence escorted Bridenstine to the ninth-floor executive suite, an entourage of Secret Service agents and aides in tow. NASA's senior staff scrambled as Pence poked around, eager to meet people, and showed Bridenstine to his new office. They were in a great mood, and in no rush, taking care to shake as many hands as possible. There was no press, no cameras. Pence just wanted to get a sense of the place.

NASA lifers had seen politicians and their appointees come and go with each election cycle. In bureaucratese, they were *Wee Bees*, as in: "We be here when you come, we be here when you go." But they had never seen anything like this—the vice president personally showing the new administrator around the executive suites.

For the past year, the White House had talked incessantly about making space a priority. But that was just rhetoric. Here, in this private moment, it was clear that they meant it. Pence's enthusiasm, however, could not change the fact that Bridenstine had been handed the monumental task of not only putting NASA back on track for the moon but injecting urgency into the agency at a time when some of its biggest programs were lagging. The Space Launch System rocket and Orion spacecraft NASA would use to fly astronauts to the moon were years behind schedule and billions of dollars over budget. The lander that would ferry the crew to the lunar surface did not exist. Neither did the space suits they would wear. If any of the White House's plans were going to come to fruition, Bridenstine would have to dramatically change the trajectory of an agency confronting the bureaucratic stiffness that comes with middle age. And given the delay in his confirmation, he would have to do it quickly.

PENCE'S WELCOME WAS warm and heartfelt; Trump's was warm and threatening. In the East Room of the White House, on June 18, 2018, two months after Bridenstine was sworn in, Trump hosted the third meeting of the National Space Council. As the president acknowledged the members of his administration, he went looking for one of its newest additions. "Administrator Bridenstine—congratulations, wherever you are. Where is he?" Then, finding Bridenstine, Trump said, "Congratulations. You better do a good job. I'll say you're fired in two minutes."

The room laughed, if somewhat uncomfortably. Given the turnover in the young, chaotic adminsitration, Trump's attempt at humor also had the sharp edge of truth. Bridenstine managed a chuckle and offered his "Yes, sir" fealty. But he was about to get a sense of what it would be like to serve in an adminstration that was looking to move fast, collapse norms, and make a mark, especially in space. The Space Council meeting was supposed to be about managing the danger posed in space by the increasing number of spacecraft, satellites, and debris. But Trump had something else in mind.

Two months earlier, while speaking to troops at Marine Corps Air Station Miramar, he went off script and starting musing about space. He started by talking about how "very soon we're going to Mars," which was complete fiction since no such program existed and his own policy was to direct NASA to the moon first. The audience, a few hundred enthusiastic Marines, eagerly hoorahing their commander in chief, was into it. Soon Trump pivoted from the Red Planet to national security, saying that space is "a warfighting domain just like the land, air, and sea. We may even have a Space Force—we have the Air Force, we'll have the Space Force. We have the Army, the Navy. You know I was saying it the other day—because we're doing a tremendous amount of work in space—I said, 'Maybe we need a new force. We'll call it the Space Force.' And I was not really serious. And then I said, 'What a great idea. Maybe we'll have to do that.'"

It sounded like a lark, and perhaps it was just another stream-of-consciousness Trumpian riff. But the idea of the Space Force, or something like it, was a serious one that had been floating in national security circles for decades. In January 2001, a commission led by Donald Rumsfeld, who would serve as the secretary of defense under George W. Bush, warned in a report of a "Space Pearl Harbor." While it didn't explicitly call for a new branch of the Armed Forces dedicated to space, it did say that the United States' increasing reliance on space demands "that U.S. national security space interests be recognized as a top national security priority."

More recently, Mike Rogers and Jim Cooper, the congressmen behind the proposal in 2017 to create a "Space Corps" that would be placed under the Air Force, agreed. But their effort faced strong resistance from inside the Pentagon. The creation of a new service would "create enormous upheaval," Deborah Lee James, the Air Force Secretary, said at the time. "Sometimes the juice is not worth the squeeze." Trump's own defense secretary, James Mattis, said he did not "wish to add a separate service that would likely present a narrower and even parochial approach to space operations."

In Trump's off-the-cuff comments to the Marines, he made it seem like he'd suddenly stumbled on the idea of the Space Force. In reality, the White House had been seriously considering the idea for months beforehand. A report issued by the Office of Management and Budget found there were "fundamental deficiencies" with the way the Pentagon organized its efforts in space. "Just as other domains—land, sea, and air—have their own departments, studies have made a strong argument that space, as the fourth warfighting domain, should be similarly organized."

With the report in hand, Trump's advisers helped raise the issue. Soon Trump was on board with the idea and eager to make his mark on history by helping create the first new branch of the Armed Forces since the creation of the Air Force in 1947. The Pentagon would push back, but he was the commander in chief. They worked for him.

Trump was running late for the meeting. And so the dignitaries on the Space Council, including Executive Secretary Scott Pace, Bridenstine, Marine Corps general James Dunford, the chairman of the Joint Chiefs of Staff, and others waited near the fireplace of the White House Green Room, making small talk under a 1767 portrait of Benjamin Franklin that captures him reading a pamphlet, deep in thought. Finally, Trump came in and walked straight up to Dunford.

"General, can you create me a Space Force?" he asked.

Dunford, a former infantry officer who had served as the commander of the 5th Marine Regiment during the Iraq War, was surprised at the request but took it in stride. "Sir, I don't know if this is the right time," he parried, and explained the difficulties of standing up a whole new branch of the military. But Trump was impatient.

"I'm asking you if you can do it," he said.

"Well," Dunford replied, "if you give the order—" At that, Trump left.

The president's speechwriters made a few tweaks to his remarks, and eventually the meeting began. Soon after congratulating Bridenstine, Trump got to his surprise announcement: "When it comes to defending America, it is not enough to merely have an American presence in space. We must have American dominance in space. So important," he said. "Very importantly, I'm hereby directing the Department of Defense and Pentagon to immediately begin the process necessary to establish a Space Force as the sixth branch of the Armed Forces. That's a big statement."

Then he said, "General Dunford, if you would carry that assignment out, I would be gretaly honored." Finding Dunford in the room, he looked at him and said, "Got it?"

"We got it," Dunford replied.

"Let's go get it, General."

And there it was—a direct order from the president. It was, as Trump said, a "big statement," one that took virtually everyone in the room by surprise. Bridenstine was impressed by the action, but also shocked. Why hadn't anyone clued him into this beforehand? And when might the president spring a surpise on him?

IN THE MEANTIME, he had more immediate concerns. Not only was Bridenstine the first NASA administrator to come from the halls of Congress, he was the first who had not been alive during the Apollo 11 moon landing. His first memory of the space program was the sickening fireball that engulfed Challenger, killing all seven people on board, including Christa McAuliffe, the teacher from New Hampshire who had been chosen by NASA to be one of the first civilians to fly on the space shuttle.

"Everyone who was alive when Neil Armstrong and Buzz Aldrin walked on the moon, they know exactly where they were when that happened," Bridenstine would later say. "I wasn't alive back then. I don't have that memory. My generation does not have that memory. Our memory is watching the Challenger explode with Christa McAuliffe on board."

It was January 28, 1986. Bridenstine was ten years old in Ms. Powers's fifth-grade language arts class at Dunn Elementary School in Arlington, Texas. As news of the disaster spread, the teachers stopped classes and huddled together. One was crying. Another brought in a television set, and the class watched as the news "kept replaying it over and over," he recalled years later. "I remember it like it was yesterday. There are those events that happen in history where everybody knows exactly where they were, and that was an event like that for me."

Now, as NASA's chief, he carried the burden of safeguarding the lives of the astronauts under his care. If anything happened, he would be the one who would have to answer to the president—and the nation. The National Space Council that Pence led was focused on crafting the policies that would set NASA back on a course to the moon, and keep the United States ahead of China and Russia. But any moon missions were years away. Bridenstine's more immediate concern was the first human spaceflight missions to the International Space Station from U.S. soil since the retirement of the Space Shuttle. Those, in all likelihood, would happen on his watch. They were his greatest responsibility—and his greatest fear.

CHALLENGER AND THEN the disaster of Columbia, which came apart as it was returning from space in 2003, were singular events that reverberated across generations. They not only led NASA to cancel the shuttle program but shifted the ethos of an agency that had, in the Apollo years, embraced the inherent risk of pushing boundaries while refusing to accept failure as an option. When NASA launched its first Space Shuttle mission in 1981, it estimated that the risk of death was between 1 in 500 and 1 in 5,000. But that was just an educated guess. Years later, after it had flown several missions, NASA recalculated the odds of losing the crew on that first voyage. They were 1 in 12.

Bridenstine was the heir to a more mature and risk-averse NASA, the one that bore the scars of loss. Under his leadership, NASA would be flying astronauts for the first time since the Space Shuttle was retired. But this time, the agency wouldn't own and operate the rockets or spacecraft. It had outsourced flights to the space station to Boeing and SpaceX in 2014.

Both contracts were behind schedule, and SpaceX had had two failures of its Falcon 9 rocket, the same model that would be flying NASA's astronauts. The first came in June 2015, when it blew up a couple of minutes into flight while carrying four thousand pounds of cargo and science experiments in its Dragon capsule to the space station. The second was in September 2016, when the rocket exploded on the pad during an engine test, the giant fireball sending thick plumes of smoke billowing over Cape Canaveral.

Both those accidents were on Bridenstine's mind as he took over an agency in the midst of profound transformation. The U.S. space effort was now not led merely by NASA but dependent on partnerships with a growing number of companies competing against one another in a new, commercial space race. If Bridenstine was going to return astronauts to the moon, he would need to harness the best of American industry, and force it to perform. He would need bipartisan support from Congress and contributions from international partners. In the years to come, he would build a coalition of nations in a diplomatic alliance designed to stay ahead of China and Russia and ensure that space would be a peaceful and propsperous sanctuary for generations.

But that would come later. For now, as NASA and SpaceX prepared to fly astronauts to the space station for the first time since the Space Shuttle was retired, astronaut safety was the priority, and he recalled a visit to SpaceX while he was still a member of Congress. As Elon Musk gave him a tour of the factory floor, Bridenstine asked him what worried him most now as the company prepared to fly people for the first time. He was expecting Musk to say it was the vessels used to keep pressure in the fuel tanks as the rocket burned through propellant. Those vessels had been the culprit of both rocket explosions, and Bridenstine assumed they remained the greatest risk.

Musk assured him that the problem had been solved. What worried him, he said frankly, was the launch abort system. In the case of a problem with the rocket booster, the Dragon spacecraft had its own engines that were designed to quickly shoot the spacecraft away to safety, a feature the Space Shuttle, mounted on the side of the rocket, had not had. In order to fire the engine fast, as would be necessary in an emergency, SpaceX used what are known as hypergolic propellants, which ignite instantaneously when combined. The fluids were highly flammable and dangerous and released under an extraordinary amount of pressure. "There's a lot that can go wrong," Bridenstine recalled Musk saying.

As he began his term as administrator, Bridenstine made it clear to his team that safety would be the hallmark of this new exploration campaign. "If there's a catastrophe," he told them, "the whole world stops. The president stops what he's doing. Leaders from other nations stop what they're doing. Programs get shut down or delayed. We can't allow that to happen."

PART II

• • •

Earth Orbit

2018–2020

CHAPTER 7

Starman

The chances of success weren't great. Some at SpaceX just wanted the untested rocket to get high enough so that when it exploded, it didn't damage the Cape Canaveral launch site. It was, after all, Launch Pad 39A, of Apollo 11 fame. With a rocket this risky, simply not destroying the historic patch of space history would count as a victory.

Officially, the rocket was called the Falcon Heavy. Internally, SpaceXers referred to it as the "Frankenrocket" since it was comprised of two Falcon 9 boosters mated to either side of a third. Musk was confident that his engineers could make the design work, and at a press conference introducing the machine in 2011, he predicted it would launch sometime the following year. "This is a rocket of truly huge scale," he said at the National Press Club in Washington, D.C.

It was also hugely complicated. The SpaceX team soon discovered that mating together three rocket boosters with a combined twenty-seven engines and getting the whole off the ground was far more difficult than Musk had made it seem. "At first it sounds real easy, you just stick two first stages on as strap-on boosters, but then everything changes," Musk said at one point. "All the loads change. Aerodynamics totally change. You've tripled the vibration and acoustics. You sort of break the qualification levels on so much of the hardware." He added, "We were pretty naïve."

Falcon Heavy didn't launch in 2012. It wouldn't be until the end of 2017 that SpaceX's engineers started to think that maybe it would fly soon, and even then, they girded themselves for a fireball. Even Musk said the odds of a successful flight were between 50 and 70 percent. "In theory, it should work," he said. "But where theory and reality collide, reality wins."

When he first conceived of Falcon Heavy, Musk thought it could be the vehicle used to get people to Mars. It would lower the cost of access to space, he felt, and push rocket technology forward with a vehicle that would be able to carry "more than a fully loaded Boeing 737 with 136 passengers, luggage, and fuel" to Earth orbit. More than its capacity to lift large amounts, it symbolized the possibility of space expansion at a time when American ambitions had retreated. The year he introduced Falcon Heavy, NASA was at a low point. In July 2011, the Space Shuttle would fly for the last time after a thirty-five-year run, leaving the United States with no way to fly astronauts anywhere. Musk believed SpaceX could take up the mantle by building a rocket with even more power. "Although the Space Shuttle is obviously retiring this year, this is something that America can be really proud of—the fact that there is actually going to be a vehicle with twice the capability of the Space Shuttle," he said.

Since the Falcon Heavy could very well explode, SpaceX's engineers didn't want to risk a payload of any value on its first flight. Certainly not an expensive satellite from one of their customers. Some had suggested launching a satellite designed by a university group, but then SpaceX could be in the position of crushing some students' dreams. They didn't want to do that, either.

"The number of things that could have gone wrong was extremely high," one engineer recalled later. "Looking under the hood, I didn't have a lot of faith in the mission succeeding. But it was a known risk, and we were okay with the idea that maybe it wouldn't succeed." This was purely a test flight, a chance to see if Frankenrocket could fly. If it didn't, SpaceX would pick up the pieces and try again. Fail fast and learn—that was the SpaceX way. It had taken the company four tries to get its first rocket, the Falcon 1, to orbit in 2008. Later, in 2015 and '16, the two Falcon 9s exploded. And it had lost several others returning from space as they attempted to land on a ship at sea. Musk called the mishaps RUDs—Rapid Unscheduled Disassemblies—and had put together a video of some of the most spectacular explosions, an anti-highlight reel of fire and smoke set to the John Philip Sousa march "The Liberty Bell" that had been the theme of one of his favorite TV shows, *Monty Python's Flying Circus*.

For the first flight of the Dragon capsule in 2010, SpaceX put a wheel of cheese on board, an homage to the Monty Python skit in which John Cleese goes to a cheese shop that has no cheese. For Falcon Heavy's payload, Musk wanted something different, and one night he talked about it over drinks with his friend Jonathan Nolan, the film-maker. They lamented the state of space exploration—how in the decades since NASA sent men to the moon, space had become uncool, even boring. In their view, space was "the wide yawning black infinity," in the words of Carl Sagan, and the site of some of humanity's greatest achievements. Musk wanted a bold object to go up with the Falcon Heavy, something that would embody the Monty Python ethos—"And now for something completely different"—but also herald a new era of space. He wanted art.

"We were talking about how do you inspire people?" Nolan later recalled. "How do you get people to talk again, how do you drive the conversation? And we're trying to think of an image—something to sort of shock people into looking and thinking about this again."

On December 1, 2017, Musk announced on Twitter what they had come up with. "Payload will be my midnight cherry Tesla Roadster," he wrote. "Destination is Mars orbit. Will be in deep space for a billion years or so if it doesn't blow up on ascent." The Tesla would separate from the rocket to fly on its own. It would not enter Mars orbit, however, but rather fly on a long looping arc around the sun in a trajectory that would bring it past Mars.

"*Why?*" one of his followers responded, asking the question that many—even some people inside SpaceX—were asking. What was the point?

"There is no point," Musk wrote. "It's just for fun and to get the public excited. Normally, when a new rocket is tested, they put something really boring on, like a block of concrete or a chunk of steel or something. The car is just the most fun thing we could think of." He added, "I love the thought of a car drifting endlessly through space and perhaps being discovered by an alien race millions of years in the future."

Some flat-out hated the idea. It felt crass, mixing two of Musk's companies in a vulgar promotional stunt. Space was a global commons, the great beyond, a place of hope and optimism. Now it would be reduced to the backdrop for a billionaire's toy. Instead of cool, this had the potential to be seriously lame. Instead of art, this could be pollution—exploration as marketing.

But it was what Musk wanted. And so SpaceX's engineers got to work figuring out how to pack a red convertible sports car into a rocket. Musk was pushing for the early part of 2018—not a lot of time to design and build a special mount to keep the vehicle safe during the violent force of launch while inside the Falcon Heavy's nose cone or fairing. Satellites were built to survive not only the supersonic flight out of Earth's atmosphere but the harsh vacuum of space. Cars were not. What would happen when a Tesla designed for Earth got propelled to orbit? They weren't sure.

Engineers began tinkering on the Tesla behind a giant curtain for secrecy, swapping out the windshield for an acrylic one they hoped wouldn't crack. Any loose parts got welded down or removed. The tires were of particular concern. No one wanted the big Tesla-in-space reveal to be marred by a quartet of tires exploding, so they put them into a vacuum chamber first to make sure they'd survive. "It was a ton of work," the engineer said. "And it was all very last-minute."

Musk didn't want to just put a Tesla into space; he wanted to film the car flying. He wanted an iconic shot, one that would mark the modern era the way the 1968 *Earthrise* photo or the image of Buzz Aldrin saluting the flag on the lunar surface had symbolized the dawn of the

Space Age. But when engineers prepped the Tesla in the processing facility on Cape Canaveral and sent test images to Musk, he hated them. "These suck," he said. The pictures were too close and made the car look fat and stupid. Musk had a specific vision for how the images should look, and these were not it.

Emily Shanklin, SpaceX's senior director for marketing and communications, tapped Sam Friedman, the company's in-house photographer and videographer, to fly to Florida and make some better images. It was an assignment with a high degree of difficulty and, since Musk was personally invested in it, peril. Shooting in space presented all sorts of technical challenges, including how to cram cameras inside the rocket fairing. When he arrived at Cape Canaveral, Friedman took some time walking around the Tesla, imagining where the cameras would be mounted on various points on the vehicle. He snapped some pictures and sent them back to SpaceX headquarters, hoping the new shots would meet Musk's vision.

They did not. Friedman tried again. And again. But nothing satisfied Musk, who was losing patience. And Friedman was getting tired of flying back and forth between SpaceX's headquarters and Cape Canaveral, only to be told his work was no good. In January 2018 he decided he'd just move to Florida until he got it right. He knew the only way to get a decent framing was if the engineers would build mounts that could attach to the car like giant selfie sticks. The engineers didn't appreciate the extra challenge, but it was what Musk wanted, so it's what they did.

At some point, Musk and Nolan came up with another idea—adding a passenger to the Tesla. They chose a mannequin, named him Starman, and outfitted him with one of the space suits SpaceX had designed for the astronauts it would soon be flying.

On one side of the car, a camera would catch Starman in profile, right hand on the steering wheel, left elbow perched on the window sill. A camera on the hood would frame him straight-on, as if he and the Roadster were flying toward the viewer. A third camera was mounted on the back of the vehicle, looking over Starman's shoulder to the dash and, hopefully, into the depths of space. Finally, Musk was pleased. But Friedman still had work to do. Just because the images looked good in a controlled environment on Earth did not mean the shot would work in space.

The lighting in particular was a problem. There was no atmosphere to filter the sun's rays, so Friedman had to adjust the camera settings to prevent the images from being overexposed. He found a production company in Orlando and rented some high-wattage lights to mimic the sorts of conditions he thought might be found in space. To get the composition right, he also needed to know where the sun and Earth would be in relation to the car. The money shot would be to get the Tesla cruising through the void with Earth in the background—an image suggesting humans would be going next, and that eventually spaceflight would be as common as driving.

Friedman worried that the cameras would be shaking so much during the voyage that every frame would be blurry. It was so high-risk that the engineers wouldn't touch Friedman's cameras. "I was the one who tightened the bolts," he recalled. "They were like, 'You're taking full responsibility for this. If it goes south, it's on you.'"

Finally he thought he had it right, but he wasn't sure. "The night before we closed it up, I was having a panic attack about the exposure," he said. "This was the most stressed I had ever been in my entire life. I wasn't sleeping. It was awful." But there was nothing he could do now. The rocket was rolling to the launch pad. Exhausted after weeks of eighteen-hour days, he went on a family vacation in Hawaii and hoped that it would all work.

ON THE DAY of the launch, February 6, 2018, the press room at the Kennedy Space Center was so packed that NASA and SpaceX had to provide an overflow room for the journalists, many of whom had traveled from outside the United States to witness the launch—or explosion—of what was then the most powerful rocket in the world.

Local officials were expecting a hundred thousand people from out of state to cram the beaches and causeways. The hotels along nearby

Cocoa Beach were sold out. Police were all-hands-on-deck to handle traffic. The scene was reminiscent of an earlier era—the heady days of Mercury, Gemini, and Apollo. It felt like the Florida Space Coast, which had been largely forgotten after the retirement of the Space Shuttle, was back.

Musk was in the launch control center at the Kennedy Space Center. Nolan, invited to observe his idea take flight, was there too. Across the country, at SpaceX's headquarters in Hawthorne, California, crowds of employees gathered early outside the glass-enclosed mission control center, as they had done for so many of the company's major milestones. The company was bigger now, with some seven thousand employees, and they were gathered shoulder-to-shoulder, cheering so loudly it was hard for the engineers on the company's live broadcast to hear one another and track what was going on.

"We're very excited, I'm sure you can hear it," Michael Hammersley, a SpaceX materials engineer and one of the hosts of the broadcast, said as his coworkers applauded behind him.

With twenty seconds left in the countdown, when the launch director called out, "SpaceX, Falcon Heavy, go for launch," they went nuts, pulsing with the energy that came with the belief that they were witnessing history. As the countdown came to the *T-minus 10 seconds* mark, they shouted in unison: "Ten, nine, eight . . ." all the way until the engines—all twenty-seven of them—ignited, pushing a huge plume of exhaust into the air. The rocket cleared the tower, and those who were worried about the launch pad breathed a sigh of relief.

I was standing on the lawn outside the press center of the Kennedy Space Center, watching from a distance of around three miles, about the closest one could get to the pad. Like other launches I had attended, I could see the white-orange flash of the engines igniting, but I knew it would take a few moments before the sound arrived. It hit me square in the chest, a wall of sound coming in unrelenting waves. As the rocket rose, I thought of Musk's prediction that it might explode. I also knew that it should be flying not just up but eventually out and away from the coast, over the Atlantic Ocean, which had been cleared of all boat traffic. But at first, it seemed like it was going straight up, meaning that if it did blow up, the shrapnel would be raining down over land, maybe right on top of us.

But it didn't explode. It kept going and going. Inside SpaceX headquarters it was pure pandemonium. They had cheered during the countdown, when the engines fired, as it continued to climb, and now the side boosters separated and began their journey back to Earth, where they would attempt to land simultaneously on a pair of pads at Cape Canaveral, a few hundred yards from the shoreline. As they got close, they reignited three of their first-stage engines, slowing them down. Their landing legs unfurled, and they touched down softly at almost exactly the same moment, triggering sonic booms that reverberated across the coast and set off car alarms in the parking lot of the press site.

It looked just like an animated video that SpaceX released before the flight had depicted it—a bit of science fiction coming to life.

The landings of the side boosters overshadowed the big Tesla-in-space reveal, which was slightly mishandled on the broadcast. David Bowie's "Life on Mars?" started to play, but viewers continued to see a map of where the rocket was in relation to Earth. When producers finally cut to the Tesla, however, the image was dramatic—a view from the camera Friedman had mounted to the arm sticking out from the front of the car, showing Starman seated in the convertible, with an expansive panorama revealing the ocean and Earth below. Then the feed switched to the camera behind Starman, showing his view of the blackness of space laid out in front of him, like a long road on the journey around the sun. The crowd at SpaceX erupted into cheers. On vacation in Hawaii, watching with his father, Friedman was ecstatic, jumping up and down, screaming, letting out weeks of frustration with each exuberant cheer. He had nailed it.

After the boosters landed, SpaceX kept the live stream going for more than four hours, an uninterrupted feed of Starman in the Tesla in space. SpaceXers stood around watching in clusters. It was a moment of brilliance, the ruby-red car set against the blackness of space, Earth glowing in blues and greens in the background. Starman's space suit was a sleek and elegant departure from the bland, bulky gear of the past, a new suit that prized form as well as function.

The image of Starman cruising through space as if on the Malibu strip went immediately viral on social media, converting doubters to believers. It was indeed art, with a few hidden gems that the company soon unveiled. Printed on the Tesla's circuit board were the words MADE ON EARTH BY HUMANS. The dashboard screen read DON'T PANIC, a nod to *The Hitchhiker's Guide to the Galaxy,* one of Musk's favorite science fiction novels. And it won over at least some of the skeptics. "I'm a believer in the stupid Tesla," one of the engineers who had looked askance at the whole idea told me later. "Elon obviously had a vision I didn't see. I take it all back. We just sat there and watched it, thinking, 'Wow, I can't believe we did this.'

"And then the alcohol started flowing."

SHORTLY AFTER THE launch, Musk told an audience at the South by Southwest conference in Austin that the point of the flight was "to get the public—you—excited about the possibility of something new happening in space, of the space frontier getting pushed forward. The goal of this was to inspire you, and to get you to believe again—just as people believed in the Apollo era—that anything is possible."

Musk was already a celebrity, but the launch made him even more so. His fan base—a strong cultish following of optimists who believed in his futuristic vision of electric vehicles and space exploration—was growing and starting to include a broader audience outside the techno-enthusiasts who had been with him for years. The live broadcast of the Falcon Heavy launch attracted 2.3 million viewers on YouTube, at the time the second-largest audience in the site's history. Musk encouraged his growing legion of fans, saying in Austin that he hoped for "general support and encouragement and good will."

He had plenty of that. People were starting to believe and were more than eager to help, to cheer him—and his hopeful vision of the future—on. Later, when he'd become controversial and divisive for his

pugilistic and unfiltered tweets, lashing out at critics, angry and puni-
tive over perceived slights, it would become difficult to remember the
Musk who served more as an inspiration than agitator as his wealth
and power grew. It'd be hard to fathom that in several years, after he
would become a White House adviser during Trump's second term,
he'd be so reviled by some segments of the population that protestors
would gather outside Telsa plants to denounce Musk's role cutting the
federal government. But for now, he was still widely beloved, flying
high, somewhere beyond gravity's reach.

The launch, however, wasn't just about art or excitement. SpaceX
had just demonstrated real technological progress, proving it could design, build, and fly a very powerful rocket, capable of delivering an
enormous amount of mass to orbit. The Falcon Heavy was on par with
one of the nation's workhorse big-boy rockets, the Delta IV Heavy,
which the National Reconnaissance Office relied on to lift its biggest
and most important spy satellites. Large, powerful rockets had long
been considered national assets, a show of might and prestige that perhaps should have been granted the "U.S.S." designation like a Navy aircraft carrier or destroyer. NASA had for years been working on its own
heavy-lift vehicle—the Space Launch System, which would be even
more powerful than the Saturn V rocket that flew the Apollo astronauts
to the moon. But its price tag was becoming unsustainable, costing
some $10 billion in development by that point, with an estimated $1 billion or more for each flight. Plus, it was expendable. After launch, the
booster would fall into the ocean, never to be used again. By contrast,
SpaceX had said that the reusable Falcon Heavy had cost some $500 mil-
lion to develop and could be launched for as little as $150 million.

Bezos was impressed. By then, Blue Origin had launched and landed
its New Shepard rocket six times. As the company moved toward offering suborbital space tourism flights, it had recently unveiled a new
crew capsule with large windows meant for Earth-gazing. New Shepard,
though, only flew to an altitude of just over sixty miles. Compared to
Falcon Heavy, it was a single-engine pipsqueak, not powerful enough
to even reach orbit. With Falcon Heavy, SpaceX was leaping so far
ahead of Blue that it was hard to even consider them rivals anymore.

Before the launch, Bezos took a conciliatory tone that seemed to show his respect, a moment of graciousness in a relationship that had been defined largely by antagonism: "Best of luck @SpaceX with the Falcon Heavy launch tomorrow—hoping for a beautiful nominal flight!"

"Thanks," Musk responded, adding a winking, kissy-face emoji.

Falcon Heavy's development had been closely tracked not only by Bezos, but by space agencies around the world, and its successful debut left them astonished. The European Space Agency was in the middle of building its own expensive, expendable rocket; the organization's chief, Jan Wörner, wrote an envious blog post titled "Europe's move," in which he worried that the continent was falling behind. "Totally new ideas are needed, and Europe must now prove it still possesses that traditional strength to surpass itself and break out beyond existing borders," he wrote. Europe, he insisted, needed to be more innovative and start developing "a launcher system that eschews traditional solutions."

China, which was also developing a heavy-lift rocket, the Long March 9, was startlingly candid in its admiration. The *Global Times,* a state-run news site, wrote that Falcon Heavy had "totally crushed all other current rockets in the world." It added, "What really shocks us Chinese is not only that our country currently doesn't have rockets of such magnitude, but the fact that we are almost 10 years behind. More importantly, what our country has to desperately catch up with is actually a private U.S. enterprise. To put it more bluntly, this time the Americans showed us Chinese with pure power why they are still the strongest country in the world and how wide the gap really is between us and them."

China didn't just rely on single-use rockets; it launched them over land, often over populated areas. Its three main launch sites were inland, a relic of the Cold War, when the regime wanted its ballistic missile sites far away from enemy lines. China's rockets littered the ground with toxic fuel and debris that sometimes hit houses. In 1996, an earlier version of the Long March family lifted off, started flying sideways, and crashed into a hillside, setting off an explosion that leveled much of a nearby village. Six people were killed and fifty-seven injured, according to Chinese authorities, though many think the casualty toll was higher.

If China wanted to catch up to the United States and become a great spacefaring nation, it would need to not only reach for the moon, but to emulate SpaceX. A few months after the Falcon Heavy launch, China announced it would begin developing its own reusable rockets.

JIM BRIDENSTINE WAS also awestruck by the launch and landings, and by the engineering feats behind them. In that moment, it became clear to him: SpaceX was for real. The images of Starman and the twin boosters landing were a revelation that stood in contrast to the images of Space Shuttle Challenger exploding into an orange fireball that he had watched over and over in Ms. Powers's fifth-grade language arts class. Here was something hopeful and optimistic that represented what space was supposed to be.

The congressman-turned-NASA-administrator's goal was to drum up support for space exploration. NASA was a powerful brand, but now so was SpaceX—enough so that he could capitalize on its success. He had photos of the booster landings and Starman framed and hung in his office. They were symbols, he told visitors, of what was to come. If SpaceX could fly Falcon Heavy, then maybe the even more powerful and improbable BFR, or whatever it was called, was not the fantasy many considered it to be.

Bridenstine believed Musk was a genius, single-handedly reinvigorating the nation's space program. But he didn't want to get carried away. Musk was a man, not a messiah. SpaceX was a government contractor, funded in large part by taxpayers. Under his watch, the company was going to be launching astronauts—not a mannequin. That would require rigorous oversight. Bridenstine's job, he told himself, was to help Musk and his team succeed, but also hold them accountable.

CHAPTER 8

Flying by Swipe

I t had been a trippy, far-out conversation well before Joe Rogan busted out the spliff. From the beginning, Elon Musk—Rogan's guest on a live-streamed episode of his popular podcast on September 7, 2018—sounded something like Hunter S. Thompson on industrial-grade psychedelics.

Here was Musk on artificial intelligence: "The percentage of intelligence that is not human is increasing, and eventually we will represent a very small percentage of intelligence."

On cyborgs: "You will be essentially snapshotted into a computer at any time. If your biological self dies, you could probably just upload it to a new unit—literally."

On his plans to dig tunnels under Los Angeles to circumvent traffic: "I'm not asserting it's going to be successful. But I don't see any other ideas for improving the traffic. So, in desperation we are going to dig a tunnel and maybe that tunnel will be successful and maybe it won't."

And if anyone was looking for evidence that Musk, even while sober, spoke like he had just exhaled a spectacular bong hit, here he was on watches: "It's kind of amazing that you can keep time mechanically on a wristwatch with these tiny little gears." Rogan tried to match his tone, gushing that the technology in self-driving cars could evolve to the point where there is "some electromagnetic field around the cars that as cars come close to each other, they automatically and radically decelerate because of magnets or something."

On it went. "I'm an alien," Musk said. He postulated that reality did not exist and that humanity was dwelling in some sort of simulation taking place in a computer-generated metaverse. It was only *after* all that meandering that Rogan pulled out the joint and basically dared Musk to toke. "Come on, man. You probably can't because of stockholders, right?" Musk, who was wearing a T-shirt that read OCCUPY MARS, took a quick hit and arched his eyebrow behind a veil of smoke—a cartoonish image that would soon be broadcast across the Internet and cable news. Rogan was soon wondering, what if there were a million Elon Musks, and Musk was saying, "I don't think you'd necessarily want to be me." Musk's phone started buzzing. And buzzing. Until he could no longer ignore it.

"You're getting text messages from chicks," Rogan said.

"No. I'm getting text messages from a friend saying, 'What the hell are you doing smoking weed?'"

"It's legal," Rogan reminded him. In the state of California, at least.

Musk, as if anticipating some blowback, said, "I'm not a regular smoker of weed. . . . I don't actually notice any effect." That was perhaps true. He had treated the joint the way he had treated the whiskey Rogan had poured for him, with polite restraint—a couple of modest sips that left the glass half full. And the conversation after the hit was no more bizarre than before, though it did get decidedly more mellow, as they moved ahead, guided by Rogan's frenetic, attention-deficit-disorder interviewing style, into the perils of social media and human relations.

"This may sound corny," Musk said at one point. "But love is the answer."

"It is your answer," Rogan replied.

"Yup."

"Yeah, it is. It sounds corny because we're all scared. You know, we're all scared of trying to love people, being rejected, or someone taking advantage of you because you're trying to be loving."

"Sure."

"What if we all could just relax and love each other?"

"It wouldn't hurt to have more love in the world," Musk said.

"It definitely wouldn't hurt."

"Yeah," Musk said, sighing. "I think people should be nicer to each other, and give more credit to others. And don't assume that they're mean until you know they're actually mean. You know, it's easy to demonize people. You're usually wrong about it. People are nicer than you think. Give people more credit."

When the two-and-a-half-hour odyssey of an interview was over, Musk checked his phone again. There weren't just texts from friends. There were countless TV clips and news headlines questioning why the CEO of Tesla and SpaceX was publicly taking a hit of marijuana.

A CNN ANCHOR noted that Tesla's "stock is down, but Mr. Musk appears to be kind of high. What's going on?" MSNBC's correspondent said it was yet another example of Musk "seeming to dare people to call him out." *The Wall Street Journal* reported that the appearance on Rogan "was another example of the CEO's unorthodox style that has won him legions of fans. But some analysts and investors say his erratic behavior is creating distractions for the company and its employees." The article quoted an analyst as saying it was "becoming more clear that Tesla needs to entertain a major change in the C-suite. The ongoing, effectively self-inflicted public relations crisis is now affecting key personnel within the organization."

Tesla's investors were already getting nervous. The Rogan podcast came a month after Musk had tweeted that he had the "funding secured" to take the company private at $420 a share. The price was a pot joke, and the Securities and Exchange Commission sued Musk, alleging he had lied to investors. Musk settled, agreeing to pay a fine of $20 million; Tesla also ended up paying $20 million, as the SEC sought to remove Musk from the chairmanship of Tesla and ban him from serving as the CEO of any publicly traded company.

Jim Bridenstine didn't really care about any of that. Tesla's stock price was not his concern. As the head of NASA, he cared that the CEO of the company that was about to launch American astronauts to the International Space Station had just taken a hit of a drug that was still federally illegal. If the chief executive could act this way, then what about the engineers under him, the ones on the production line working on the rocket and the spacecraft? Would they smoke weed and show up at work high? Soon after being shown footage of Musk's appearance on the podcast, Bridenstine imagined a worst-case scenario—a Challenger-like explosion—and the subsequent investigation that would seek to determine if he had done everything in his power to keep the astronauts safe.

Musk's behavior was even more galling to Bridenstine because of his personal values. A conservative and a Christian, he didn't drink—not even in college—and never experimented with recreational drugs. In Congress, he voted against a bill that would have allowed buying marijuana with food stamps and was given a "D" rating for his "hard-on-drugs" stance by the National Organization for the Reform of Marijuana Laws. As a Navy pilot, he had flown the E-2C Hawkeye, a Navy reconnaissance plane, on missions over Central and South America. His job was to take intelligence reports on the movement of drugs and weapons, and even human trafficking, and then validate the information before working with the Coast Guard and local authorities on the ground. It was satisfying to know he was helping prevent drugs from getting to American streets. "It's also about preventing the cash from getting into the hands of really bad people," as he told me. His squadron alone was responsible for helping seize about $2 billion worth of cocaine every year.

Bridenstine knew Musk liked to flaunt rules and regulations he felt were unjust. At the time, Musk was warring with the SEC over the "funding secured" tweet, mocking the agency on Twitter as the "Shortseller Enrichment Commission." Bridenstine didn't want to let Musk intimidate him, or to give him special treatment. It didn't matter that SpaceX was one of the most innovative companies since Apple, or that he had been fanboying over the Falcon Heavy launch and decorating his office with images of it just a few months earlier.

"You can't be doing that," Bridenstine told Musk when they spoke by phone. "The whole world is watching you."

Musk let loose an awkward giggle and assured him that it had been a spur-of-the-moment thing. "I don't normally do that," he said.

The explanation was not reassuring. NASA had a zero-tolerance policy on drugs, and that applied to contractors as well. Members of Congress called Bridenstine, asking to know how he was going to handle this. SpaceX's competitors also applied pressure, hoping to capitalize. Even within NASA, people were asking, "How does Elon get away with this?"

The NASA administrator had to do something more than slapping Musk on the wrist. To outsiders, SpaceX could appear a bit like a freewheeling college campus. Young people. Tattoos. Piercings. Long hair. T-shirts and jeans everywhere. It was not out of the question—perhaps even likely—that at least some of them were dabbling in illicit substances. As Gabe Sherman, Bridenstine's chief of staff, counseled him, "If anything goes wrong with one of SpaceX's human spaceflight missions, the questions are going to come back to NASA leadership: 'You saw these things and did nothing?'"

Bridenstine convened a meeting among NASA leadership to ask what tools were at his disposal to hold SpaceX accountable. The answer was a review of SpaceX's safety culture, called an Organizational Safety Assessment, a provision allowed under the contract. NASA investigators at the Office of Safety and Mission Assurance would descend on the company, inspecting its facilities, interviewing hundreds of employees at every level and at multiple sites. It would be, in the words of William Gerstenmaier, then the head of NASA's human spaceflight division, "pretty invasive." If there were any problems, Bridenstine said, he wanted to find them.

In the weeks that followed, I heard rumblings about the review from sources and broke the story in *The Washington Post*. By then, NASA leaders convinced Bridenstine that he couldn't just pick on SpaceX. If he was going to order a review of its culture, he would also have to do it for Boeing, the other company under contract to fly NASA's astronauts. The space agency did not want to seem as if it was favoring one contractor over another—and SpaceX also complained that it was being singled out. Adding Boeing, a trusted partner of NASA's since the dawn of the Space Age, was merely for show; in an interview with me at the time, Bridenstine confirmed it was Musk's behavior that had triggered the investigations.

Later, speaking to reporters at NASA headquarters, he said the decision was influenced by past tragedies, including the Apollo 1 capsule fire in 1967 and the two Space Shuttle disasters. "Every single one of those accidents had a number of complications," Bridenstine said. "Of course, the technological piece was a big piece of it. But the other question that always came up was, 'What was the culture of NASA? What was the culture of our contractors, and were there people that were raising a red flag that we didn't listen to, and ultimately did that culture contribute to the failure and, in those cases, to disaster?'"

THE SAFETY REVIEW added to the tension between NASA, the stodgy bureaucracy, and SpaceX, the innovative start-up. This wasn't just a clash over culture, but an attempt by the arm of the federal government that had landed men on the moon to assert itself. SpaceX had exceeded everyone's expectations, and the commercial sector was on the rise. But space exploration remained a national priority, especially amid a space race with China and Russia. NASA was still the world's preeminent space agency. It wasn't ready to cede the space enterprise to any private company, especially one led by a mercurial billionaire increasingly courting controversy.

Critics from traditional aerospace contractors seized the moment and pointed to the fact that SpaceX had had two catastrophic failures of its Falcon 9 rocket, the same one that was to be used for human spaceflight missions. One of the accidents had happened on the pad when the rocket, fully loaded with propellant, ignited into a bright orange mushroom cloud that sent a shockwave across Cape Canaveral. That had sparked criticism of the way SpaceX fueled its rockets.

To get more power out of the Falcon 9 booster, SpaceX "super-chilled" its propellants, keeping the liquid oxygen, for example, at minus

340 degrees Fahrenheit. As a result, they became denser, allowing SpaceX to pack more fuel into the same space and get better performance. But they also quickly boiled off, which is why SpaceX would load its rocket at the last minute before lifting off—a process known as "load and go." The problem with applying the method to human spaceflight was that it required astronauts to be on board the spacecraft during fueling, which stood in contrast to decades of NASA protocol.

For years, the process was done in the reverse order: Fuel the rocket, make sure it's stable, then allow the astronauts on board. That would limit their exposure to a disaster. That's how Boeing planned to fuel the rocket that would launch its spacecraft. Thomas Stafford, the legendary astronaut who flew during Gemini, Apollo, and the Apollo-Soyuz mission, slammed SpaceX's plan, saying the company was going to get people killed. Separately, in a letter to NASA leadership in 2015, the retired Air Force lieutenant general who was then the chair of the agency's space-station advisory committee wrote that "there is a unanimous, and strong, feeling by the committee that scheduling the crew to be on board the Dragon spacecraft prior to loading oxidizer into the rocket is contrary to booster safety criteria that has been in place for over 50 years, both in this country and internationally."

NASA and Boeing had similar cultures: formal and conservative, with a strict hierarchy and a military-like deference to authority. SpaceX could hardly have been more different. Employees were brash and iconoclastic. Their culture encouraged dissent. Hurt feelings mattered less than correct answers. Musk encouraged employees to stand up and walk out of meetings that were a waste of time. Shotwell told her troops, "If we're throwing a bunch of shit in your way, you need to be mouthy about it. That's not a quality that's widely accepted elsewhere, but it is at SpaceX." It certainly wasn't widely accepted at NASA, where many of the top officials felt that the company still could not be trusted.

IT HAD BEEN that way for years, well before NASA decided to assign its astronauts to fly with SpaceX. When NASA's top human spaceflight officials had gathered back in August 2014 to select one or more companies that would build the next spacecraft to fly NASA astronauts to and from the International Space Station as part of what was called the Commercial Crew program, Boeing had been the overwhelming favorite. For years, NASA had used the Space Shuttle to transport its crews to the orbiting laboratory. Now, with the Shuttle retired, the space agency would look to the private sector. Within the agency, there was a clear favorite. "If it ain't Boeing, I ain't going," as one saying went.

The decision was up to William Gerstenmaier, head of the human spaceflight division. An elder statesman who'd started his NASA career in 1977, "Gerst," as he was known, seemed to wear the burden of safeguarding the lives of astronauts on his shoulders, which perhaps explained his tendency to slouch. He spoke softly, in a lulling monotone, and looked like he could doze off at any minute, snoring through his bushy gray mustache. But at NASA, he was perhaps even more powerful than the politically appointed administrator, and his stolid, unflappable demeanor was well suited for running the division and its various factions. Sitting at the head of the table in a ninth-floor conference room at NASA headquarters in Washington, D.C., he started polling senior staff on who should be awarded a contract.

"Boeing."

"Boeing."

"Boeing."

"Boeing."

At the far end of the table, Phil McAlister was growing alarmed. He was director of NASA's commercial space division and felt strongly that NASA should award two contracts, not just one. Yes, Boeing had scored higher on its technical and management approaches, earning Excellent to SpaceX's Very Good. But SpaceX's bid was 60 percent less expensive than Boeing's: $2.6 billion compared to $4.2 billion. No one at this meeting was talking about that. In McAlister's view, everyone was just voting for Boeing as a matter of routine. At one point, one of the officials said their vote was influenced in part because "SpaceX is a pain in the ass to work with."

McAlister hoped his concern didn't show on his face. "I was a deer in the headlights," he told me later. "I'm thinking, 'How do I turn this around? I mean, this is the paragon of human spaceflight all voting one way. But it does not appear to be reasonable from an external standpoint.'"

When it was his turn, McAlister started making his case methodically, like a defense attorney trying to persuade a jury—or in this case, Gerst, the one-man judge. Publicly, NASA had said it was open to awarding multiple contracts, but the consensus inside the agency was that it could only afford to pick a single company. McAlister argued that this was a mistake. Two providers would ensure that if one had a failure, there would be a backup. When the Space Shuttle had its two deadly disasters—Challenger in 1986 and Columbia in 2003—it took more than two years for NASA to return to flight.

Then there was the cost factor. McAlister turned to Bill McNally, NASA's lead procurement official, and asked if could remember when a federal agency "picked a company that was 60 percent more expensive?"

"No, I've never seen that," McNally said. "If you do pick Boeing, we will be in uncharted territory."

SpaceX was litigious, McAlister reminded Gerst and the panel. Musk had just sued the Air Force over the right to compete for national security launch contracts. And here NASA's chief procurement officer was essentially saying that SpaceX would have valid grounds to make a complaint. "You're risking a protest, which is going to add a year to this whole development process," McAlister said.

Next, he targeted the deputy head of safety and mission assurance, who had cast her vote for Boeing with a caveat: that Boeing perform a test of the capsule's emergency abort system while the vehicle was in flight.

"But that proposal doesn't exist," McAlister said. "You can't just add in an in-flight abort test." Rules were rules, he was saying: Federal procurement law prohibited Boeing from revising its bid to include the test after the fact. "So, what you're telling me," he continued, "is that the head of safety and mission assurance believes that the Boeing proposal was unsatisfactory?"

"I didn't say that."

"Well, yes, you did," McAlister said. "You just didn't say it the way I said it."

He had made his case, but it wasn't good enough. After the meeting, NASA's procurement staff drafted a source selection document for Gerst's signature. It listed Boeing as the sole winner.

Undeterred, McAlister kept up his campaign. "The thing that was in my favor is I think everybody left the selection meeting thinking, 'Oh, for sure Bill's going to pick Boeing; I don't have to do anything,'" he said. "But I was in his office almost every day trying to convince him to pick two."

One day, McAlister went too far. "If you only pick one," he told Gerst, "in my opinion, that would be reckless." That was a trigger word for the NASA veteran. He had worked closely with the astronaut corps for decades and had gotten to know them and their families. He had given so much of his professional life to ensuring their safety. He had lived through both the Challenger and Columbia disasters. *Reckless?* The implication made him furious. Gerst was not one to ever lose his cool, but now he couldn't help it. His face turned red, and he started yelling.

McAlister backpedaled fast. He had overstepped. He was sorry. Still, as he walked away, chastened, hoping he wouldn't get fired, he felt as if he were finally getting through. Indeed, he was. Many at NASA had serious concerns about the design of the Dragon capsule that would fly NASA's astronauts to the space station, which in some key respects was a marked departure from the way NASA had done things for decades. But in September 2014 the space agency awarded two contracts: one to Boeing, and another to SpaceX.

"This was not an easy choice," Charlie Bolden, then the NASA Administrator said at the time.

IF NASA'S TOP executives were skeptical of SpaceX, some of the astronauts were outwardly hostile. Few believed SpaceX would ever fly. At least not anytime soon. Some assumed that the astronauts assigned to

SpaceX were in purgatory. They'd spend years training on a spacecraft that would never lift off, while their colleagues got to actually go to space with Boeing.

"There was nobody, absolutely nobody, who had any confidence SpaceX could fly people," Doug Hurley, one of the astronauts chosen to fly in the Commercial Crew program, told me. "Everyone thought Boeing would fly first by a long shot."

Some in the astronaut corps felt it was their job to tell SpaceX how the spacecraft should be designed. People from NASA would visit SpaceX and "start dictating we need this, we need that, without really understanding how the vehicle works," recalled Garrett Reisman, a former NASA astronaut then working for SpaceX. "In some cases, it didn't make any sense, or they were just applying lessons learned and assumptions based on the Space Shuttle or the Soyuz or something in the past, not understanding this vehicle operated very, very differently."

The two sides plainly distrusted each other. "The SpaceX people were generally very dismissive of these NASA types. In the beginning, there were preconceived notions on the part of SpaceX that these guys are dinosaurs and stupid," Reisman recalled. "We were constantly challenging them and had interns telling them that they were stupid."

One had even tried to change the way SpaceX talked about its systems. The company's engineers called the displays GUIs, pronounced "Gooeys," an acronym for graphical user interface. Pilots, the astronaut informed them, called them "displays," and he wanted everyone to use that term from then on. "GUIs," he said, was out. Reisman lost it and ordered the astronaut outside, where he berated him. "What the fuck are you doing?" he said. "You're destroying your credibility over the fucking nomenclature. I don't give a shit if we call it, Fred, as long as it works."

Shortly after SpaceX won the Commercial Crew contract, astronauts started to train on the Dragon spacecraft. Almost immediately, many started to complain about its design.

One of the main points of contention was that the capsule didn't have a control stick. Flying a spacecraft without one was like trying to drive a car without a steering wheel, they groused. How would they be able to reach the space station without one? For the astronauts who had flown fighter jets in the military, which was a good number of them, the stick was their connection to the aircraft, the main feature of the cockpit. NASA's fleet of T-38 training jets had sticks. When Neil Armstrong took over manual control of the lunar module, he grabbed the stick to land Eagle on the moon. The Space Shuttle had a stick. Boeing's Starliner spacecraft had a stick.

After years of training in the military and then at NASA, many of the astronauts had built up muscle memory for flying with a stick. In the agency's male-dominated culture, it had become a phallic appendage, an extension of themselves.

Where the hell was the stick? The lack of one in SpaceX's Dragon wasn't a design edit; it was castration. In the face of the astronauts' uniform dissent, however, SpaceX stayed steadfast: no stick. If Musk had his way, there'd probably be no switches or buttons, either, though there were several below the touchscreens. Such manual controls were old and analog, as antiquated as rotary phones. Musk was not just the founder and CEO of SpaceX, but also its chief designer, and he was building an iPhone of a spacecraft. Instead of a stick, the Dragon had touchscreens. The astronauts would fly by swipe.

To the pilots, it felt like a slight, because it was. In the hierarchy Musk was building, astronauts would no longer be the lone stars of space exploration. They'd have to share the spotlight with engineers, software designers, and programmers who were building a new generation of rockets and spacecraft capable of flying themselves. John Glenn and the rest of his cohorts may have been extraordinary pilots, with skills honed after years of training. But they couldn't have written the code to program a 140-foot-tall rocket booster to turn around in space, fly back to Earth, and land accurately on a ship that was also guided autonomously by algorithms, a couple hundred miles out in the Atlantic Ocean. Engineers, Musk believed, were the top of the chain, the leaders who would build the future. He had such faith in them—and such a low regard for some other fields of expertise—that he'd often put engineers in all sorts of other roles at the company. SpaceX's head of human resources, for example, was a former avionics engineer with degrees in aerospace and astronautical engineering.

If humanity were truly to become a spacefaring civilization, the whole enterprise would have to change. Control sticks were just the beginning. Fortunately for SpaceX, it had some backers at NASA, who believed in the future the company was trying to create. "If you want to talk more about the commercial paradigm in general, we need to stop listening to astronauts," one NASA official once told me. "They represent a minority of users if we're able to expand this universe. And if the hardware flies itself, then I'm sorry, but I'm going to have to treat what you say with skepticism because you're a pilot. And you're no longer objective about this conversation when I'm eliminating pilots."

SpaceX was out to dismantle the very definition of the word "astronaut," what it meant and who got to be one. In the new space age, everyday people would be able to fly. Forget "the Right Stuff." Poor-to-mediocre stuff would suffice. As Musk liked to say, if you could handle a roller coaster at Disney, you could fly Dragon. The astronauts would take control of the spacecraft only in the event of an emergency. Otherwise, the spacecraft would fly itself.

REISMAN KNEW THE astronauts were going to hate the touchscreens. He'd felt the same way when he first tried them. *Oh, man, this is going to be a mistake,* was his first thought. But the more time Reisman spent with the system, the more he grew convinced it could work. The problem wasn't so much getting used to the touchscreen but to the way the vehicle handled, which could be "squirrely," as he put it. "There was a steep learning curve, but I felt they could become proficient at it given practice," he said. Docking with the space station required small bursts from the spacecraft's Draco thrusters to keep it on a steady, level course. It required precision and patience—the ideal speed at docking was 0.2 meters per second, a snail's pace. Flying a fighter jet with your hair on fire this was not.

"You're not dogfighting or doing aerobatics," he recalled. "You're doing slow, small corrections, trying to keep it going right down the middle."

But with the astronauts, he was pushing against a deeply embedded grain. A few of them, he knew, "hated it before they ever walked in the building," Reisman recalled. Their reaction: "No way. Over my dead body."

A couple of astronauts were particularly condescending toward the SpaceX engineers, continuing to say the touchscreens were a terrible idea and that they needed a stick and manual switches. Their attitude was, according to one of the people involved in the discussions, "I've flown in space—this is what you need to do."

Two of the most senior astronauts were at least respectful enough to conceal their doubts. Doug Hurley and Bob Behnken both had been military test pilots—Hurley in the Marine Corps, Behnken in the Air Force. Together they had thousands of hours flying some of the most sophisticated jets in the world, not to mention their time in the Space Shuttle. Flying by stick had been so embedded in them that it was "down to the DNA level. It was instinct," Hurley recalled.

When one of the first crews of astronauts showed up at SpaceX to test the touchscreen system in a simulator in 2016—flying from Houston direct to Long Beach in their T-38s, with their crew cuts and sunglasses and blue flight suits—Reisman asked them to keep an open mind. If the technology wasn't there yet, it would be. Plus, he had grown increasingly impressed with the level of talent at SpaceX. They were kids, yes. They could be brash and unpolished. Some of them looked weird. But they were smart and dedicated and really, really good at building software.

Hurley was first to try out the simulator. The goal was to manually fly the spacecraft and dock with the International Space Station, lining up what looked like the sight of a rifle to hit a bull's-eye that marked the docking port. For someone as accomplished as Hurley, it should have been easy. During his NASA career he had done about a thousand simulated dockings of the Space Shuttle. In the Marines, he had logged more than 5,500 hours—the equivalent of more than seven months—in twenty-five different aircraft. If NASA could get a washing machine to fly, Hurley would be able to dock it.

But within moments, he was struggling to gain control. He'd push the input to raise the spacecraft, but the nose of the capsule would suddenly veer off to the side. When he tried to straighten it out, the spacecraft would suddenly accelerate, bringing him in way too hot. Trying to slow down would only spin the craft off course. He was all over the place, thinking that if he was scrambling so badly, there was no way the rookies with him were going to succeed. "The thing was just hard to fly," he said. "It was a big mess."

In the end, he was able to dock on the first try. But nobody else could. Astronaut after astronaut, NASA's best and brightest, crashed Dragon into the space station. Frustrated, Hurley confronted Reisman. "Dude, this is unflyable," he said. "I mean, if you have six pilots and five of them can't dock the vehicle, we're in trouble. We're in big trouble." It was, he said, using military lexicon, "unsat." Unsatisfactory. And that was putting it lightly.

Over at Boeing, the simulator for its Starliner spacecraft had a stick. Astronauts who'd flown in the Space Shuttle found it familiar, and even the rookies didn't have much trouble. Every one of them could dock it.

CHAPTER 9

Question Everything

Touchscreens weren't the only SpaceX innovation NASA had been skeptical about. The company had assembled a prototype of the docking system it wanted its Dragon spacecraft to use to park at the International Space Station. If the touchscreens were sleek and modern, the docking system looked like it had been built by a hobbyist in a garage with spare parts. Which was because it essentially had been.

Leading up to the award of the Commercial Crew contract, NASA's engineers were perplexed that SpaceX would even go through the trouble of building a docking system on its own. NASA had been working to design one for more than a decade, in partnership with Boeing, and NASA had offered it to SpaceX for free. All the company had to do was install it. This was no small offer. The docking system was one of the most complicated and vital components of the whole spacecraft: It would hold the capsule to the space station with an airtight seal, something that had to work 100 percent of the time.

Boeing, of course, was installing NASA's apparatus on its Starliner. But SpaceX wasn't so sure. Its engineers had designed virtually the entire spacecraft in-house and were skeptical about outsourcing such an important element. The person tasked in 2013 with assessing NASA's docking system was Jaret Matthews, then a thirty-six-year-old engineer who had worked for nearly a decade at NASA's Jet Propulsion Laboratory before coming to SpaceX. It was clear, upon first inspection, that the NASA docking system was robust, as it had to be since it was designed to be used on an array of spacecraft. But Matthews thought he could build something tailor-made for Dragon with a simpler, more elegant design that, importantly, wouldn't add too much mass to the vehicle.

"There was an opportunity to simplify it considerably," said Matthews, who at the time was one of the lead engineers on Dragon.

At other companies, such a thought would likely have been laughed at. *A young engineer is going to improve on NASA's design? Why even bother?* But SpaceX operated by a "first principles" approach—a philosophy around innovation that had become trendy among many Silicon Valley types. The idea, as Musk has said, was to take a physics approach to "boil things down to their fundamental truths and reason up from there, as opposed to reasoning by analogy." To Musk, reasoning by analogy meant essentially, "copying what other people do with slight variations." This was especially important, he said, "when you want to do something new."

Matthews had been at SpaceX just a few months when tasked to evaluate the docking system. But in that short amount of time, he had understood his job was not to merely take the easy path. His job was to make the best docking system in the world. At SpaceX, he was given the freedom to start fresh, to be creative. Even if that meant rejecting the space agency's system as it was going up against Boeing for the Commercial Crew contract to fly NASA's astronauts.

"Question everything. Everything," as Gwynne Shotwell once said. "That's what innovation is. . . . Nothing is sacrosanct. Nothing."

The rule had the weight of doctrine and applied up and down the company, from interns all the way up to Musk. "Obviously, we want to hire people from the best schools. But where people think you're so smart, and you're so good, you must never be wrong? That's bullshit," Shotwell said. "Elon is wrong. Not very often, by the way, it's super irritating. He's not wrong very often. But he is wrong on occasion, and he wants to be questioned. He's like, 'Don't let me go jump off the stupid cliff. Please tell me if I'm doing something that's really dumb.'"

Matthews imagined how NASA's design could be improved, and working with an intern, he began to sketch out a prototype. The main deviation from NASA's system was in the "soft capture" design, the part that first extends a ring with latches that slides into the space station's port and is flexible enough to adjust any small misalignments. Even though the spacecraft is moving very slowly as it parks itself, the ring has to absorb the initial contact and then be able to make slight adjustments. Once the soft capture is complete, twelve latches clamp into place and hold the spacecraft firmly to the station in a process known as "hard capture."

In NASA's design, there were six mechanical arms that were used to maneuver the soft-capture ring into place. The arms were programmed to act like springs, to be able to absorb the impact when the spacecraft came in contact with the station. But they were complex, heavy, and required a lot of power. If the software or electronics controlling the arms failed, the whole docking could go awry. Matthews and his intern, Craig Western, developed a simpler design using springs and dampers, which required no software or electronics.

The design was so simple that for the prototype they purchased mountain bike shocks from an online cycling shop. They bought the rest of the parts from McMaster-Carr, a sort of Home Depot for industrial tools and materials, including a chain connected to a hand crank that was used to deploy the capture ring. Their prototype, which they dubbed the "McDocker," looked like something out of a nineteenth-century steampunk factory—clunky and rudimentary. But it was also a symbol of first-principles thinking and a manifestation of SpaceX's approach to innovation.

Engineers were not only allowed to question the fundamentals of design but to make them simpler through revision and editing. The policy was part of what Kiko Dontchev, SpaceX's vice president of launch, called the company's "algorithm." "A common misconception in engineering is something that's complex is more reliable or better," he said during a conference. "False. Oftentimes, the most simple thing is the best design, and the most reliable. Complexity, we like to say, is the devil. Spending a lot of time optimizing, simplifying and deleting parts from your product is critical."

Not only had Matthews and Western done that, they had done it quickly, in about two months, fulfilling another SpaceX mantra. As Dontchev said, "There's always time to gain. There's always efficiencies to be brought. You think you went fast enough? Bullshit. You can go faster. Every time my team is like, 'Dude, we can't go faster.' You're like, 'Yes, you can.' Move the goalposts. You'll be surprised what happens when you challenge people."

When their prototype was finished by the fall of 2013, Matthews and Western showed it to Mark Juncosa, one of Musk's most trusted engineers. Unlike some at the company, who shied away from dealing with Musk directly, Juncosa was unafraid of the boss. He told Matthews this was something Musk would want to see, and that they should go show him the prototype that instant. Without an appointment, they rolled the McDocker over to Musk's cubicle and asked him to take a look.

Musk studied it intensely, pulling and pushing on the docking ring, while rubbing his chin. After just a few minutes, he said, "Yep, let's do this." There were no deliberations. No consultations with other engineers. No memos or meetings. Musk liked what he saw and simply made the decision to go.

"At SpaceX, once you have the green light you can just run a million miles an hour in that direction," Matthews said. "There's no hand-wringing about it. You just start going."

At first, NASA was incredulous that SpaceX was rejecting its soft capture design and attempting to build something on its own. When Matthews and his boss flew to Houston to meet with engineers from the agency and present their design as part of SpaceX's Commercial Crew bid, they were met with "blank stares," he recalled. "You could see it behind their eyes," he said. "They're thinking, 'These guys are idiots. They have no idea what they're getting into.' I mean, these people had spent two decades working on this problem. And you can see all that kind of going on in the back of their heads, like, 'You guys are going to do *what* now?'"

If he couldn't overcome NASA's skepticism, SpaceX could lose the contract, Matthews feared. And it would be his fault.

In the months leading up to the contract decision, the agency still wasn't convinced SpaceX was a reliable partner. If anything, it was even more skeptical, citing the new docking system as a "weakness" in SpaceX's proposal. Matthews responded to the space agency with written comments, but still NASA was suspicious, and called for a meeting to discuss the spacecraft at the Kennedy Space Center.

In a windowless room that seemed to have been furnished during the Apollo era, the NASA contract evaluation team sat across a large table from the SpaceX engineers, who squirmed uncomfortably, like PhD candidates having to defend their theses. Musk began the meeting by acknowledging the paradigm shift NASA was undergoing by outsourcing human spaceflight to the private sector. But he assured the board that SpaceX would take that sacred trust seriously. In his typical grandiose fashion, he said the mission was important not just to the nation but to humanity as a whole, a path to becoming a truly spacefaring species.

Then, as Musk looked on, Matthews anxiously began to detail his docking system. The whole contract, worth some $2.6 billion, could come down to his presentation. But being able to show charts and graphs made his task easier—as did the fact that he passionately believed in the system. It *was* a simpler design. Musk thought so too, which is why he had signed off on it in the first place. Matthews left the meeting feeling confident that he had finally changed NASA's mind.

Still, when NASA announced that SpaceX had won the Commercial Crew contract alongside Boeing in September 2014, it came as a shock to many at NASA and even to some at the company.

Despite the victory, the astronauts remained adamant about wanting a stick in the spacecraft. It didn't help their cause when one of

Musk's young sons, a proficient gamer, sat down at the simulator and was able to dock the spacecraft on the first try. (Musk himself was not so adept. Docking was "a very slow flying task, and he would just spam the buttons and try to ram her home, of course to disastrous results," one former SpaceXer recalled. "He didn't have the patience for it.")

"These guys are supposed to be the world's best pilots, and my kid could do it just fine," Musk groused. SpaceX put months into perfecting the software. The engineers followed what's known in Silicon Valley as an "agile" development process, working simultaneously on different parts of the system, instead of the more step-by-step linear approach of fixing one problem before moving on to the next. The programmers at SpaceX weren't afraid to try new ideas. If they worked, great. If not, they'd try something else until they got it right. They iterated, working out all the kinks in the system, until it finally came together.

Bob Behnken and Doug Hurley continued to keep an open mind about the SpaceX approach, however unusual it seemed to NASA. And over the next few years, the more that Hurley and Behnken watched the SpaceX engineers, the more they were impressed with their intelligence, dedication, and speed. They worked hard, and eventually the simulator was flyable. The inputs on the screen now controlled the vehicle in a way the astronauts were used to. It was simpler and easier to fly. But, yes, it was different and took some getting used to. Instead of flying by feel, astronauts had to look at the screen, reach across, and press virtual buttons, in a small but deliberate act that in time became the new normal.

They appreciated that the setup was convenient, even if it was only to be used in an emergency; otherwise, the spacecraft would fly itself. "Everything was right there on the screen that I needed in order to accomplish the task," Hurley said. "From the standpoint of human-vehicle interface, they made it about as good as they can make it." During the day, Hurley, Behnken, and other astronauts would fly in the simulator, offering insights and suggestions to the developers, who would take notes and then work that evening or even overnight to make upgrades.

"And the next morning we would fly the sim with those particular changes," Hurley recalled. "And you would say, 'Oh, yeah, you fixed all that.' It was really cool because you just saw how quickly SpaceX could get something done."

EVENTUALLY THE SOFTWARE was good enough that SpaceX started allowing VIPs to test their skills in the simulator. On a visit to SpaceX's headquarters, soon after being confirmed as NASA administrator in 2017, Bridenstine tried it out. As a Navy pilot, he had flown the F/A-18 Hornet, playing the enemy for the Top Gun pilots. (And yes, it had a stick.) Docking the SpaceX Dragon was easy, he bragged—nothing like landing on the deck of an aircraft carrier. (Eventually SpaceX would even release a version of the simulator online, allowing anyone to test their astronaut skills.)

Surrounded by astronauts in their one-piece flight suits, not too different from the kind Bridenstine had worn in the Navy, he felt close to the astronaut corps, who reminded him of the men and women he'd served with. These were American heroes who were about to accomplish something extraordinary—launching into space from U.S. territory for the first time in years, and on a pair of brand-new spacecraft, no less.

It was a thrilling, momentous time for NASA and for the country, he believed. After Mercury, Gemini, Apollo, and then the Space Shuttle, the agency was now about to write the next chapter in the history of human spaceflight. After the Space Shuttle was retired in 2011, many people mistakenly thought NASA had closed its doors. Astronauts flew on Russian rockets from a Soviet-era Baikonur Cosmodrome in Kazakhstan, a world away from Cape Canaveral. It had been decades since they'd received ticker-tape parades down Broadway after their missions, but now they were truly out of sight and out of the public's mind. Flying with the Russians meant engaging in Russian rituals, no matter how bizarre. Before flights they watched the 1970 Russian film *White Sun of the Desert,* though no one is entirely sure why. They were blessed by an Eastern Orthodox priest wearing a golden robe, who splashed them in the face with water in a disconcerting spurt that could feel like it came from a water gun. Perhaps the most well-known ritual consists of the astronauts urinating on the right rear tire of the bus that transports them to the launch site, because that's what Yuri Gagarin did in 1961, when he became the first person to fly to space.

The Soyuz spacecraft isn't built for comfort. All three astronauts are wedged in together so tightly they can barely move. And mission control doesn't bother cluing them in to the countdown. There's no *ten, nine, eight* . . . Instead, the astronauts get a five-minute warning, then another at one minute, and then the engines begin to roar.

Landing in the Soyuz has been described as a "train wreck followed by a car crash followed by falling off your bike." Former NASA astronaut Scott Kelly described it this way in his memoir, *Endurance:* "When the parachute opened, the capsule spun and twisted and turned violently in every direction. If you can get in the right frame of mind, if you can experience it like an adventure ride, this can be great fun. On the other hand, some astronauts and cosmonauts, after their first Soyuz landing, have said that they were being thrown around so violently they became convinced something had gone wrong and they were going to die."

The Space Shuttle landed like an airplane on the tarmac at the Kennedy Space Center, where the astronauts were given a heroes' welcome. The Soyuz crashed in the frozen Kazakh desert, hundreds of miles from civilization, where yak herders sometimes showed up on horseback.

Bridenstine wanted the astronauts to become heroes again. Those entering service today, he believed, were every bit as accomplished as Neil Armstrong, John Glenn, and the explorers who seized hearts and headlines at the dawn of the Space Age—maybe even more so. Take Jonny Kim, a first-generation Korean American, whose father operated a liquor store and whose mother worked as a substitute elementary school teacher. He'd enlisted in the Navy and become a SEAL; while serving in Iraq in 2006, he witnessed two squad-mates being shot and killed. Unable to save them, he decided to become a doctor. A few years after graduating from Harvard Medical School he was selected as a NASA astronaut, one out of eighteen thousand applicants. It was a stunning career path that meant, as Senator Ted Cruz of Texas once said, "He can kill you and then bring you back to life. And do it all in space."

This was the real Right Stuff. And that was exactly what Bridenstine wanted to highlight. "I'd like to see kids growing up, instead of maybe wanting to be like a professional sports star, I'd like to see them grow up wanting to be a NASA astronaut, or a NASA scientist," he said during a meeting in which he urged NASA to better promote its stars. "I'd like to see, maybe one day, NASA astronauts on the cover of a cereal box, embedded into the American culture."

The Astronaut Office at the Johnson Space Center in Houston was a problem, however. It was its own secretive realm, walled off from the rest of the space agency. How it operated was a mystery. Even the astronauts groused that the selection process for crews—who got to fly and when—was as unknown to them as it was to the general public. The only thing more mysterious than crew selections was how they were chosen to be astronauts in the first place.

The culture was steeped in military tradition, promoting selflessness, the group over the individual. It was an ethos that existed from the beginning of the Space Age. When it came time to design a mission patch for the Apollo 11 moon landing, Neil Armstrong, Buzz Aldrin, and Michael Collins forbade NASA from putting their names on the patch, as had been custom. The moon landing was a team effort, the result of countless hours, from thousands of people toiling to make the mission a success. So, no names, the astronauts insisted. Today, that story is legend in the halls of the Johnson Space Center. Even now, when it comes to offering interviews to reporters, NASA will often make the entire mission crew available—or no one at all—so one astronaut won't get singled out over the others.

As administrator of NASA, Bridenstine ran into this wall head-on. He respected the self-effacing culture and, coming from the military, was familiar with it. Still, he had a job to do—promote the agency and what it was trying to accomplish. The more the public saw the astronauts—their stoicism, their probity, their optimism—the more he could garner support from the country as a whole, he felt. More importantly, as the politician in him knew, he'd get support from Congress, which he desperately needed to fund the human missions to the space station, as well as the administration's renewed effort to return to the moon.

When Bridenstine said he wanted to put on a show to announce the crews that would lift off from U.S. soil, the Astronaut Office was a hard no. They hated the idea of a spectacle and did not want astronauts used as pawns in the political game Bridenstine and his Trump-appointed staff were playing. But Bridenstine's aides were insistent. The answer from Houston was always *No* at first. The first *N* in NASA, some liked to say, did not stand for *National*. Bridenstine and his team would work to find a way to yes. That included the promise that the astronauts would be introduced and celebrated as teams—one group assigned to fly on SpaceX's Dragon capsule, the other on Boeing's Starliner.

It worked. The event was held in August 2018. Bridenstine brought the astronauts on stage at the Johnson Space Center auditorium and recited their impressive biographies. NASA packed the room with employees and VIPs, including members of Congress, whom Bridenstine singled out with praise and gratitude. And as the astronauts took their seats on the stage, Bridenstine asked them questions about themselves and the missions they were about to undertake.

To Behnken, who along with Hurley was assigned to fly on SpaceX's first human spaceflight mission to the space station, he said, "For some of you, the last spacecraft you flew was the Space Shuttle, which was designed in the 1970s. Bob, can you tell us about the advancements in the technology and design of the SpaceX Crew Dragon?"

"The way we described the Space Shuttle was there are about 3,000 switches inside," Behnken said. "There was no situation that the astronauts couldn't make worse by touching the wrong switch at the wrong time. . . . We're grateful the next vehicle we're going to fly on is going to be a little bit more automated and doesn't have as many switches."

"It's like flying an iPhone," Bridenstine said.

OUTWARDLY, SPACEX AND BOEING executives professed that there was no competition between them over who would fly first. The astronauts displayed nothing but comity and teamwork, rooting for one another. Soon, both groups would fly, and that would be good for the country, giving it two spacecraft capable of flying astronauts.

Privately, however, it was cutthroat. Boeing was still the favorite to fly first and win the honor of restoring human spaceflight to American shores. Everyone thought so. Four years after it had won its contract, SpaceX was still seen as a backup to Boeing in some quarters of NASA. It didn't matter that by late 2018, SpaceX had completed fourteen successful flights to the space station, delivering cargo and supplies, or that the Falcon 9 was quickly becoming the industry's workhorse rocket. "The prevailing opinion within NASA in general was, 'Hey, we got a sure thing with Boeing, because of their heritage and history, you know, their big aerospace blah, blah, blah,'" Hurley recalled. "'And if we get the second provider with SpaceX, it's a bonus.'"

To be assigned to fly on SpaceX, then, was at best a dead end—because those astronauts would never fly. It might even be a death sentence—because if SpaceX did fly, the thinking was, "If they don't kill you, you'll be lucky," Hurley said. "There were probably five people in the entire agency that thought we'd be successful." SpaceX would treat you like cargo, Hurley was told. You'd be little more than a "biological payload."

That only strengthened Hurley's resolve. He had now spent a few years getting to know the SpaceX team. Yes, they were young and did things differently. But as he watched them build the spacecraft from the ground up, they had not only earned his admiration, but his trust.

"Once I was assigned to SpaceX, it was like, hey, I'm on the team. You guys have embraced us, and we're embracing you. And we're going to get through this together, even though we certainly had some well-documented issues getting to the launch pad. I was as emotionally invested as I have ever been about anything I've done in my life. It's because you had to be. We were the underdog. But we're going to beat you, and we're going to beat you bad."

Moving with urgency, however, didn't mean rushing. Being first meant you crossed the finish line alive in the safest and most reliable spacecraft. "I'm not in this business to be posthumously awarded anything," he said.

To make sure of that, SpaceX would fly Dragon in a test flight to the International Space Station called Demo-1 without anyone on board. Only if that went well would the company then proceed to launch astronauts there. For the uncrewed flight, the autonomous spacecraft would separate from the rocket, catch up to the space station traveling in orbit at 17,500 miles per hour, and dock itself to the orbiting laboratory. It would stay there for a few days, so NASA could see how it fared in the vacuum of space. Then it would fly home, crashing through Earth's atmosphere and testing its heat shield as the capsule generated temperatures of about 3,500 degrees Fahrenheit. Finally, it would splash down in the sea under four parachutes that needed to deploy reliably and at the right time. If everything went successfully, it would help to cement the feeling that the entire endeavor was actually going to happen—the first new major human space campaign in more than forty years.

Trump's first State of the Union address was coming up in February 2019, and Bridenstine was hoping that NASA would be mentioned. He reached out to Scott Pace, the executive secretary of the National Space Council, and urged him to push the president's speechwriters to at least include a line about the progress the agency was making. In his own speeches, Bridenstine had been saying there would soon be "American astronauts flying on American rockets from American soil." It was a ready-made sound bite, and Trump's speechwriters grabbed it, noting also that 2019 marked the fiftieth anniversary of the Apollo 11 moon landing.

The White House invited Buzz Aldrin to the Capitol for the speech. He was sitting in the VIP balcony next to Karen Pence, the vice president's wife, and when Trump singled him out, Aldrin waved and gave two thumbs-up to a standing ovation of Democrats and Republicans alike. Then the president added: "This year, American astronauts will go back to space on American rockets."

"*This year.*" The words took Bridenstine completely by surprise.

There was no way he could guarantee the flights would happen on that schedule. Delays and setbacks were common in development programs, especially one as complicated as building new spacecraft. Just a few months earlier, Boeing had a major setback when it suffered a propellant leak during a test of the emergency abort system of its Starliner spacecraft.

If disaster struck, or if there were more delays, Bridenstine now would be in violation of the president's timeline, one announced on the most public of stages. The reference to 2019 may have been a Trump ad-lib, but for Bridenstine it carried the weight of a direct order from his commander in chief.

And now the clock was ticking.

BY MARCH 2019, NASA and SpaceX were ready for the Demo-1 test flight. Both organizations were happy with the software used by the spacecraft to dock autonomously. And if some at NASA had been initially skeptical about SpaceX's ability to develop its own soft-capture docking system, their doubts melted away the more they watched Matthews and his team working. NASA engineers who visited SpaceX headquarters studied the McDocker prototype Matthews and his crew had built from scratch, and were impressed. It also helped that they would embed themselves for weeks at a time at NASA's Johnson Space Center in Houston, working alongside NASA engineers at a mock-up of the space station to simulate docking in more than 450 tests, as well as thousands more computer simulations.

"They saw how the system behaved, and we started to win fans," Matthews said. "They always gave us anything we asked for, and were so super-rational in our discussions. They were essential to our ultimate success in so many ways."

But now there was a new hiccup. A couple of weeks before the Demo-1 test flight, Roscosmos, the Russian space agency, started complaining. NASA's main partner on the International Space Station was concerned about Dragon's ability to dock itself, and worried it might crash instead. Privately, some at NASA and SpaceX suspected Russia was playing games. Flying American astronauts to the station since the Space Shuttle had retired in 2011 was a lucrative business. With a monopoly on human spaceflight, Russia had jacked up the price and was now charging NASA more than $80 million a seat. Now SpaceX was threatening that revenue stream.

Still, NASA worked to allay the Russians' fears, while also downplaying them. "I don't think it'll be a problem once we go through the details of why it's safe, and we can explain the details of why we're moving forward," William Gerstenmaier, the head of NASA's human spaceflight division, said at the time. This was a test flight. The whole purpose was to see how the rocket and spacecraft performed before putting astronauts on board. If SpaceX encountered problems—and most at NASA believed they would—that was the best way to find weaknesses in the design and fix them. "I fully expect we're going to learn something on this flight," Gerstenmaier said. "I guarantee everything will not work exactly right, and that's cool."

Still, the Russians ordered the crew already on board the station, including a NASA astronaut, Anne McClain, to be ready to evacuate and take refuge in their Russian Soyuz spacecraft, which would be used as a lifeboat in the event that SpaceX's innovations triggered a disaster.

THE LAUNCH, ON March 2, 2019, went smoothly, but the autonomous docking was still a huge hurdle. "Paranoid" was how Benji Reed, SpaceX's director of crew mission management, described his emotional state. SpaceX had never done anything remotely like it before. The version of Dragon that delivered cargo to the station didn't carry people or dock itself to the station. Rather, it pulled up alongside the ISS and an astronaut inside would use the station's robotic arm to grab it and pull it in. But for human spaceflight missions, NASA wanted visiting spacecraft to be able to park themselves.

"If something goes wrong, you're just in the void of space," Musk once told me. "You're in a vacuum—with nothing. You have the space suits and a lot of backup systems. But it's still a dicey situation."

Matthews, too, was "absolutely terrified." Docking Dragon amounted to "crashing into the space station. It's a controlled crash, but it's a crash, nonetheless." Essentially, he said, it was "a 20,000-pound vehicle running into a $100 billion national lab."

The SpaceX and NASA teams were already on edge. When the spacecraft attempted to open its nosecone, which would reveal the docking system, one of the Dragon's sensors reported that it had opened successfully. But the other said it did not.

It was a tense few moments, but soon cameras on the space station showed Dragon floating through the blackness of space with its nose cone open, passing through several predetermined waypoints, inching closer and closer. Soon it loomed large.

From his console in SpaceX's mission control center in California, Matthews watched with his arms folded over his head, as nervous as he had ever been about anything in his life. Next to him sat his wife, Aarti Matthews, the Demo-1 mission manager. If the docking didn't go well, he'd not only let down his employer, but his partner too. Dragon continued to approach the station smoothly and steadily, until finally it eased into place gently, like a nice, soft kiss.

"Soft capture confirmed," the mission control director called out, and Matthews was on his feet, high-fiving his colleagues.

But while his portion of the mission was essentially over, Dragon's return a few days later, on March 8, was another harrowing leg of the journey. That, too, would prove the doubters wrong. The spacecraft undocked, then plummeted back toward Earth and slammed into the atmosphere. Its speed generated a fireball that engulfed the capsule and charred the outside like a marshmallow. But everything worked as designed. The heat shield. The chutes. The recovery. Splashdown was scheduled for 8:35 A.M. And, after decelerating from 17,500 miles per hour, the capsule hit the water at exactly 8:35 A.M.

SpaceX had put a mannequin—named Ripley, after Sigourney Weaver's character in the 1979 movie *Alien*—in one of the crew cabin seats and outfitted it in a space suit laden with sensors. Their data showed that any real astronauts would have had a very comfortable ride. "I can't believe how well the whole mission has gone," Reed said, at a press conference after the splashdown. "Everything happened just perfectly—down to the second."

Steve Stich, who oversaw the Commercial Crew program for NASA, said the agency was getting close to finally putting astronauts on board. "I don't think we really saw anything in the mission so far that would preclude us from having the crewed mission later this year," he said.

Bridenstine, not one to show much emotion, was elated—and relieved. He had flipped out when Musk took the hit of marijuana on the Joe Rogan podcast, and the safety review was going to start soon. But now he could have hugged Musk, and all the SpaceX engineers, who were making spaceflight look easy. SpaceX not only had won the contract, but it was now continuing to earn NASA's trust. Blue Origin could not say the same. It had decided years earlier to sit out the competition and, as was now becoming clear, had ceded precious ground to one of the fastest-moving companies on the planet.

Even though SpaceX's test flight had been a resounding success, Bridenstine wouldn't allow himself to get too emotional. There was still a long road ahead. While he was hopeful that Stich was right—that NASA would fulfill Trump's directive to fly astronauts by the end of the year—he knew setbacks could pop up at any time.

CHAPTER 10

By Any Means Necessary

On March 12, 2019, a few days after Dragon splashed down successfully, Jim Bridenstine was once again struggling to keep his emotions in check. This time, though, it was anger the NASA administrator was trying to repress. A pair of Boeing executives had arrived at his office to deliver some very bad news. Quietly seething, Bridenstine tried to make sense of what they were saying and what it would mean.

Leanne Caret, the president and CEO of Boeing's Defense, Space, and Security division, and Jim Chilton, who oversaw the company's Space and Launch division, had come to NASA headquarters to tell Bridenstine that the gargantuan rocket they were building to return the agency to the moon was going to be delayed yet again.

The unimaginatively named Space Launch System had been in development for nearly a decade, more time than had elapsed between John F. Kennedy's declaration that the United States would put a man on the moon and when Neil Armstrong left his footprints there. Not only had the rocket been delayed repeatedly, but its costs had ballooned to $20 billion and counting. The longer it stayed on the ground, the more it appeared to be not a vehicle for deep space exploration, but a symbol of government waste and inefficiency. Or, worse, that NASA had withered to the point of obsolescence. In the years since Apollo, many asked why NASA could not repeat the feat. The SLS was one of the reasons.

The White House had been trying to jump-start NASA's lackluster programs, particularly the development of the SLS rocket and Orion spacecraft, which had been languishing since their inception in 2011 and had still yet to fly together. In December 2017, President Trump had signed Space Policy Directive-1, officially making a "return of humans to the moon for long-term exploration and utilization" national policy. The memorandum also directed NASA to ensure the moon missions would be "followed by human missions to Mars and other destinations," though the locations of those other destinations were not named, and a plan and schedule for Mars was left undefined.

The first step in achieving those goals was to be in June 2020, in the months leading up to the November presidential election, when the SLS rocket would fly Orion around the moon. The flight, known as Exploration Mission-1 (yet another example of NASA's lack of imagination), would not have any astronauts on board. Still, after years of failed deep-space human exploration efforts, it would be a moment of triumph the administration could claim as its own—and no doubt use in campaign ads.

For months, Bridenstine had been trumpeting the upcoming launch as proof that NASA was on the right track. Now, in his office, Caret and Chilton were telling him that EM-1 would not launch in June 2020. The rocket simply would not be ready. Boeing's current schedule, they said, called for a delay of more than a year.

Bridenstine couldn't believe what he was hearing. The new date, November 2021, wasn't just well past the election; it was so far into the future that Bridenstine might not even be the NASA administrator by then. This was unacceptable. How was he going to explain it to the White House? To Trump?

The timing was awful. In 2018, China had launched thirty-nine rockets, more than any other nation, showcasing the arc of its ambitions. It sent up the beginnings of its BeiDou satellite constellation, a global navigation system designed to compete with GPS and bolster the Chinese military. The country's most impressive feat came when China launched a rocket that landed a spacecraft on the far side of the moon in January 2019, something that had never been accomplished by any other country.

Landing there was an enormously difficult feat that demonstrated just how far China's space program had come. The China National Space Administration said the Chang'e-4 mission "marked a new chapter in the human race's lunar and space exploration," and indeed it did. The moon rotates around its axis. But since it takes precisely as long for it to rotate as it does to complete its monthly orbit around Earth, it shows us only one side, "like a dancer circling, but always facing, its partner," as NASA has put it. In order to communicate with its lander on the moon's far side, China had separately launched a relay satellite to a spot where the Earth and the moon's gravity balance each other out, known as a Lagrange point—another first.

If NASA's engineers looked on in awe, American scientists did so with envy. The spot on the moon that China was now exploring with its rover, the South Pole–Aitken Basin, was believed to be the site of an ancient meteor strike that could help solve the mysteries of the formation of the solar system. The basin had been one of the top priorities for the U.S. National Academies of Sciences, Engineering, and Medicine for decades. In China, the *Global Times,* a news organization run by the Communist Party, ran an exuberant editorial. It parroted the language of John F. Kennedy and Neil Armstrong while also asserting that the new era of space exploration would hew to Chinese values. "Unlike mankind's mania in the past, the Chinese people ultimately harbor the dream of shared human destiny and practices open cooperation," it read. "We choose to go back to the moon not because of the unique glory it brings, but because this difficult step of destiny is also a forward step for human civilization!"

There were also signs that its ambitions in space were not just about furthering human civilization, but territorial. The year before, Ye Peijian, the leader of the country's lunar program, invoked China's sovereign claims to islands in the South China Sea as an analogy for what it hoped to achieve on the lunar surface. "The universe is an ocean, the moon is the Diaoyu Islands, Mars is Huangyan Island," he said. "If we don't go there now, even though we're capable of doing so, then we will be blamed by our descendants. If others go there, then they will take over, and you won't be able to go even if you want to. This is reason enough." He also predicted that "in the foreseeable future, we will be in a leading position in lunar exploration."

China was planning another moon mission within a couple of years, one that would return samples. A space station to rival the aging International Space Station was in the works. China was also gearing up to send a rover to Mars, making it only the second country, after the United States, to operate a spacecraft on the Red Planet. More was coming in a new Age of Discovery that resembled Europe's quest in the fifteenth and sixteenth centuries to cross oceans to the New World.

And here NASA's moon rocket was being delayed yet again.

BRIDENSTINE WAS ANGRY, but he didn't yell or pound the table. That wasn't the style of the former Navy pilot. He informed Caret and Chilton that this was a stunning disappointment, and that they needed to get the SLS program on track, and fast. After they left, offering their apologies and vows that they were working diligently, Bridenstine called James Mazol, one of his aides when he had served in Congress but who was now working for Senator Robert Wicker, the Mississippi Republican who was chairman of the Senate Commerce Committee.

The following day, Bridenstine was going to testify before the committee during a hearing titled "The New Space Race: Ensuring U.S. Global Leadership on the Final Frontier." Bridenstine told his former aide that he wanted Wicker to ask him a pointed question about the status of the SLS rocket. He had a bombshell he was ready to drop.

WICKER OPENED THE hearing with his first question for Bridenstine teed up for him. "Last week," he began, "NASA informed Congress of yet another delay in EM-1. NASA had planned to launch no later than June of 2020. However, NASA now says that further delays are anticipated. What about that? What are your plans to address this situation?" Peering down from the dais, Wicker looked up from his notes at Bridenstine. "And have you considered alternatives?"

"Yes, sir," Bridenstine responded, grateful that his planted question had surfaced so early in the hearing. "We have considered alternatives."

He took a moment to praise the potential of the SLS rocket, which was supposed to be the most powerful ever launched. But then he threatened the very vehicle that was at the center of NASA's moon exploration plans.

"I want to be really clear," he said. "I think we as an agency need to stick to our commitments. Sir, if we tell you and others that we're going to launch in June of 2020 around the moon, which is what EM-1 is, I think we should launch around the moon in June of 2020. And I think it can be done. We need to consider as an agency all options to accomplish that objective."

Among those options, he said, was to sideline the SLS and instead fly the Orion capsule on another rocket. In other words, he was saying NASA might outsource one of the most important missions in years to SpaceX's Falcon Heavy rocket or the United Launch Alliance's Delta IV Heavy. Both of these rockets could theoretically replace SLS. This was a radical departure from the way NASA had long done business. Despite its troubled history, the enormous costs and repeated delays, the SLS was so untouchable that it transcended politics.

The SLS had survived as long as it had for one reason: Senator Richard Shelby, the powerful Republican chairman of the Appropriations Committee, which was responsible for doling out money to the federal government's various agencies. For nearly a decade, one of Shelby's main pet projects had been the SLS, which had been headquartered in his home state of Alabama. The fact that it had never flown was of no concern. It provided thirteen thousand jobs and pumped $2.4 billion into the state's economy.

While Boeing was the SLS's prime contractor, the rocket was also supported by other heavy hitters in the military-industrial complex. Aerojet Rocketdyne supplied the engines. Northrop Grumman built the side boosters. And the United Launch Alliance built the rocket's upper stage. (Which also meant it would push back at attempts to have its Delta IV Heavy replace SLS.) Manufacturing was spread across several key congressional districts, from Alabama to Texas to Florida, so that it enjoyed widespread—and bipartisan—support in Congress, which every annual budget cycle pumped about another $2 billion into the program.

After all, it was Congress, not NASA, that gave rise to the SLS in the first place, amid some of the biggest tumult the space agency had ever endured. In 2010, the Obama administration soured on NASA's flagship human exploration effort at the time, called Constellation. That program would have built the Orion spacecraft and a pair of rockets: one to replace the aging Space Shuttle and fly astronauts to the International Space Station, and the other a large, heavy-lift rocket to fly astronauts to deep space. But after years of cost overruns and delays, the Obama White House determined that Constellation was on "an unsustainable trajectory" and "perpetuating the perilous practices of pursuing goals that do not match allocated resources." It moved to kill the program.

Congress had other ideas. Led by Senators Shelby (who had NASA's Marshall Space Flight Center in his state), Nelson (of Florida and the Kennedy Space Center), and Kay Bailey Hutchison (of Texas and the Johnson Space Center), Congress passed the NASA Authorization Act of 2010, reversing the White House's decision by reinstating Orion and rebranding the Ares V heavy-launch vehicle as the Space Launch System. Not only that, the law directed NASA to build the rocket using Space Shuttle and Constellation contracts so that the engines used on the shuttle would power the SLS. In other words, NASA's new rocket would use technology designed in the 1970s, while preserving jobs and keeping money and contracts flowing into key states.

Not everyone went along with the effort to keep the SLS, however. Dissenting in a speech on the House floor, Gabby Giffords, then the Democratic Congresswoman from Arizona, who happened to be married to Mark Kelly, the NASA astronaut, charged that the bill "will force NASA to build a rocket designed by senators and not engineers."

Critics started referring to it as the *Senate* Launch System, or even the *Shelby* Launch System. As the SLS's protector, Shelby had shielded the rocket against numerous government watchdog reports that highlighted its cost overruns, schedule delays, and NASA's lax oversight. He proudly ensured that every year, the SLS would be fully funded whether it flew or not. When it came to SLS and parochial interests involving his home state of Alabama, he played rough. As one longtime lobbyist told me, "Shelby doesn't go after the quarterback. He goes after the quarterback's family." And no NASA administrator, not even one serving a Republican administration, was going to threaten his rocket.

PUBLICLY, SHELBY ISSUED a polite rebuttal to Bridenstine's congressional testimony. "While I agree that the delay in the SLS launch schedule is unacceptable, I firmly believe that SLS should launch Orion," he said in a statement.

Privately, he and his staff were less courteous. After the hearing, Bridenstine dispatched Jim Morhard, the deputy NASA administrator, to call Shelby's office to try to smooth things over. Morhard was a longtime Senate staffer who'd served as the chamber's deputy chief protocol officer—the perfect person to try to make peace. It didn't go well. He was only able to reach an aide in Shelby's office, who refused to put the senator on the line and instead berated Morhard with a string of invectives and expletives so loud that Morhard had to hold the phone away from his ear while Bridenstine, sitting nearby, looked on nervously.

"Would Senator Shelby speak with Bridenstine?" Morhard asked gently.

"Is Bridenstine going to submit his letter of resignation?" the aide said.

"No."

"Well then, we have nothing to talk about," the aide said. "You just tried to cancel the senator's program."

"Well, no," Morhard said again, trying to stay calm. "That's not what we did."

Bridenstine was astonished that a high-ranking Senate aide would act this way. When the call was over, he turned to his deputy and said, "Well, we sure made them angry."

Morhard, recovering from the broadside, gave a deadpan response: "You think?"

SHELBY WAS NOT someone you wanted as an enemy in Washington. Within a couple of days, Brindenstine secured a meeting and visited Shelby's office in the Capitol to try to patch things up. But when he arrived, Shelby's secretary apologized and said that the senator was terribly busy and Bridenstine was just going to have to wait. So he did—for an hour—a bit of purgatory, he thought, for crossing the powerful Appropriations chairman.

"You know, I like NASA, but I'd just as soon sell it," Shelby said when he finally met with Bridenstine. The SLS dispute wasn't about NASA, he said. "This is about northern Alabama." If Bridenstine didn't change course, Shelby threatened to call for his resignation.

Bridenstine made it clear that the SLS was a national asset, a key part of NASA's plans, and that it could very well be the best option for the mission. But he also held firm. NASA was going to do a review and look at alternatives. He reiterated what he'd said in public: "Sir, we're going to look at all options to get to the moon."

Bridenstine's testimony had created shockwaves beyond just the office of the senator from Alabama. The space press corps had rarely seen a NASA administrator make such a daring move, even if it was destined to fail. Trying to sideline SLS in an effort to get it going demonstrated that either Bridenstine was crazy or that he and the White House really were serious about creating a sense of urgency for getting to the moon. Either way, it was a great story.

"In a Major Shift, NASA May Use Commercial Rockets for Next Moon Mission," read the headline from *Axios*. *Forbes,* in a story titled "Surprise NASA announcement puts future of new mega-rocket in doubt," wrote that Bridenstine's announcement "shocked the space world."

If Bridenstine was worried about the political fallout and Shelby's wrath, his concerns were soon allayed. Pence was thrilled by the attention Bridenstine, and, as a result, the White House's moon effort, was getting. This was just what NASA needed, he told Bridenstine a few days after the hearing, during a meeting in the vice president's West Wing office. It would also shake up Boeing. Pence liked the idea of threatening to bench the SLS so much that he vowed to borrow it and use it himself. In a couple of weeks, Pence had a big speech coming up at the Marshall Space Flight Center in Alabama. He was going to do some shaking up of his own.

PENCE'S SPEECHES USUALLY went through several drafts and were often revised on Air Force Two, where he would dictate changes. To his speechwriters' chagrin, Pence was sometimes known to edit a speech even after the text had been loaded into the teleprompter, creating a bit of a scramble. But as the vice president liked to say, "Remember, there's no such thing as good writing, only good rewriting." A former talk-show radio host in Indiana, he had developed a folksy style, heavy on biblical references and a tendency to go on too long. When he would ask his wife how a show went, she would tell him, "It was really good—once you started taking calls."

The address he was going to deliver at the fifth meeting of the National Space Council in Huntsville, Alabama, on March 26, 2019, would also go through a series of last-minute revisions as speechwriter Matt Grinney channeled the vice president's thoughts. The most prominent edit didn't come from Pence, but rather his boss. Trump didn't like surprises, and Pence had survived as long as he had under the mercurial and volatile president in part because he diligently kept Trump in the loop. Before heading to Huntsville, he informed the president that he intended to use the speech to light a fire under NASA and get it moving on the White House's chief space policy goal—a return to the moon.

Nearly a year and a half after Trump had signed Space Policy Directive-1, the moon effort was lagging. NASA was aiming to send astronauts to the lunar surface by 2028, but in reality it was no closer to achieving that than it had been under any of the presidential administrations since the Apollo era. Despite all the Trumpian hype about "making space great again," the signature effort of the president's space policy was headed nowhere.

Even though his own policy called for a return to the moon, Trump had thought it was a lame destination. The United States had already done it, fifty years ago. He wanted NASA to send people to Mars—now, *that* would be new and exciting. Never mind the fact that NASA didn't have the budget or capabilities for a Mars mission. If the moon was going to be the goal, Trump wanted NASA to move faster; 2028 was too far away. He wanted a win now, something he could take credit for. Just before the Huntsville speech, he told Pence to move up the timeline.

Not 2028, Trump said: 2024.

WARY OF LEAKS, the vice president's office didn't share the speech with anyone outside the White House. But the night before the National Space Council session, while staying on the outskirts of Huntsville in a run-down motel with holes in the ceiling, Bridenstine's staff got word about the expedited timeline. They also learned that the vice president was going to single out NASA's leadership, saying, "In Space Policy Directive-1, the president directed NASA to create a lunar exploration plan. But as of today, more than fifteen months later, we still don't have a plan in place."

A public scolding from the vice president was not good in any administration. That was especially true in the Trump White House, where cabinet members and political appointees were being shown the door at an unprecedented rate. Bridenstine's aides told him that he had to somehow buttonhole the vice president before the speech and assure him that 2024 was doable and that there really was a plan. The next day, shortly before Pence took the stage, Bridenstine was able to pull him aside and assure him that NASA would meet whatever timeline the White House set out.

"Sir, I want you to know we're making this happen," he told Pence. "We have a plan. We're going to get it done."

Moving the date from 2028 to 2024 was going to come as a shock to the NASA workforce, many of whom weren't fans of Trump and would see the accelerated timeline as a political stunt. And despite all the talk of sidelining SLS, it was clear there were no alternatives. NASA was stuck with SLS, a fact that would doom Pence's ambitious schedule before he even made it public.

EVERY SPACE SPEECH gets compared to John F. Kennedy's call to send astronauts to the moon "not because it is easy, but because it is hard." None of them are its equal. That's in large part because in the years after he delivered it, Kennedy's speech morphed from rhetoric to prophecy, as NASA, receiving wartime levels of funding, as if it were an adjunct of the Pentagon, made the impossible happen in order to beat the Soviet Union.

At the time of Kennedy's speech at Rice University, in 1962, NASA was four years old. It had sent to space a total of four men who, combined, had spent a grand total of ten hours and twenty-two minutes in space flight. The idea of going to the moon was initially thought laughable, especially by the end of the decade, as Kennedy had vowed.

Speaking before a packed house that included several cabinet members, space executives, and luminaries such as Buzz Aldrin, Pence, by contrast, wasn't pushing NASA to do anything in 2019 that it hadn't already accomplished fifty years earlier, and with far less computing power. He wanted NASA to rid itself of the detritus that had encrusted the agency since Apollo. Pence wasn't going to Huntsville to inspire a nation; he was going there to kick some ass.

No one was exempt. Not NASA, not its workforce or contractors. Not past administrations or Congresses that were responsible for the fact that "NASA's exploration efforts were left adrift with no clear direction, focus, or mission," as he said. He took aim at Boeing and the SLS rocket, a mismanaged program that he lamented "has been plagued by bureaucratic inertia, by what some call the 'paralysis of analysis.'"

"After years of cost overruns and slipped deadlines," he continued, delivering his remarks at the U.S. Space & Rocket Center, the Alabama museum home to Space Camp, "we're actually being told that the earliest we can get back to the moon is 2028. Now, that would be eighteen years after the SLS program was started and eleven years after the president of the United States directed NASA to return American astronauts to the moon. Ladies and gentlemen, that's just not good enough. We're better than that. It took us eight years to get to the moon the first time, fifty years ago, when we had never done it before, and it shouldn't take eleven years to get back."

He repeated the line Bridenstine's aides feared but, thanks to Bridenstine's pre-speech intervention, with an ad-libbed caveat: "In Space Policy Directive-1, the president directed NASA to create a lunar exploration plan. But as of today, more than fifteen months later, we still don't have a plan in place. But Administrator Bridenstine told me, five minutes ago, we now have a plan to return to the moon."

What NASA needed now, he said, was "urgency," a word Pence used six times in the speech, as he raised the specter of China's growing prowess in space and cast their ambitions as another Sputnik moment.

"Make no mistake about it: We're in a space race today, just as we were in the 1960s, and the stakes are even higher." In January, he noted, "China became the first nation to land on the far side of the moon and revealed their ambition to seize the lunar strategic high ground and become the world's preeminent spacefaring nation."

Then Pence dropped the bombshell. Saying he was speaking on behalf of President Trump, he said the "stated policy" of the administration was to get astronauts back on the moon in five years—by 2024.

"The first woman and the next man on the moon will both be American astronauts, launched by American rockets, from American soil," Pence proclaimed. And NASA and Bridenstine were "to accomplish this goal by any means necessary."

By any means necessary. Bridenstine took that as more than a directive; this was an order. And it was clear Pence meant it. He made good on his vow to steal Bridenstine's line about sidelining SLS. In the speech, Pence said, "If our current contractors can't meet this objective, then we'll find ones that will. If American industry can provide critical commercial services without government development, then we'll buy them. And if commercial rockets are the only way to get American astronauts to the moon in the next five years, then commercial rockets it will be."

He even took aim at the space agency itself, saying NASA needed to "transform itself into a leaner, more accountable and more agile organization." As Pence saw it, the mission or the destination wasn't the problem. Organizational dysfunction was.

The destination was the moon's South Pole. China, too, was aiming for the same place, Pence knew, and whoever got there first could end up claiming a key component to sustaining life on the moon—water—which could also be used to create rocket fuel on-site.

Despite Bridenstine's assurances that NASA would make 2024 happen, in reality he had no idea how the agency was going to accomplish it. There was no real plan. Pence's speech had sent shockwaves through the industry, and private companies hoped the accelerated timeline would come with an infusion of cash that would pad their bottom line. Lockheed Martin, which was building the Orion spacecraft for NASA, said a 2024 landing could be achieved "with the right level of commitment, urgency, and resources." Blue Origin saw opportunity as well and immediately started lobbying the White House, saying it was poised to meet the 2024 timeline. And Boeing, chastened, said it had "implemented changes in both processes and technologies to accelerate production" of the beleaguered SLS rocket.

Musk responded to a tweet from NASA about Pence's speech, writing, "It would be so inspiring for humanity to see humanity return to the moon!" But when asked if his BFR system—which Musk had rechristened as Starship in late 2018—would be ready in time to meet Pence's goal, his response was noncommittal; he said only that it would be "for sure worth giving it our best shot!"

Those without a financial stake in the outcome, however, believed a landing in 2024 was not possible. Not even Aldrin, who was sitting in the audience next to Homer Hickam, the former NASA engineer and author. Aldrin's hearing was going, and Hickam had been whispering into his ear what the vice president was saying. "When the vice president said, 'We're going to go back to the moon in 2024,' I translated that to Buzz, and he shook his head," Hickam recalled. He said, "Too early. Way too early."

The heads of the space agencies from other countries felt the same way. At a space conference the same day of Pence's speech, David Parker, the director of human and robotic exploration at the European Space Agency, pulled up a chart from the month before that showed how NASA was planning for a human landing in 2028. "This was NASA's architecture for getting to the moon when I woke up this morning," he quipped. "Maybe it's evolved over the course of the day."

AS BRIDENSTINE WAS reeling from Pence's announcement of the 2024 timeline, his mind wandered to another phrase in the speech, one that had taken him completely by surprise: "First woman." He'd had no idea the vice president was going to call for a woman to land on the moon. Politicians were not usually in the position of making crew assignments. But Pence had been struck by the growing diversity of the astronaut corps, which had changed dramatically since the days of Mercury, Gemini, and Apollo, when the only diversity among the white male astronauts was whether they had flown fighter jets in the Navy or the Air Force. While preparing for his Huntsville speech, he'd heard Scott Pace, the executive secretary of the National Space Council, matter-of-factly note that a woman on the moon was an inevitability. Pence had grown excited and stopped Pace. "Do you realize how monumental that would be?" he said. "That's got to go in the speech."

Just two weeks before the Huntsville speech, Pence had gone to NASA headquarters to join Bridenstine in speaking with Anne McClain on the International Space Station, who was about to perform a feat that would make history: the first all-female spacewalk. McClain had all the Right Stuff: She was a West Point graduate and a Marshall Scholar, with a master's degree in aerospace engineering; a lieutenant colonel in the Army, where she flew Blackhawks and Lakotas; a combat veteran who spent fifteen months in Iraq and earned a Bronze Star; an athlete who played on the national women's rugby team and earned the nickname Animal, which she also used as her NASA call sign. And now she was going to make history by venturing outside the space station with another female astronaut, Christina Koch. Both were members of the NASA astronaut class of 2013, the first time the space agency had selected as many women as men for its prestigious training program.

Pence, whose son was a Marine Corps pilot, was starstruck with McClain and eager to speak with her about her time in space. "I know you have an important spacewalk coming up later this month. I understand it is going to be the first all-female EVA. Can you talk to us about your preparations?" he said, using the NASA acronym for "extravehicular activity."

McClain deftly deflected the question, talking not about herself, but rather about the mission and the team, and their work on the "technical preparations to make sure everything is ready." She didn't mention what would make her spacewalk special. She didn't want to be treated differently in any way because of her gender. Sally Ride had been the same way. After she became the first woman in space in 1983, a NASA official handed her a bouquet of flowers to welcome her home. None of the other astronauts on the mission, all of them men, were offered flowers. So Ride refused them.

McClain's gracious but purposefully banal answer carried that selfless tradition forward. It was too bad, then, that in the end she didn't actually get to complete the all-female spacewalk. The space suit that she was supposed to wear didn't fit. During training, back on the ground, she had selected a relatively large size for the torso section of her suit, because she thought that in the weightless environment of space her body would elongate, perhaps by as much as two inches. But once she was actually in orbit, the suit felt too big. The problem was that there was only one size-medium on board the station, which was reserved for Koch. McClain pulled herself from the mission, saying, "This decision was based on my recommendation. Leaders must make tough calls, and I am fortunate to work with a team who trusts my judgment."

Even though she didn't complete the spacewalk, it was still easy to imagine someone of McClain's expertise and character on the moon's surface one day. The astronaut corps had any number of exceptional women. It could be Koch, a scientist who worked for a year as a researcher in the United States' Antarctic program. It could be Nicole Mann, a Marine Corps colonel and fighter jet pilot, who flew combat missions in Iraq from the deck of an aircraft carrier. It could be Jessica Meir, who had a doctorate in marine biology and had worked as an associate professor at Harvard Medical School and at Massachusetts General Hospital. If it wasn't Meir, it could be Jessica Watkins, who had a doctorate in geology. Or it could be Kayla Barron, a Navy submarine warfare officer with a master's in nuclear engineering from Cambridge. Or Jasmin Moghbeli, a Marine Corps helicopter pilot who served in Afghanistan and had a degree in aerospace engineering and information technology from MIT. Or it could be any of the other highly accomplished women in NASA's astronaut corps, who were raising the bar on what "the Right Stuff" meant.

In the vice president's "first woman" announcement, Bridenstine saw a master stroke of political calculation. If he had to get astronauts on the moon within five years, he would need a lot more funding from Congress. To do that, he'd need support from Democrats, generally hostile to all of Trump's plans and policies. But now he could use the fact that NASA intended to make history. This was something even House Speaker Nancy Pelosi would get behind, Bridenstine thought. Who would vote against funding a program that would send the first woman to the moon?

CHAPTER 11

Artemis

By the time of Vice President Pence's galvanizing speech in March 2019, which promised to return U.S. astronauts to the moon within five years, Blue Origin was starting to feel like the corporate behemoth Jeff Bezos had wanted it to become. It had grown to two thousand employees, and its footprint was expanding rapidly. Two months before Pence's speech, Blue Origin had broken ground on a 200,000-square-foot engine manufacturing facility in Huntsville, Alabama, a not-so-subtle ploy to cozy up to Senator Richard Shelby. The company also started work on expanding its headquarters in Kent, Washington, just south of Seattle. In Florida, it announced plans to further enlarge the sprawling campus it was building on Cape Canaveral for the development of its New Glenn rocket. And it had been rehabbing a historic launch site there, Launch Complex 36, just down the road from where SpaceX operated.

This frenzy of building reflected Bezos's determination to make Blue Origin a formidable player in the burgeoning private space industry. Since becoming CEO in 2017, Bob Smith's mandate had been to transform Blue Origin into a functioning company that would be able to compete for, and win, government contracts and start taking in the one thing that had eluded the company for most of its nearly two decades of existence: revenue. NASA, of course, was a main target. So was the Pentagon, which was about to open another round of contracts to launch national security satellites to orbit, which could be worth billions. SpaceX and the United Launch Alliance were the only companies certified to fly the missions. In the next round, the Pentagon was planning again to award just two contracts, making them the clear favorites. Blue Origin might be shut out for years to come.

If Blue really were to compete, Smith thought, it needed better intelligence on its competitors, especially SpaceX. Blue Origin had taken note of Musk's update on Starship, then still called BFR, during his 2017 presentation in Adelaide. Musk had claimed that BFR/Starship would not only be the biggest and most powerful rocket ever built; it would be fully reusable—both its rocket booster and second-stage spacecraft would be able to land and fly again. It would even be able to be refueled in space, he had said, a feature that could allow the vehicle to traverse the solar system.

Musk, of course, was known to make wild predictions, but he also had an impressive track record of making the "impossible merely late," as he liked to say. At a meeting in April 2018, Greg Allen, then Blue's senior manager of market strategy, presented a memo he had crafted for the company's senior leadership. Titled "SpaceX BFR Competitive Analysis and Strategic Implications," it reported that the rocket could not only dominate the U.S. launch market but "plausibly achieve a monopoly" even if it didn't accomplish all of the initial goals that Musk had set out to accomplish.

SpaceX was moving remarkably fast and was "farther along than would normally be expected," Allen wrote. SpaceX had already "developed and tested large composite tank structures for BFR, whereas Blue Origin has yet to build a test tank for New Glenn. . . . Blue Origin should therefore consider SpaceX's BFR as a potentially serious medium term competitive threat and undertake efforts to meet the challenge by a BFR-class competitor."

One way to do that could be to undercut SpaceX's attempts to win government investment. Mimicking the strategy Bezos himself had laid out a couple of years earlier—"From now on, we go after everything

SpaceX bids on"—Allen wrote that Blue could "seek to deny, delay, and diversify government development funding for BFR." That would require warfare in Washington, D.C., where, he warned, SpaceX's lobbyists were already pushing for the cancellation of NASA's SLS rocket and Orion spacecraft, thinking it represented "the best path to receiving billions in BFR development funding."

Among the other options to make Blue competitive would be to start funding several ambitious initiatives, he wrote, including developing a reusable second stage, the ability to refuel its spacecraft in orbit, and eventually the means to store propellants in space—all at a time when it had yet to launch a single rocket to orbit.

Bezos would later tell me that he wasn't interested in copying competitors, but rather looking to their success for inspiration. "My view is you look at what people are doing right and try to be inspired by it," he said. "You don't copy it straight ahead because if you do, you're not pushing the state of the art forward. It just doesn't work as well. Close following is not an easy business strategy. There are some companies that have succeeded with a close-following business strategy. I actually think it's much harder than it seems. You're really much better off pioneering and pushing forward."

IN ADDITION TO better competitive intelligence, Smith felt the company needed access to the ultimate center of power: the White House. But before it could ingratiate itself with the Trump administration, there was a bit of smoothing-over it would need to do. Tension between Bezos and Trump had been simmering for years, heightened by Bezos's ownership of *The Washington Post*. During the 2016 campaign, Trump repeatedly trashed the *Post* for its coverage of him, and he also went after Bezos and Amazon. Bezos, in turn, made a flip joke about shooting Trump off to space, followed by a more serious warning of what he feared a Trump presidency would portend.

Trump's behavior "erodes our democracy around the edges," Bezos said during a conference sponsored by *Vanity Fair* a month before the election. He added that Trump's attempt "to try and chill the media and threaten retribution, retaliation—which is what he has done in a number of cases to people in the media—is not appropriate."

When he bought the *Post* in 2013, Bezos knew the media business would be no walk in the park. But he said he admired the leadership and toughness of Katharine Graham, the former owner who'd stood up to the Nixon White House during Watergate, even when Attorney General John Mitchell pushed back on a story and told reporter Carl Bernstein that Graham was going "to get her tit caught in a big, fat wringer if that's published."

In a note to the *Post* staff shortly after buying the paper, Bezos wrote, "While I hope no one ever threatens to put one of my body parts through a wringer, if they do, thanks to Mrs. Graham's example, I'll be ready." After Trump won the election, however, Bezos grew more conciliatory, offering his congratulations and saying, "I for one give him my most open mind and wish him great success in his service to the country."

Seven months later, in June 2017, Trump invited Bezos and the leadership of the *Post*—including publisher Fred Ryan, editorial page editor Fred Hiatt, and executive editor Marty Baron—to a private dinner at the White House. Over cheese soufflé, pan-roasted Dover sole, and chocolate cream tart, the president "sought to be charming," Baron recounted in his book, *Collision of Power: Trump, Bezos, and The Washington Post*. "It was a superficial charm, entirely without warmth or authenticity." Bezos also made an attempt at comity. When Trump mentioned that his wife, Melania, enjoyed shopping on Amazon, Bezos joked that he should "consider me your personal customer service rep." Despite the jokes and small talk, it was clear that Trump "saw all of us at that table as his foes," Baron wrote. "Perhaps most especially Bezos because he owned the *Post* and, in Trump's mind, was pulling the strings—or could pull them if he wished."

Through the course of his presidency, Trump continued to rant against Amazon, Bezos, and the *Post*, which he started referring to as the #AmazonWashingtonPost—along with the press in general, which he derided as the "enemy of the people." By contrast, he seemed to have a soft spot for the commercial space sector. Maybe he saw himself in the billionaire entrepreneurs driving it. Maybe he liked the big phallic rockets Bezos and Musk were shooting off to space, or saw similarities to the real estate business, skyscrapers shooting into the clouds. Maybe he just admired the press they generated.

Whatever the reason, in 2018 Trump sat in a cabinet meeting in which Vice President Mike Pence was to give an update on the work of the National Space Council. "Rich guys, they love rocket ships. That's good. That's better than us paying for them," Trump said, referring to the federal government, with a trio of model rockets, including SpaceX's Falcon 9, on the table before him.

He praised SpaceX in particular, which had just flown the Falcon Heavy rocket and landed both side boosters gently back on their pads. "I don't know if you saw Elon with the rocket boosters where they are coming back down," he said. "To me that was more amazing than watching the rocket go up because I had never seen that before. Nobody has seen that before. They are saving the boosters and they came back without wings, without anything."

SpaceX had not only disrupted the industry, it had dazzled the president. Now Smith wanted to as well. But he needed the kind of White House access his competitors enjoyed. As two of the largest defense contractors in the world, Lockheed and Boeing were embedded in all levels of government. Once an outsider, SpaceX now also had a line to the White House as one of NASA's premier contractors and had become a player in Washington.

In 2014, SpaceX spent more than $1.5 million on lobbying. By 2019, that had become nearly $2.5 million. Blue Origin, by contrast, spent a mere $70,000 in 2014—chump change in D.C. circles. But as the election started to heat up, and Trump became the front-runner and then the president, Bezos started doling out more. By 2019, the company would spend almost $1.4 million.

To open doors with the White House, Blue Origin turned to the lobbying firm of Barnes & Thornburg, which had close ties to Pence and the Trump campaign—one of its managing partners was a member of Trump's inaugural committee. The expenditure quickly paid off. A few weeks after Pence's Huntsville speech, Smith was invited to the

White House to meet with Marc Short, Pence's chief of staff, and Scott Pace, the head of the National Space Council.

Though Pence was not in attendance, the gathering, in April 2019, was in his West Wing office. It was Smith's first White House meeting as Blue Origin's CEO, and he was hoping to put Blue Origin on the map, or at least in the consciousness of the Trump administration. He wanted to show that Bezos was serious about space exploration and willing to personally fund many of the projects himself. Smith started his pitch on why the Pentagon's launch contracts program should be expanded to allow for a third provider, and how Blue Origin was positioning itself to serve those missions.

He didn't get far before Short interrupted him. "You have a *Washington Post* problem," he said.

The problem was that the *Post*'s editorial board had trashed Pence's Huntsville speech as a political ploy. Under the headline "Mike Pence, Boldly Sending America Back to Where Man Has Gone Before," the editorial writers opined: "Sticking another American flag in the ground just to watch it wave will accomplish little. Anyone calling for the 21st-century lunar renaissance should be thoughtful about the space program's goals and honest about the costs. Under this administration, unfortunately, that may be too much to ask."

It took a moment for Smith to register what Short was saying: He wanted Blue Origin's leadership to press Bezos to have the *Post* write something positive about Pence and his lunar ambitions. This was, of course, an absurd request. Yes, Bezos owned both the *Post* and Blue Origin. But the two companies were wholly separate, and Bezos had vowed at the time not to interfere with the *Post*'s editorial decisions.

Smith was taken aback by this jarring welcome to Trump's Washington. This was not how he had expected the meeting in the White House to go. No, Bezos would not lean on the *Post* to tilt its coverage, in the news pages or in editorials. Later, in 2025, Bezos would reverse that decision, at least in part, writing in an email to the *Post* newsroom that the opinion pages would stand "in support and defense of two pillars: personal liberties and free markets. We'll cover other topics too of course, but viewpoints opposing those pillars will be left to be published by others." The move was largely seen as a way to curry favor with Trump during his second term. But it wasn't the first time Bezos sought to mollify the Trump White House. As it turned out, after the 2019 meeting with Short, Bezos would find himself in a position where he felt he needed to appease the president.

INTERNALLY AT BLUE Origin, they had been calling it the "vision speech," and it had been in the works for months. The talk, set for May 9, 2019, at the Washington, D.C., Convention Center, would be a sort of coming-out for Bezos the Space Nerd. Bezos rehearsed big presentations like this again and again until he was intimately familiar with the material. Every line, every pause, every hand gesture was scripted. The autobiographical talk would trace the arc of his fascination with space, from his childhood to Blue Origin's present-day work, before turning to the company's future and how it hoped to achieve its goal of "millions of people living and working in space."

I had witnessed Bezos's passion for space firsthand in 2017 when I was invited to the annual spring conference he put on at a stylish Palm Springs, California, resort. Called MARS—for Machine learning, Automation, Robotics, and Space—it was an intimate, eclectic gathering of business executives. The CEOs of Ford, Intel, and Bose were there the year I was, as well as thirteen scientists and engineers from MIT and the MIT Media Lab, a few NASA officials, and three astronauts.

Each morning, Bezos sat front and center during a series of TED-like talks with titles such as "Eating Digital," "The New Era of Extreme Bionics," and "Juno's Exploration of Jupiter." In the afternoons, there were demos of bionic dogs prancing about the resort's gardens, virtual reality tours of Mars, drones the size of pennies. In one of the courtyards stood the New Shepard rocket and capsule Blue Origin had flown to the edge of space and back, as the company prepared to finally fly people on it.

On our first night, Bezos entertained us by climbing up a ladder into a nearly fourteen-foot-tall robot that resembled the manually controlled attack robots in the movies *Avatar* and *Alien*. "Why am I feeling so much like Sigourney Weaver?" Bezos joked while inside the cockpit. Andy Weir, the author, read from his forthcoming science fiction novel, *Artemis*. The next night, Intel CEO Brian Krzanich showed off a swarm of dozens of autonomous drones flying in formation to light up the night sky in the form of an American flag, then a three-dimensional sphere, and finally a rocket ship flying to Mars.

MARS was a glimpse into the future, or at least the future Bezos and the tech titans there were hoping to build. So, two years later, when Bezos began his vision speech at the D.C. Convention Center by talking about how in a few hundred years people would be mining asteroids for humanity's energy needs, I was not surprised by a topic that, coming from the lips of some other business executive, might have seemed downright weird.

IN SOME WAYS, the "vision" speech built on another talk Bezos had delivered thirty-seven years earlier.

For most of his life, Bezos had been obsessed with space, not only for the awe of exploration, but for how it could benefit Earth and humanity. As a child, he came across a book called *The High Frontier*, written by Gerard O'Neill, a Princeton physics professor, that envisioned a future where humanity moved into the cosmos by building massive spaceship colonies in orbit around the Earth. As a high school senior at Palmetto High in Miami, Bezos channeled O'Neill's vision, saying during his speech as his class's valedictorian, "The Earth is finite, and if the world economy and population is to keep expanding, space is the only way to go." All heavy industry, he predicted, would move into space so that Earth could be preserved as if it were "a national park."

Now, almost four decades later in May 2019, he invoked O'Neill again. Standing on a stage, with purple-tinted lighting and two enormous display screens behind him, Bezos showed a newspaper article that covered his high school speech with that quote highlighted. "I still believe that," he said, beaming. If humanity spread out into space, "we can have a trillion humans in the solar system," he said. "Which means we'd have a thousand Mozarts and a thousand Einsteins. This would be an incredible civilization."

Where would they live? Not, as Musk was planning, on Mars or any other celestial body. O'Neill had investigated this, Bezos said, by asking the question: "Is a planetary surface the best place for humans to expand into the solar system?" The answer O'Neill came up with was no. Bezos agreed. Mars was too far away. "Round-trip times to Mars are on the order of years," he said. Another problem was that different planets have different gravitational forces. "You're going to be stuck with whatever gravitational field they have, and in the case of Mars that's one-third G," he said, referring to the fact that Mars's gravity is one-third of Earth's.

Instead, O'Neill came up with the idea of "manufactured worlds, rotated to create artificial gravity with centrifugal force," Bezos continued. Like a real estate agent, leafing through a catalog of apartments, he went through a series of short videos depicting what the space colonies could look like.

"These are very large structures, miles on end," he began. "They hold a million or more people each." Inside the first colony, there was a pastoral scene: a red barn, silos, fields full of crops. Then, in the background, the skyline of a futuristic city.

Next video: a mountain scene with an elk standing on a ledge overlooking a river and a verdant valley below. "You could have recreational ones that keep zero G so you could go flying," he continued. "Some would be national parks."

Next: a European village, maybe in Austria or Switzerland, next to a mountain range. "These are really pleasant places to live," he said. "Some of these O'Neill colonies might choose to replicate Earth cities. They might pick historical cities and replicate them in some way."

Next: a futuristic city with tall, spiraling skyscrapers that resembled sculptures more than residences. "There'd be whole new kinds of architecture," he said. "These are ideal climates. These are shirt-sleeve environments. This is Maui on its best day, all year long. No rain. No storms. No earthquakes. What does architecture even look like when it no longer has as its primary purpose shelter? We'll find out."

He was laying it on thick, going for the close. "These are beautiful. People will want to live here. And they'll be close to Earth so that you can return. Which is important because people are going to want to return to Earth. They're not going to want to leave Earth forever."

The audience for Bezos's speech, numbering several dozen, mostly consisted of a handpicked crowd of space enthusiasts. His talk came across as completely sincere, and it was enthusiastically welcomed by the attendees, who seemed eager to believe in Bezos and his vision, no matter how far out it was. What's not to love about an uber-rich guy publicly indulging his dream of recreational space colonies that allow you to fly around in man-made environments where not only the temperature is regulated but the amount of gravity as well?

But I wondered how it would be received by, well, everyone else. Especially at a time when Amazon had been under attack for invading people's privacy with its eavesdropping Alexa, creeping into every nook and cranny of modern life, and developing a stranglehold on e-commerce—all while reaping enormous windfalls. Bezos, too, had been criticized for his extreme wealth, for giving relatively little to charity, for building an empire on the backs of low-paid workers with meager benefits and reports of horrible working conditions. Notoriously, Amazon had once parked private ambulances outside warehouses during heatwaves instead of paying for air-conditioning.

In his speech, Bezos acknowledged the disconnect, perhaps foreseeing the criticism that would likely come. "We have to realize there are immediate problems, things we have to work on. They're urgent. I'm talking about poverty, hunger, homelessness, pollution, overfishing in the oceans," he said. But there were long-range problems as well that could only be solved by building the infrastructure to go to space. On Earth, resources are finite and dwindling rapidly. In space, he said, they are abundant, only hard to access. He wanted to change that. "You either go out to space or you need to control population on Earth," as he once told me. "You need to control energy usage on Earth. These things are totally at odds with a free society. And it's going to be dull. I want my grandchildren to be using more energy per capita than I do. And the only way they can be using more energy per capita to me is if we expand out into the solar system. And then we can keep Earth as this incredible gem that it is."

The colonies in space were all still hundreds of years away, he conceded during his presentation. His purpose in the here and now was to begin to open space to the masses. That, he said, started with bringing down the cost of launching rockets and opening up a new path to orbit, like an early trade route across the Atlantic or a railroad to the West.

"We're going to build a road to space and then amazing things will happen," he said. The first stop, he said, should be the moon. On screen, where the images of the space colonies had been, were the words Pence had said in Huntsville. Bezos read them aloud for emphasis: "It is the stated policy of this administration and the United States of America to return American astronauts to the moon within the next five years."

Bezos paused and then gave his unequivocal endorsement. Even if the editorial board of his newspaper had issues with the expedited schedule, he did not. "I love this," Bezos said. "It's the right thing to do. For those of you doing the arithmetic at home, that's 2024. And we can help meet that timeline."

Earlier in the presentation, he had unveiled a towering, life-sized mockup of the lunar lander spacecraft the company had been working on. Though it was still called Blue Moon, it had evolved quite a bit since Bezos first pitched it at NASA headquarters two years earlier. Speaking fluent engineering, he described the "deep throttle capability" of its engines, its ability to land up to 6.5 metric tons on the moon's surface, and the way it would touch down using machine-learning algorithms that had been programmed to understand the terrain of the moon.

"We were given a gift, this nearby body called the moon," he said. "We know a lot now about the moon that we didn't know during the Apollo days, or even twenty years ago." The most important discovery was the abundant reserves of water in the form of ice nestled in the permanently shadowed craters at the lunar poles. Water was not just necessary to sustain human life, but its component parts, hydrogen and oxygen, could be separated by electrolysis and turned into rocket propellent. Only a true space nerd got excited about rocket fuel. The discovery of water on the South Pole of the moon was precisely why Blue Origin had designed the engines of Blue Moon to run on liquid hydrogen and liquid oxygen.

Behind the scenes, the company was thinking deeply about how to use the resources of the moon to build solar cells and habitats. The settlers going west in the nineteenth century were able to live off the land, and the first space settlers would have to as well. In hundreds of years, space settlers might mine asteroids to build the massive space colonies Bezos fantasized about. But first, he wanted his company to start practicing on the moon.

"It's time to go back to the moon," he said, a statement that put him in lockstep with the White House. "This time to stay."

Throughout his presidency, Trump continued to rail against the *Post* and Bezos. But Bezos saw Pence for the space advocate that he was. And despite the pressure the White House was putting on the *Post* and Amazon, the administration could be good—very good—for Blue Origin. It would not be the last time Bezos would seek to align himself with Pence and defend his space policies against criticism from the *Post*'s editorial board.

THAT NIGHT, BEZOS celebrated his speech by hosting a small group of people at a Washington restaurant. In order to attend, I agreed to keep the gathering off the record. But another guest, Lori Garver, the former NASA deputy administrator, described the evening in her memoir, *Escaping Gravity*. As she wrote:

> [Bezos] gathered about a dozen of us for a relaxed sit-down meal and discussion after his presentation at the DC Convention Center. He held court throughout the feast, asking and answering questions about space and politics—two of my favorite topics. Caroline Kennedy was seated between Jeff and me, and she told us about meeting John Glenn in the Oval Office when she was five years old. After saying hello to the astronaut, she explained, she turned to her father and expressed her

disappointment, saying she thought she was going to get to meet the monkey. I'd heard John Glenn tell the same story, but Jeff's loud and infectious laugh signaled he hadn't heard it before.

Less than twelve hours after the dinner broke up, Bezos was at NASA headquarters again. It was May 10, 2019, and Bezos's goal was the same as it was on his last visit: to pitch the Blue Moon lander to NASA leadership. The previous meeting, in March 2017, had been so early in the Trump administration that Bridenstine had not yet been confirmed as NASA administrator, and the space agency's moon program was still in its infancy. Now, Bridenstine had been given a mandate by the vice president and was eager to listen to everyone who could help him meet it.

Bezos was more than happy to oblige. He again came armed with poster boards mounted on easels. One, titled CARGO DELIVERY, detailed the amount of mass the lander would be able to deliver to the lunar surface. Another vowed that Blue Origin would be ready to land 2 CREW BY 2024, and that the company was working at developing LOX/LH2 EXPERTISE—shorthand for liquid oxygen and liquid hydrogen.

At the bottom, the last line of one of the poster boards, SUBSTANTIAL COST SHARING, which was Bezos's way of saying he was going to pay for most of the development of Blue Moon himself. Bridenstine liked the sound of that. Since the vice president's speech, his team was not only working on how to get astronauts to the moon in five years, but on how much the effort would cost. The numbers were going to be enormous, and Bridenstine was not sure how he was going to get Congress to fund the effort. Having the richest man in the world promise to write some checks certainly wouldn't hurt.

That month, Bridenstine convened a meeting with his top communications aides. It was bothering him that the project was known clunkily as Exploration Mission-1. Internally, the programs were named for their component parts, the SLS rocket and the Orion spacecraft. There was nothing lofty, like Mercury, Gemini, or Apollo. "There's no cohesive story," he told his staff. "I'm trying to mobilize people, but we don't have a program. We have parts and pieces." He paused for a moment. "We need to name it. We need a brand for this."

Some on the team were against naming the program. Give it a name and Congress can kill it, longtime NASA staffers warned him. That's what happened to the Constellation program. Bridenstine didn't want to hear it. The right name, he believed, would bring national pride and support. China knew the power of a good name. Its moon program, Chang'e, was named for the moon goddess celebrated in Chinese folklore and culture. Its Long March rocket program got its name from the historic six-thousand-mile trek by Chinese communists in the 1930s that led to the emergence of Mao Zedong.

Suddenly, Bridenstine popped out of his chair and left the room, telling his puzzled team, "I'll be back." He started roaming the hallways of NASA, asking everyone he ran into if they had any ideas for what to call the moon program. He poked his head into offices and cubicles. Finally, a few days later, he ran into Alex MacDonald in a snack area. MacDonald was NASA's chief economist and the author of *The Long Space Age: The Economic Origins of Space Exploration from Colonial America to the Cold War*. He was also something of a Renaissance man—a historian and academic, with an appreciation for modern art. His father had been a fan of the classics, and when MacDonald was a child, he recounted the *Iliad* and the *Odyssey* from memory as bedtime stories.

"We could always go back to Artemis," MacDonald told Bridenstine.

"Artemis?" Bridenstine said, looking curious but also a bit confused. He was aware that Andy Weir had just published a novel with that title, but he couldn't quite place the name.

Artemis was the twin sister of Apollo, MacDonald explained, and also the goddess of the moon. The only problem was that NASA had already used the name for a mission years earlier, involving a pair of spacecraft used to study the moon and solar wind.

No one had ever heard of the earlier mission, Bridenstine thought. And no one outside of the scientific community cared. Artemis was perfect.

"That's it," Bridenstine said. "That's the name."

. . .

BRIDENSTINE WAS IN love with Artemis and wanted to announce it as soon as possible. Normally, such a decision would need sign-off from the White House and the National Space Council. NASA's Capitol Hill liaisons would be dispatched to give a heads-up to the appropriate congressional committees and make them feel like they were a part of the plan. But Bridenstine didn't have time for a bunch of meetings on the subject. That would only delay the inevitable.

On May 13, he decided to hold a press conference ostensibly to announce that the White House had agreed to boost NASA's budget by $1.6 billion. But he also wanted to drop the Artemis name as well. Late in the day, as his team scrambled to prepare for the teleconference, which was to start at 7:00 P.M., Bridenstine called Gabe Sherman, his chief of staff, who was already headed for his train ride home, for a gut check. Sherman didn't answer, so Bridenstine went running after him and was out of breath by the time he caught up to him.

Was he making a mistake to name the program? Bridenstine wanted to know. Or should he go for it? "Let's do it," Sherman said.

Shortly before the teleconference started, Trump scooped Bridenstine in part by crowing on Twitter: "Under my administration, we are restoring @NASA to greatness and we are going back to the Moon, and then Mars. I am updating my budget to include an additional $1.6 billion so that we can return to Space in a BIG WAY!"

On the media call, Bridenstine said the funds would "accelerate our return to the lunar surface." It was an attempt to demonstrate momentum, that NASA did in fact have a plan—and the money, or at least the first deposit—to meet the 2024 deadline. But he also sought to manage expectations, acknowledging that "in the coming years, we will need additional funds." At the end of the call, after fielding questions from reporters skeptical about the timeline and whether Congress would really pay up, Bridenstine said he had some closing remarks. It was time for his own surprise announcement.

"The first time humanity went to the Moon, it was under the name

Apollo," he said. "The Apollo program forever changed history, and I know all of us here in this room and on the phone are very proud of the Apollo program. It turns out that Apollo had a twin sister, Artemis. She happens to be the goddess of the Moon. Our astronaut office is very diverse and highly qualified. I think it is very beautiful that 50 years after Apollo, the Artemis program will carry the next man and the first woman to the Moon. I have a daughter who is 11 years old, and I want her to be able to see herself in the same role that the next women that go to the moon see themselves in today. This is really a beautiful moment in American history, and I'm very proud to be a part of it."

The White House was not happy with the breach of protocol and let Bridenstine and his staff know they should have given them a heads-up. But it was too late. The name was out there. There was no taking it back now.

NASA'S FLAGSHIP EXPLORATION mission may have had a name. But that didn't ensure anything, especially not with Trump in the White House. Two years earlier, he had asked Robert Lightfoot, the acting NASA administrator, what the plan was for Mars. Lightfoot said NASA was planning for sometime in the 2030s. Trump wasn't pleased with the answer and asked what it would take to get humans to Mars during his first term. He was annoyed when Lightfoot said it was not possible.

"But what if I gave you all the money you could ever need to do it?" Trump asked. "What if we sent NASA's budget through the roof, but focused entirely on that instead of whatever elese you're doing now. Could it work then?"

Getting to Mars was exceedingly difficult. On average, it's 140 million miles from Earth, though they come to within about 40 million miles every twenty-six months when their orbits align them on the same side of the sun. But even then, it would take months to get there. Even if they did survive the trip, because of the alignment of the planets the astronauts would likely have to stay there another two years before the long journey back, meaning NASA would have to pre-position tons of cargo, food, water, shelter, and communications equipment ahead of time. Not to mention that the radiation environment on Mars is extreme, the air is toxic, and the average temperature is a balmy minus 85 degrees Fahrenheit.

The answer was still no, which annoyed the president even further.

Trump was still thinking about Mars three weeks after Bridenstine christened the Trump administration's moon program Artemis, when the president caught Jeff DeWit, NASA's chief financial officer, speaking on the Fox Business channel.

Fox host Neil Cavuto wondered why NASA was "refocusing on the moon," and asked, "Didn't we do this moon thing quite a few decades ago?" He added: "I thought we would advance beyond that, and I thought we would target Mars."

DeWit was a political operative who had served as Trump's campaign chairman for Arizona. But he understood NASA's logic in using the moon as a stepping stone to Mars.

"What we're doing now is enabling a sustainable presence on the lunar surface," he said. It was a necessary step before going to Mars sometime in the 2030s, he explained. It was a coherent and reasonable answer that accurately summarized NASA's plans. But it did not sit well with Trump, who fired off a tweet.

"For all the money we are spending, NASA should NOT be talking about going to the Moon. We did that 50 years ago. They should be focused on the much bigger things we are doing, including Mars (of which the Moon is part), Defense and Science!"

It was an impulsive, stream-of-consciousness post that seemed to contradict his very own space policy directive. And it sent NASA officials scrambling. A few hours later, Bridenstine put out a tweet of his own, attempting to clarify the president's remarks. "As @POTUS said, @NASA is using the Moon to send humans to Mars!" he wrote.

To get to either destination, it was becoming clear to him, NASA wouldn't just have to survive the perils of deep space exploration, but the unpredictable whims of politics as well.

CHAPTER 12

"We Just Blew It to Smithereens"

As Saturday-duty assignments go, Craig Bailey's wasn't a bad one: Go shoot a surfing festival down on Cocoa Beach. It was April 20, 2019, a lovely afternoon on Florida's Space Coast, with a high of 78 degrees, a slight breeze, and bright blue skies mirrored by the ocean. Bailey, a veteran photographer at *Florida Today* with nearly forty years of experience, focused his lens on the shoreline, snapping shots of the surfers prancing in the waves like seals. Then, in the distance, he noticed a plume of smoke rising over Cape Canaveral.

That was not unusual. There were often prescribed burns over there to weed out brush and prevent wildfires, so Bailey didn't think anything of it at first. "But then I looked at it again, and that's where things got weird," he recalled later. The smoke color was "off," he said, an ominous orange with a faint greenish tint, unlike anything he had ever seen. No one on the beach seemed to notice, but Bailey's journalistic instincts kicked in and he started snapping photos of the bizarre orange cloud, whatever it was.

He texted Emre Kelly, *Florida Today*'s excellent space reporter, to let him know something might be up out at the cape. Kelly said he'd look into it. Not long later, Kelly texted back: Something blew up out there.

Well, Bailey responded, we have pictures.

Astronaut Doug Hurley was home in Houston when his phone rang that afternoon. It was Lee Rosen, SpaceX's vice president of mission and launch operations. Like Hurley, he was a retired military officer and, officer to officer, gave Hurley the courtesy of telling him directly: "The Dragon capsule that was going to take you and Bob Behnken to space just blew up."

"What?" Hurley said, incredulous.

"We blew it up," Rosen repeated. The team had been testing the SuperDraco engines used in the spacecraft's emergency abort system when the whole thing exploded.

"You've got to be shitting me."

"We just wanted to let you know," Rosen said. "We're going to work through this."

Hurley hung up, trying to process what he'd just been told. This was the spacecraft that had just completed a flawless test flight to the International Space Station without any crew on board. It was going to be the ship that carried humans—him and Bob Behnken—to space from American soil for the first time in years. It was going to make history and perhaps be destined for the Smithsonian. Now it was in pieces.

Holy shit, Hurley thought. *What the hell happened?*

JIM BRIDENSTINE WAS also in Houston—at a national robotics championship, encouraging the fifteen thousand students at the event to consider a career at NASA—when his aides pulled him aside and told him the news.

They found a side room and started taking calls. Details were murky. In the first, chaotic moments, all Bridenstine could glean were the basics: Coming just six weeks after the successful Demo-1 flight, the explosion had occurred as SpaceX was testing the engines that would shoot the astronauts' capsule away from the rocket in the event of an emergency, like a cork popping from a bottle of champagne.

For the abort system to work, the SuperDraco engines had to be able to fire quickly. SpaceX used hypergolic propellants—a fuel and an oxidizer that ignite instantaneously when mixed. For the capsule's eight SuperDracos, that was nitrogen tetroxide as the oxidizer and hydrazine as the fuel. When they combined as intended, they would give the Dragon capsule some real giddyup: propelling the spacecraft to 100 miles per hour in 1.2 seconds, ultimately breaking the sound barrier and exerting high G-forces on the astronauts inside.

Huddled with his aides, Bridenstine demanded answers from his staff. After *Florida Today*'s story posted online with Bailey's photos, he started fielding calls from members of Congress, who were pushing for information he didn't have. "Nobody knew what the deal was," he recalled.

He was grateful that no one had been hurt. It wasn't like the 1967 Apollo 1 disaster, when astronauts Gus Grissom, Ed White, and Roger Chaffee were killed in a fire while trapped inside the command module during a preflight test. Still, it was a jarring reminder of the perils inherent in human spaceflight. And Bridenstine's mind quickly went to the day a few years earlier when he had toured SpaceX headquarters with Musk. He had asked Musk what his greatest fear was as SpaceX worked to fly astronauts for the first time, and Musk had immediately named the emergency abort system. He even remembered what Musk had said: "There's a lot that can go wrong." And now it had.

BEFORE THE ACCIDENT, some at NASA had been concerned with another issue—the parachutes SpaceX was using to slow the 21,200-pound capsule as it fell through Earth's atmosphere. During the Demo-1 test flight without astronauts, the chutes worked perfectly, all four red-and-white canopies inflating like giant jellyfish before the capsule splashed into the water. But during a series of tests designed to see how the parachutes performed under more extreme conditions, SpaceX was discovering problems. It would drop test articles—often giant sleds meant to mimic the mass of the capsule—from helicopters at various altitudes over the Nevada desert. A few ended up as lawn darts, plunging into the ground, including one that crashed in April 2019, the same month the Dragon capsule exploded.

Word of the parachute problems was spreading. At one congressional hearing, William Gerstenmaier, the head of NASA's human spaceflight division, was asked about it by Mo Brooks, a Republican congressman from Alabama. Brooks was no fan of SpaceX, which was threatening the traditional space companies that worked out of his state.

"The test was not satisfactory," Gerstenmaier said, deadpan. "We did not get the results we wanted. The parachutes did not work as designed."

"Can you get more specific when you say, 'It wasn't what we wanted?'" Brooks said.

Gerstenmaier allowed that the test sled was "damaged upon impact with the ground." He explained that SpaceX was conducting what was known as a "one-out" test, meaning one of the chutes would purposefully not open. "The three remaining chutes did not operate properly," he said. But failure, he added, "is part of the learning process. This is why we test."

NASA's chief engineer, Ralph Roe, had briefed Bridenstine on the problems and said he had recommended to SpaceX that it upgrade the parachutes it bought from its contractor, Airborne Systems. SpaceX was using the Mark 2 version, but the company had a Mark 3 design with stronger lines and a more robust canopy. That, Roe believed, would solve the problem they were seeing when the chutes opened in an asymmetrical way. When that happened, it put additional stress on the lines instead of the weight being distributed equally. Stronger lines seemed like a no-brainer.

SpaceX, though, felt it wasn't necessary, as did some in NASA's own leadership. They had flown the Mark 2 chutes on numerous cargo flights, without any people on board. They knew them and their limits. Switching so late in the game seemed risky, since they wouldn't have as much data to analyze. Like the fights over touchscreens and the docking system, here was another flashpoint inherent in the decision to outsource human spaceflight to the private sector.

Elon Musk celebrates SpaceX's first successful human spaceflight mission, which launched NASA astronauts Bob Behnken and Doug Hurley to the International Space Station on May 30, 2020. *(Courtesy of* The Washington Post *and Jonathan Newton)*

Jeff Bezos unveils the Blue Moon lunar lander during his "vision speech" in Washington, D.C., on May 9, 2019. *(Courtesy of* The Washington Post *and Jonathan Newton)*

For the first flight of SpaceX's Falcon Heavy rocket on February 6, 2018, the company launched Elon Musk's Tesla Roadster with a mannequin named Starman at the wheel. *(Courtesy of SpaceX)*

Both of Falcon Heavy's side boosters land back at Cape Canaveral, Florida, as part of the rocket's first test flight. *(Courtesy of SpaceX)*

Vice President Mike Pence and NASA administrator Jim Bridenstine walk through NASA headquarters after Bridenstine was sworn in as head of the agency on April, 23, 2018.

(Courtesy of NASA)

NASA administrator Jim Bridenstine participates in a crew Dragon flight simulation in 2019 with NASA astronaut Doug Hurley at the SpaceX headquarters in Hawthorne, California.

(Courtesy of NASA)

NASA administrator Jim Bridenstine and SpaceX CEO Elon Musk address the media after resolving their differences during a meeting at SpaceX headquarters on October 10, 2019. *(Courtesy of NASA)*

NASA astronauts Doug Hurley (right) and Bob Behnken (left) train on the SpaceX Dragon capsule's touchscreens leading up to their flight in 2020. *(Courtesy of SpaceX)*

NASA astronauts Doug Hurley (right) and Bob Behnken (left) in their SpaceX space suits ahead of their spaceflight mission to the International Space Station.

(Courtesy of SpaceX)

Jeff Bezos celebrates his July 20, 2021, flight on New Shepard's first human spaceflight mission, which flew past 60 miles on a suborbital trajectory. *(Courtesy of Blue Origin)*

Blue Origin took over Launch Complex 36 at Cape Canaveral, Florida, and rebuilt it for its New Glenn rocket. *(Courtesy of Blue Origin)*

A rendering of Blue Origin's Blue Moon lander on the lunar surface. Blue Origin revised the design after it lost the initial NASA competition to SpaceX in 2021. It won a NASA contract in 2023. *(Courtesy of Blue Origin)*

SpaceX's Starship Super Heavy booster separates from the spacecraft during the fifth test flight of the vehicle on October 13, 2024. *(Courtesy of SpaceX)*

The chopstick-like arms of Starship launch tower catch the Starship Super Heavy booster on its fifth test flight. *(Courtesy of SpaceX)*

Catching the booster is a key step toward SpaceX's goal of rapidly reflying the world's most powerful rocket. *(Courtesy of SpaceX)*

NASA's Orion spacecraft flies by the moon November 21, 2022, during the Artemis I mission. *(Courtesy of NASA)*

Blue Origin's powerful New Glenn rocket on its launch pad ahead of its first successful flight on January 16, 2025. *(Courtesy of Blue Origin)*

"The NASA team felt like they weren't getting responses," Bridenstine recalled. "They were identifying problems and SpaceX was not responding."

WORSE, MUSK SEEMED distracted. He was spending much of his time in South Texas working on Starship, his next-generation rocket, a vision that had now become his muse. But Musk wasn't happy with the pace of progress—on the rocket, or the site in Boca Chica, Texas, that was slowly transforming into a manufacturing facility and a private spaceport. The man tasked with building the site was John Muratore, who had spent twenty-nine years in the Air Force and then at NASA before moving over to SpaceX, helping the company build its launch pads at Cape Canaveral. Seeking another challenge, he eagerly volunteered to go to Boca Chica, which reminded him of the early days of Cape Canaveral: "Isolated and alone with lots of potential."

"My first impression was, it was a place where we could make history," he said. He and a skeleton team had been building up the site, but by the fall of 2018, Musk was getting restless, always wanting to know, "Why aren't you moving faster?" On October 19, during a 7:00 P.M. conference call, Musk was "very unhappy with the rate of progress," Muratore said. And he had some news. No longer was the rocket going to be built out of carbon composites, a state-of-the-art material that is very strong but also very light, an ideal combination for rockets. Instead, Musk wanted to use a stainless-steel alloy.

Musk said that this was something he had been "contemplating for a while." It was a novel and somewhat crazy idea, or, as he said, "counterintuitive." No rocket or spacecraft had ever been made out of stainless steel, a fact his engineers reminded him of. Still, Musk was convinced it would work. "It took me quite a bit of effort to convince the team to go in this direction," he said. This was no small decision. The material used for the rocket would need to be able to withstand extreme temperatures—from the minus 272 degree Fahrenheit liquid methane fuel, to the aproximately 5,000 degrees it would generate as it reentered the Earth's atmosphere.

Steel could handle both extremes much better than carbon fiber, which could withstand 300 degrees, maybe 400 tops. Steel was good up to about 1,500 degrees, meaning SpaceX wouldn't have to cover the entire body with heat shield tiles, saving weight and mass. Steel also was far less expensive. Carbon fiber cost $135 per kilogram, compared to $3 per kilogram for steel.

On the conference call, Musk announced his decision with typical bluntness. "Fuck composites. I want to start building Starship out of stainless steel at Boca Chica." And he added, for emphasis, "Starting tomorrow."

At the time, there were just a few dozen people at Boca Chica. The facilities consisted of "a double-wide trailer and a couple small buildings," Muratore recalled. Turn it into a manufacturing site? Tomorrow? There was no way. Muratore reached for the unmute button to tell Musk it was impossible.

Then he thought about how his boss would react—how he'd insist and tell them to find a way, no matter what it took—and retracted his finger. He had led the team that transformed two pads at Cape Canaveral: the famed 39A as well as 40, which SpaceX had refurbished and then rebuilt after a Falcon 9 rocket exploded in 2016, destroying the pad. If SpaceX could do that, it could build the rocket "field of dreams" Musk was envisioning.

What the hell, Muratore thought. When he clicked unmute, he heard himself telling Musk they'd get it done: "Yes, sir. We'll be ready."

The next day, he took a picture of the site. "It was literally a green field," he recalled. But he started prepping the ground so that the company could start building.

Musk started coming down to Boca Chica at least once a month to chart the team's progress, usually on weekends. He approved every significant decision and signed off on every purchase over $50,000. At the time, SpaceX had started ramping up an entirely new business line based in Redmond, Washington—Starlink satellites, which would beam Internet signals to stations on the ground, allowing broadband access from remote areas of the globe. The plan was to build, in-house, hundreds, then thousands, then tens of thousands of the small satellites, and hoist them to orbit. If it was going to work, SpaceX would need to get them up quickly. The program was draining enormous amounts of capital, and while Musk believed SpaceX could pull it off, he was also aware that several other companies, including Teledesic, Iridium, and Globalstar, had tried in the past and failed spectacularly. Making matters more complicated, SpaceX was trying to build a new rocket and a new engine, the Raptor, all at the same time, and none of it was guaranteed to work.

"He was really worried that taking on Starlink and Starship was going to kill the company," Muratore said. "I think we all were worried too, because what he would say was, 'Building large rockets has broken rocket companies, and trying to build satellite networks has broken satellite companies. We're going to take on both of these at the same time, and we're doing it all internally.'"

Not only did Muratore have to build a production facility from scratch, he had to do it on the cheap. That meant he had to be creative. When he wanted to hire a crane to lift a small Starship prototype called Starhopper, he couldn't. The crane was too expensive, Musk said. Find another way. Muratore did. He came up with a plan to lift Starhopper with hydraulics and move it to a large trailer. It was a perhaps risky improvisation, but it saved tens of thousands of dollars.

Over the next several months, Musk's visits prompted whirlwinds of activity. "He'd look at everything, climb around, constantly had great suggestions," Muratore said. His team worked sixteen-hour days, seven days a week, taking "naps in whoever's car was closest," as one former SpaceXer told me. "The level of dedication was special; I would dare you to tell me you've been tighter with a group of people." SpaceX was full of believers, dedicated to the mission. The first group to locate to South Texas were among the most die-hard, a band willing to follow Musk virtually anywhere. "At the time we never had more than a hundred people," Muratore recalled. "It was an amazing few months. People were sleeping on-site."

Still, the constant refrain from Musk was: "Why aren't we moving faster?"

By July 2019, Muratore and his crew had built a rudimentary pad for Starhopper. Short and stubby, it looked more like a water tower with three legs than a spacecraft. But it was a tough little rocket that had survived a fireball during an engine test at the pad that went awry. "Big advantage of being made of high strength stainless steel: not bothered by a little heat!" Musk tweeted. Less than ten days later, on July 25, it took off to an altitude of about sixty-five feet, then touched back down in a soft landing. "Water towers *can* fly haha!!" Musk wrote on Twitter.

Muratore, who served as the flight director, marked the occasion by passing out Beeman's chewing gum to his team. The obscure brand, with an odd taste, had been favored by military pilots and was featured in *The Right Stuff* when Chuck Yeager asked for a stick before his flight breaking the sound barrier.

A month later, SpaceX flew it again, this time to nearly five hundred feet. Starhopper not only flew up but over, from one concrete pad to a second about a hundred yards away, landing softly and with precision. That gave Musk confidence to move to the next stage of Starship's development, testing even larger prototypes that could fly to even higher altitudes. There would be no rest after Starhopper's flight. There would be no pausing to reflect on what they had built out of the mud in South Texas. There would only be progress—relentless, head-down, plow-through-the-line progress.

"He was always wanting to go further and faster and higher," Muratore said. "Whereas most people might want to stop and breathe and savor the moment, he just wants to go for the next ride. He wants to get back on the roller coaster as fast as possible. He just wanted to go again."

As the Boca Chica site developed, it became something of a tourist attraction. A public road to the beach ran right alongside SpaceX's property, and space fans and Musk acolytes made pilgrimages to see the prototypes. Towering and stainless-steel shiny, they looked like surreal sculptures set amid the cactuses and yucca. Eventually, at the end of September, Musk decided to show off all the work his teams had accomplished in transforming this little nook of South Texas wasteland into a rocket garden. For the first time, Musk was going to open the site to the media and select VIPs.

He arrived in his private jet a few days early. Muratore and his wife, Mary, greeted him at 11:00 P.M., bringing barbecue in case he was hungry, because there was no food on-site. Muratore showed Musk around, detailing all the preparations they had made moving from Starhopper, which had one engine, to the larger Starship prototypes, which would have three. Musk was inspired. Starship, he said, was starting to seem real. And if Starship was real, then maybe, just maybe, an eventual trip to Mars could be real as well.

Their minds brimming with visions of the future, Musk and Muratore stayed up talking until two in the morning, musing about the form of government that humans might install on Mars, the kinds of tools and technologies they would need, and problems like radiation and long-term flight. "We talked for three hours. I mean, it was till late."

BRIDENSTINE WASN'T INVITED to the Starship presentation in Boca Chica and wouldn't have gone if he had been. All this attention on Starship was distracting from what should be SpaceX's top priority, he felt: NASA's Commercial Crew program to launch American astronauts from American soil.

The president had said in his State of the Union address that NASA would restore astronaut flights to the International Space Station by the end of 2019. But here it was, late September 2019, and neither SpaceX nor Boeing was anywhere close to being ready. Boeing hadn't even flown its test flight without astronauts. SpaceX had, but then its Dragon capsule had exploded. The safety review, triggered by Musk's appearance on Joe Rogan, was finally underway. Musk's preoccupation with Boca Chica and Starship was grating.

"Am I missing something?" Bridenstine said to his senior leaders during a meeting after Musk had invited the press to Boca Chica. "Why is he announcing all this stuff about Starship, when taxpayers are funding him billions of dollars to fly crew?"

"Hit him," Jim Morhard, the deputy NASA administrator, advised.

"We have all the federal funding. What's he going to do? We have to get him focused."

Bridenstine agreed. He needed to rebuke Musk—again. This time he wasn't going to hide behind something like an Organizational Safety Assessment as he did after the Joe Rogan incident. This time he'd do it publicly and on the platform of Musk's choice: Twitter. He had his staff draft a tweet to post on the eve of Musk's presentation, a sharply worded shot across the bow.

"I am looking forward to the SpaceX announcement tomorrow," it began. "In the meantime, Commercial Crew is years behind schedule. NASA expects to see the same level of enthusiasm focused on the investments of the American taxpayer. It's time to deliver. No more shiny objects." A previous draft of the tweet had gone even further, calling the rocket a "flying trash can."

Bridenstine wanted the tone to be tough and no-nonsense. The goal was to keep Musk focused on the NASA mission, not to disparage his other endeavors. He told his staff to cut the "shiny objects" and "flying trash can" lines, then post. Still, Matt Rydin, the NASA press secretary, was nervous. Musk had legions of rabid fans who would no doubt go after them. It was also a little rich for NASA, an often sluggish bureaucracy, to chastise SpaceX for being slow and distracted when it had proven it was anything but. The tweet carried some real risk, and Rydin kept his hand hovering over the mouse, afraid to click it into existence.

The risk was worth it, Bridenstine said. "As long as what we're doing is based in truth and in taxpayers' best interest. Push the button," he said.

BRIDENSTINE'S PHONE SOON rang. It was Musk. He was not happy.

Bridenstine had no business telling him how to run his company, Musk groused. He *was* focused on the Falcon 9 rocket and Dragon spacecraft that would fly NASA's astronauts to the space station, he continued. In fact, most of the company's resources were dedicated to that mission, not to Starship.

"You have no right to do this," he said.

Bridenstine stood his ground and reminded Musk of how SpaceX's Dragon capsule had just exploded. "Elon, you told me when we met in Hawthorne that the risk was going to be the launch abort system," he said. "You knew that years ago. And now the challenge is we need to make sure that risk goes away."

After his presentation the next day, Musk was still annoyed. When asked about Bridenstine's tweet by CNN, he jabbed back, pointing out that NASA's SLS moon rocket was even more delayed. "Did he say Commercial Crew or SLS?" he said, laughing into the camera as if he were looking directly at Bridenstine.

Before he sent the tweet, Bridenstine had planned to follow up with an in-person visit, a summit that would put NASA and SpaceX in a room to talk through the remaining problems and get the program on track.

"I'm coming to Hawthorne," Bridenstine told Musk. He'd bring the astronauts Doug Hurley and Bob Behnken. He'd bring NASA's chief engineer, Ralph Roe, who had been pushing for the upgrade in parachutes. And he'd bring his senior leadership, who had been overseeing the program from the beginning. He'd also bring the media.

"We could come to all kinds of agreements behind closed doors, but unless people knew about them, it doesn't matter," he recalled later. "We needed the press to document any agreements and hold them accountable."

THE MEETING, ON October 10, 2019, did not get off to a good start.

The NASA team showed up at SpaceX headquarters dressed down, in polos and khakis, trying hard not to stick out as stodgy government bureaucrats at a company where T-shirts and jeans were standard. But Musk and the SpaceX leadership had made the effort to dress up, in dark suits and crisp shirts, assuming their NASA counterparts would arrive from Washington in business attire. The sartorial mix-up only widened the chasm between the two sides. Musk was surly and impatient. He didn't make eye contact or small talk. He didn't want to be told what to do, especially in his own house. "It was not 'Hey, great to see you,'" as Bridenstine said later.

"You could have cut the tension with a knife," was how Hurley recalled the meeting.

After a short tour, the two teams assembled in a conference room. Musk sat at the head of the table. Bridenstine and Morhard, the deputy NASA administrator, flanked him along with Gwynne Shotwell, the president of SpaceX; Kathy Lueders, the head of NASA's Commercial Crew program; and Roe. Lueders and Roe disagreed about whether the Mark 2 parachutes were safe enough for human spaceflight. Roe kept saying he needed repeatable successful test results instead of the failures that could imperil astronauts' lives.

Parachutes might look like one of the simplest aspects of the spacecraft. But they're actually one of the most complex. There are two sets of parachutes. First, a pair of smaller "drogue parachutes" deploy when the capsule, weighing approximately twenty thousand pounds, reaches 18,000 feet and is traveling at 350 miles per hour. They slow the spacecraft down to 119 miles per hour and then the four main chutes are deployed at an altitude of about 6,000 feet. Fully inflated, the mains make a beautiful sight: big, bright, jellyfish-like creatures floating ever so softly. But when they are initially released, it takes a few moments for them to unfurl. And when several of them are trying to unfold themselves in close proximity, they tend to crowd one another out and unfold at different rates, meaning the stress on the suspension lines is uneven. Unfortunately, there is no way to calculate that asymmetry, or how long it is going to last. So the engineers have to make some assumptions about the conditions the parachutes will have to endure.

When it comes to safeguarding the lives of astronauts, NASA's engineers did not like making assumptions. Neither did the astronauts. Listening to the debate, Hurley wondered why there even was one: *There are things on spacecraft that can break,* he thought. *The parachutes ain't one of them, especially with my soft pink body in the vehicle.*

Musk listened with an intense, unwavering glare. SpaceX had already made significant progress in identifying the cause of the Dragon explosion, tracing it to a small amount of nitrogen tetroxide fuel that leaked past a valve into a line used to pressurize the propellant tanks. As an emergency system, it was brought rapidly to an extreme pressure—2,400 pounds per square inch. When the fuel hit the titanium valve under those conditions, it "basically destroyed the check valve and caused an explosion," as Hans Koenigsmann, SpaceX's vice president of build and reliability, told reporters at a separate event. The test, then, was a gift that revealed a fundamental flaw in the system. A change in the valve, from titanium to a material called Inconel, would solve the issue; but making a change could lead to another delay.

In the conference room at SpaceX headquarters, Musk had been thinking intently, as if he had disappeared into his own mind. When he reemerged, he stunned the room. SpaceX would make the switch to Inconel, he declared. And it would upgrade to the Mark 3 parachutes. *And* it would complete ten tests of the upgraded parachutes within a few months, and therefore not cause too much of a delay.

Bridenstine was floored. This was all news to him. He had come to Hawthorne expecting a fight. But now the NASA and SpaceX teams were aligned, seemingly as a formidable team. As if to seal the deal, Musk and Bridenstine stepped outside to greet the media. They stood side by side, shaking hands as cameras snapped away. Musk told the throng of reporters that SpaceX was committed to making sure "we've done everything possible to ensure the astronauts will be safe. Only at that point will we launch." Addressing the concerns that he'd been spending too much time on Starship, he said developing Dragon "is absolutely the overwhelming priority" and that the company has "a lot of people working super hard."

"I think probably a couple of weeks ago we were not on the same page," Bridenstine told me after the event. "But now we are, 100 percent."

MUSK WAS TRUE to his word. The company went all-in on moving to the Mark 3 parachutes and the rigorous drop testing campaign they'd require.

NASA officials had told him "there was no way you're going to do ten tests by the end of the year," Bridenstine recalled. "Elon said, 'We're going to make it happen.'" Inside SpaceX, the switch further stressed a team that was already working around the clock.

"Parachutes are way harder than they look," Musk told me at the time. "The Apollo program actually had a real morale issue with the parachutes because they were so damn hard. They had people quitting over how hard the parachutes were. And then you know we almost had people quit at SpaceX over how hard the parachutes were. I mean, they soldiered through, but, man, the parachutes are hard."

Musk peppered his parachute team with questions, asking what they needed to get the job done, offering to help personally. Musk visited the parachute manufacturer's factory to make sure SpaceX would be given top priority. But it had other, bigger customers, including the United States Army. SpaceX's order was relatively small. Finally, Musk grew frustrated and called Bridenstine. It was Christmas Eve. Bridenstine was home in Tulsa taking his kids to visit Santa Claus at the mall when he took the call.

Musk said he needed Bridenstine's help and wanted him to push Airborne to prioritize SpaceX's parachutes. Bridenstine said he would see what he could do. He hung up, impressed by Musk's tenacity, and by the fact that SpaceX had worked to quickly complete the drop tests as Musk had promised. He had been skeptical of the billionaire Silicon Valley celebrity and upset by his behavior on the Joe Rogan show. But now he was starting to see him in a different light, as someone who would get things done. As someone he could trust.

A FEW WEEKS later, in January 2020, that point was reinforced when the team assigned to assess SpaceX's culture gathered to brief Bridenstine and the senior NASA leadership on the results of their investigation.

Several months earlier, about ten NASA officials had descended on SpaceX headquarters to conduct the Organizational Safety Assessment. At SpaceX, it was viewed as an intrusive disruption and, in some quarters of the company, an insulting form of punishment. There was an air of Puritan, East Coast institutional morality attempting to impose its will. They were West Coasters, young and different, and they felt like they were being picked on for having an eccentric, impulsive boss.

"Do you really imagine because we're Californians and we work for Elon, all of us are smoking marijuana and putting the astronauts in danger?" asked Abhi Tripathi, a veteran at the company who was overseeing the certification of the Dragon spacecraft that would fly astronauts to the space station. "What do you think happens here? We work crazy hours, doing good work. Do you think we have time to smoke up on the job?"

Despite the internal grumbling, SpaceX's leadership knew it was best to accept the lashing and move on. NASA was their customer. And the customer wanted to perform the federal government's version of a proctology exam. To lead the group representing SpaceX, Gwynne Shotwell chose Tripathi, who had worked at NASA for nearly a decade as a lead engineer. He was fluent in NASA's language and understood its culture. Plus, he was one of the older employees at SpaceX, a Gen Xer at a company dominated by millennials.

Like his colleagues, Tripathi felt the oversight was an overreaction by an agency looking to preemptively cover its ass. But there was no way around it. The best thing to do was embrace it, he thought, and to do it the SpaceX way—that is, fully, all in, working as hard on the safety assessment as they had on building rockets that could fly to space, turn around, and land. As Tripathi saw it, NASA's safety review could even be an opportunity to improve. SpaceX had been hiring steadily and now stood at about seven thousand employees. This was a moment that could help ensure that all the new workers adhered to SpaceX's high standards. "I took great pains to make sure we got usable and actionable intelligence that would make SpaceX better," he told me later.

The review was something of a collaboration. Working together, NASA and SpaceX chose three hundred employees at random from all across the company, from senior executives to workers on the production lines. They filled out questionnaires about the company and how it handled safety concerns, and then each of them sat for an interview with one NASA official and one SpaceX higher-up. To encourage candor, the employees' answers were kept anonymous.

If some at NASA initially saw the company's presence in the room as a conflict of interest—SpaceXers interviewing SpaceXers—they quickly came around. Tripathi urged the SpaceX employees who served on the review not to hold back. They knew the company better than NASA. They could tell when someone wasn't telling the whole truth and knew when to ask the tough follow-up.

"NASA didn't know when somebody gave an answer that should have raised eyebrows," Tripathi said. "They didn't know what the SpaceX people knew, and they followed up in front of NASA. They wouldn't just say, 'Oh, shit, that person just said something. Let me gloss over it. Let me go to the next question and figure it out in detail after this interview is over.' No. We showed NASA that we were like homicide detectives in the meetings with them and really dug in deep."

Then again, the whole culture at SpaceX was designed to create an environment where employees openly challenged one another for the best outcome—even if that meant an intern pushing back against someone far more senior. As a result, the safety review wasn't entirely foreign; it was something that happened on a daily basis in Hawthorne. Only now, NASA had a front row seat. The report its team delivered to Bridenstine in January 2020 was positive and seemed thorough. SpaceX "opened the kimono," as Bridenstine told me later. "The feedback was great. They gave us full access to everyone. There's not a drug culture at SpaceX. That did not come out at all, which was what I was worried about, quite frankly. People felt they could report problems. They could raise the red flag."

SPACEX WAS NOW rolling. The new Mark 3 parachutes had satisfied even the most squeamish at NASA, even when pushed at the margins. At the same time, SpaceX's engineers had been working to remedy the other problem plaguing Dragon: the emergency abort system designed to ferry astronauts to safety in the event of a catastrophe. Now, in January 2020, SpaceX was ready for its next major milestone, the last before

SpaceX would fly crew: an in-flight test of the capsule escape system without anyone on board. On the day of the mission, a sunny Sunday, Musk showed up at the launch control center wearing a dark suit and tie, and a headset that allowed him to listen in on the chatter between his various engineers. Behnken and Hurley were there as well, alongside SpaceX and NASA engineers who sat side by side in a mix of suits and sweats.

As the Falcon 9 lifted off, Musk stood, as did everyone else in the room not assigned to monitor the data flowing across their computer screens. After eighty-four seconds, ground controllers issued the command to fire the abort system, and Dragon popped off the rocket, as expected, at more than twice the speed of sound.

"Dragon's away," SpaceX's John Insprucker said, during a live broadcast of the test.

"Whoa," said Musk.

Before the test, Benji Reed, SpaceX's director of crew mission management, said that the force of the ejection could send the booster reeling, possibly to the point of destruction. "We expect there to be some sort of ignition and probably a fireball of some kind," he said. "I wouldn't be surprised, and that wouldn't be a bad outcome if that's what we saw."

That's what they saw. The fireball was enormous.

"Holy shit," Musk said. "We just blew it to smithereens."

But not Dragon. By then, the capsule was safely away, reorienting itself for landing. As the parachutes deployed, Kiko Dontchev, SpaceX's vice president of launch and one of Musk's most trusted engineers, started clapping. He gave a slap on the back to Musk, who seemed momentarily shocked that the abort system had worked so well.

"So, everything's good?" he said after the capsule splashed down a few miles off the coast.

"Everything's good," Dontchev said, smiling.

Finally, Musk seemed relieved. He high-fived Dontchev and then, like a ballplayer entering the dugout after hitting a home run, the engineers in the row behind him.

The test had not only proved that Dragon had a capability NASA's

Space Shuttle did not—but also that SpaceX had fixed the problem with the valves. The same system that had once caused the capsule to explode had just shown that it could be relied on to save astronauts' lives.

Behnken and Hurley watched closely. So did their wives and children. They were all relieved to know that if the worst should happen, the Dragon capsule came equipped with an escape system. "Obviously, they are keenly interested," Hurley said after the test.

Behnken and Hurley's flight was just a few months away, so close that the deadline to submit their guest lists to NASA was coming due.

CHAPTER 13

"Thank You for Flying SpaceX"

N early a decade after the last Space Shuttle flight, after a long, tortu-
ous journey to get American astronauts back to an American
launch pad, NASA now had a date: May 27, 2020. For Hurley,
there was a particular significance. He had flown on the last Space
Shuttle mission, ending that chapter of American spaceflight; now, he
would open a new era of exploration. There was an added bonus. Hurley would be able to claim the American flag that had been left on the
International Space Station for retrieval by the first crew of Americans
to restore flight from U.S. soil. Hurley's commander on that final shuttle flight had been an astronaut named Chris Ferguson, who was now
an executive at Boeing and scheduled to fly on the company's first
crewed Starliner flight. Ferguson had made it clear in public statements
that he wanted to be the one to claim the flag. He had brought it to the
station, as he once told me, and so "it probably has a lot more significance to me than it does to, say, somebody from our competition."

That somebody was Hurley, who wanted the flag and wanted it bad.
He was in a position to win it in a dramatic, come-from-behind victory. SpaceX was the underdog. NASA's top officials hadn't even wanted
to award it a contract and had been content to relegate America's

human spaceflight program to Boeing. Now it was clear that it was a good thing they hadn't done that.

Boeing's program had been a disaster almost from the beginning. Initially, NASA excused the company's problems as the normal growing pains of any development program. Building a spacecraft was hard, they said. Just look at SpaceX, which had blown up its Dragon capsule. But then Boeing almost lost Starliner—*twice*—on its first test flight in December 2019. The capsule, without any people on board, was hoisted successfully to orbit. But once there, it immediately encountered errors. Its onboard computers were off by eleven hours, making the autonomous spacecraft think it was at an entirely different portion of the mission.

As a result, the engines that were supposed to put Starliner on a trajectory to catch up with the space station failed to fire. Other thrusters, designed to keep the capsule stable, burned precious fuel needed for other portions of the mission. Once they realized the spacecraft was acting wildly, ground controllers tried to communicate with it but couldn't. Soon, they decided docking with the station was out of the question. They had to focus on getting the capsule down safely. Given the software problems, Boeing's leaders ordered a real-time review of the rest of the vehicle's systems. Incredibly, they uncovered a second major flaw. This one could have been even more dire, causing the capsule's service module to collide with the capsule at a moment when the two segments of the spacecraft were supposed to separate before reentering Earth's atmosphere.

The company was able to send up a software patch to fix the issue and the capsule successfully landed in the New Mexico desert. But NASA's confidence in the company was badly shaken. As was the nation's. Boeing had recently suffered two major crashes of its 737 Max airplanes, killing 346 people. The plane was derided as a "flying coffin" by one congressman. Nobody had been hurt in the uncrewed Starliner fiasco, but there was no way NASA was going to allow Boeing to fly its astronauts anytime soon. During an investigation into what went wrong, Boeing admitted that it had cut short a software test before the service module separation. If they had simply kept it going, "we would have caught it," a

Boeing official said. It was a stunning admission of incompetence that rattled NASA's leadership. The space agency had largely given Boeing a pass on the safety review, focusing almost entirely on SpaceX, which had passed easily. Now, NASA realized the problem wasn't with SpaceX.

It was with Boeing.

Now that SpaceX had a launch date for Hurley and Behnken, Boeing's combativeness had melted into bitterness—and regret. The company's executives hated that SpaceX was winning and enjoying the warm embrace of NASA's leadership. Leading up to SpaceX's flight with Hurley and Behnken, a person close to Boeing told me that the company simply "can't accept" that SpaceX was flying first. "People are annoyed at Elon. How does this guy who smokes pot beat us? We have a lot of humble pie to eat here."

EVEN THOUGH NASA had announced a launch date, few in the space industry thought the mission would actually go that day. Launches are always delayed. It's one of the open secrets of spaceflight. The Space Shuttle launched on schedule only 40 percent of the time, making commercial airlines, which take off with a 78 percent on-time success rate, look like models of efficiency. The shuttle was delayed so frequently that the astronauts started telling their visiting families to "prepare for a vacation in Florida and maybe you'll see a launch."

With a new vehicle that had never flown astronauts before, NASA and SpaceX were going to proceed with utmost caution. If there was even a hint of something off, they would delay, leaning on an old mantra: *It's better to be on the ground wishing you were in space, than in space wishing you were on the ground.* In the days leading up to launch, Kathy Lueders, who oversaw the Commercial Crew program for NASA, said more than once that the space agency would not rush: "We'll fly when we're ready."

Sending astronauts to orbit was risky business that required solemnity and careful deliberation, especially given NASA's history of tragedy. When NASA learned that President Trump wanted to attend the launch along with Vice President Pence, they not only groaned, but had serious concerns. Coming at the height of his reelection campaign against Joe Biden, some at NASA feared he'd turn the launch into a red-capped Make America Great Again campaign rally, using NASA and the astronauts as his props.

If Trump were there, would NASA get "launch fever," pushing to go even if there were safety concerns? Bridenstine was peppered with this question repeatedly. As a Trump appointee, who served at the pleasure of the president, would he—could he—hold off an impetuous and mercurial commander in chief from interfering? Bridenstine vowed that Trump's presence would have no effect on NASA's decision-making, or his own. NASA had a culture in which "we want people to feel free to say no, and not feel any pressure to go with this launch," he said during a briefing with the media. But really the intended audience was Trump and his advisers, whom he hoped would hear his comments and be forewarned. He was so worried about the pressure to go that as the launch day got closer, he texted Hurley and Behnken: "If you want me to stop this thing for any reason, say so. I'll stop it in a heartbeat if you want me to."

As Bridenstine told the press, "Part of my job as the NASA administrator is to make sure people understand their safety is our highest priority." NASA would understand if there was a delay. They were used to it. The American public, he was sure, would also understand. But would Trump? If he flew all the way to Florida only to have the launch delayed, he might explode and fire his NASA chief as he had so many other members of his administration.

As word spread that Trump would attend the launch, a young West Wing staffer who had been assigned to the National Space Council walked through the White House, explaining to anyone who would listen that an astronaut launch was a sacred ritual, something out of the control of even the president. This was not a train leaving on a schedule. The White House was used to bending everything to its needs. But at the Kennedy Space Center, the president would be relegated to little more than a bystander. If there was a fuel leak or a balky valve or even a cloud that looked suspicious, NASA and SpaceX would scrub and come back another day. No question. No matter Trump's mood or dictates. The president could bomb another country, pardon his friends, veto laws, nominate Supreme Court justices. But if there was a launch scrub, he'd have to deal with it and get back on his plane.

ON THE DAY of launch, Air Force Once circled the launch pad amid gray and menacing skies that did not bode well. The National Weather Service had issued a tornado warning for the area north of the launch site, and by the time Trump stepped off his jet, the low, thick clouds were unleashing sheets of rain.

The presence of the president would not, as Bridenstine promised, force NASA to launch in less-than-ideal conditions. But it did change the tone of the event. Even before he arrived, Secret Service agents were crawling all over the facilities. One afternoon, a few days before the launch, Hurley and Behnken were hanging out at the beach house—a secluded cottage on the Cocoa Beach shoreline that had served as a gathering place for NASA astronauts since the Mercury era—when a vanload of agents showed up unannounced.

"They had no reason to come to the beach house," Hurley said. "They were just coming to see it. Didn't check with anybody. I don't know how they got through security, because typically security has a person stationed there at the beach house, making sure that you're not bothered. But they showed up anyway." NASA astronaut Drew Feustel, who had been assigned to support Hurley and Behnken during the flight, had to go out and tell the Secret Service agents to get lost.

DESPITE THE WEATHER, NASA pressed on. Near the launch pad the sky was dark, but just off to the side were bright blue skies. Koenigsmann wanted to make sure that on the screens in the launch control center showing the rocket on the pad, there were as many images of the dark clouds as blue skies so that Musk would see the bad weather and temper his expectations. If he knew the chances of a scrub were high, he'd be in a better mood if the mission suddenly had to be delayed.

"Usually, I'm the one that is more conservative when it comes to these things," Koenigsmann said. "But in this case, I did not need to be conservative with him. Frankly, he was already pretty paranoid on that. So I didn't really have to remind him. He did usually seem overly confident on many launches—very, very hawkish when it comes to 'Let's just try and fly. It's going to work.' But in this case, it was clear that he would not want to risk anything."

Musk was stressed. A few weeks before the flight, in the spring of 2020, I'd interviewed him about the mission. I'd expected him to project confidence, his usual swagger. Instead, he went dark right away. "Can't mess it up," he said, as if to himself. He was quiet for a moment, looking at his lap, as if he was pondering the consequences of failure. "Can't mess it up," he said again.

SpaceX was attempting to join the space programs of the United States, Russia, and China as the only entities to successfully launch astronauts to orbit, a fact that seemed overwhelming.

"To be totally frank, I kind of have to block it out a little bit. I mean, it's a scary thing to be launching people," he said after another pause. "We've done everything we can to make sure that the rocket is safe, and the spacecraft is safe, you know? But the risk is never zero, you know, where you're going twenty-five times the speed of sound, you're circling the Earth every ninety minutes. It's just a speed that's difficult for people to even comprehend. It must get close to thirty times faster than a bullet from a handgun."

He put the responsibility on his own shoulders. "I'm the chief engineer," he told CBS News. "So I'd just like to say, if it goes right, it's a credit to the SpaceX–NASA team. If it goes wrong, it's my fault."

Everyone at SpaceX was feeling the pressure. Gwynne Shotwell said it felt like her heart was in her throat. "I think it's going to stay here until we get Bob and Doug safely back from the International Space Station," she said. To drive home the point that SpaceX was flying real people, who were not just astronauts but husbands and fathers to young boys, she started referring to Hurley and Behnken as "the Dads." She had their photos put on work orders associated with the mission, so that her engineers would see their faces multiple times a day.

Benji Reed, SpaceX's senior director for Human Spaceflight Programs, called flying Behnken and Hurley "a sacred honor" that the company was taking seriously. "We always ask ourselves: Would you fly on this, and more importantly, would you put your family on this vehicle?" He led what he called "paranoia reviews," designed to look "under every rock" to find any hidden problem that might pop up at the worst moment.

Once while down at Cape Canaveral, he came across a room where the astronauts' families from the Space Shuttle program used to wait before launches. He was stunned to see that a whiteboard with drawings made by the children of the crew lost in the Columbia disaster was still there, preserved as a reminder.

"That really drives it home," he once told me. "This isn't just the people that we're flying—these are all of their families. So we take this extremely seriously, and we understand that our job is to fly people safely and to bring them back safely. To do that you have to humanize it. You have to see them as your friends and as your colleagues."

Not everyone felt that way. "I did not really want to talk to them, frankly," Koenigsmann told me, referring to Behnken and Hurley. "I wasn't really that excited about that. Because I just knew that nothing good can come if I know them too well."

As part of the Commercial Crew program, NASA dictated that Boeing and SpaceX meet a new metric: The chances of losing the crew could be no lower than 1 in 270. But even with some of the best engineering minds at NASA, calculating risk was an imperfect science. There were too many unknowns hidden among the millions of parts that all have to work perfectly in order to escape death.

"Identifying all of the risks is impossible," William Gerstenmaier once said during a speech. "Risk cannot be boiled down to a single statistic."

EVERYONE WAS STRESSED, it seemed, except Hurley and Behnken. At the pinnacle of their extraordinary careers, they were having fun.

They'd each get their third trip to space, perhaps their last, and they wanted to make the most of it. They were getting to fly in a new vehicle that they had helped design. For many, flying in a spacecraft that had never flown humans before might be too risky. But both Hurley and Behnken had spent the better part of their time in the military as pilots assigned to test new jets—Behnken the F-22 Raptor, Hurley the Super Hornet. Breaking in a new spacecraft would be a fitting next chapter.

Over years of training, they had become best friends. "I don't know my life before Karen and Jack," Hurley told me, referring to his wife and son. "And it's the same way with Bob. I don't really remember a friend prior to Bob. We've been through all those good and bad parts of each other's lives for the last twenty years, all the important parts."

They were members of the same astronaut class, The Bugs, so named because they were selected in 2000, the year many feared computers would all shut down because of the Y2K bug. Behnken was Air Force, Hurley a Marine, but still their careers had moved in uncommon lockstep. Eventually, both would even marry fellow astronauts—Hurley to Karen Nyberg, Behnken to Megan McArthur, and the four of them would gel into something similar to family. It made sense, then, that when Hurley was assigned to support the crew of Columbia's flight in 2003, Behnken volunteered to go with him. There was nothing like seeing a Space Shuttle launch in person, especially knowing that one day it would be their turn.

On launch day, Hurley's job was to escort the crew of seven to the shuttle and help strap them into their seats. But as he did so, Hurley started to worry that Rick Husband, the mission's commander, had forgotten a small but significant bit of protocol. Husband was supposed to take Hurley's name patch with him on the flight.

Finally, with a smile, Husband reached over to Hurley's chest and ripped off the patch affixed to his flight suit and slapped it to a bit of Velcro on the left side of the cockpit, just above the controls. Just before the engineers closed the hatch, Willie McCool, the mission's pilot, looked up from his seat and caught Hurley's eye.

"I can't wait for you to be sitting here like I am," McCool said. Hurley was grateful for the gracious gesture, the commander's calmness just before liftoff. Hurley had no idea that he would be one of the last people to see the crew alive.

Sixteen days later, Hurley and Behnken were on the ground at the shuttle landing facility, waiting for Columbia to touch down. When it disintegrated over Texas, killing everyone aboard, the pair followed Bob Cabana, a former astronaut who was then the director of flight operations, to meet the families and give them the awful news. Hurley and Behnken stood there feeling helpless, while the room erupted all around them.

Afterward, Hurley and Behnken were told to go to the families' hotel rooms and pack up their things. As they gathered clothes and toiletries, Hurley came across something he wasn't expecting: a welcome-home letter written by one of the astronauts' children, saying how glad he was to have his father home, how much he missed him.

"It took me a long time to get through all that stuff," he recalled. "And it wasn't because I hadn't flown yet and was scared of flying. It was just the emotional toll and the raw experience of what we went through. All that had what I would say was a long-term effect on me psychologically."

But at least he wasn't going through it alone. His best friend was, too. They had bonded through their shared experience as military fighter pilots and astronauts—and, like combat veterans, through tragedy and grief. After some fifteen years, as they prepared to fly together on SpaceX, they reminded themselves to stay vigilant and urged the company's engineers to do the same. One small mistake—like a leaky titanium valve under enormous pressure—could be the difference between life and death.

"We emphasized it with SpaceX folks many times," he said. "It's like, look, we lived this—*intimately.*"

A FEW HOURS before liftoff, Hurley and Behnken rode to the pad in a Tesla, listening to the playlist they had chosen. It was an eclectic mix: AC/DC's *Back in Black* ("Bob likes the real heavy metal stuff," Hurley said. "Gets him fired up"); "The Star-Spangled Banner"; and "The Girl from Ipanema" ("It's like hold music for when you're waiting to launch"). They had gone through simulations and even a full dress rehearsal, but they had never been on board as the booster was actually being fueled. Now, inside, they listened to the sounds of it being loaded with cryogenic propellants. The "pops and creaks and expansion sounds" made it feel like they were sitting on top of a living beast.

As instructed, Hurley pushed the button to arm the launch abort system. "There was this very definitive thud," he recalled. "And Bob and I looked at each other, like, *Are we going to get popped off this thing? Is this normal?* And it was one of the many times when we looked at each other and knew exactly what the other guy was thinking, though we never said a word."

Everything was proceeding smoothly for the launch, until it wasn't. At 4:16 P.M., some seventeen minutes before the scheduled liftoff, with storm clouds enveloping the site, NASA and SpaceX decided to call it off. There was just too much electricity in the atmosphere. The risk of a lightning strike was too high. Pence understood that the launch couldn't go and told Bridenstine that the delay "was the right decision." Still, Bridenstine dreaded having to tell the president. With Pence at his side, he walked into the room where Trump was waiting.

The president seemed to immediately sense what they were about to tell him. "We aren't launching, are we?" Trump said.

"Well, sir, we have too much electricity in the air," Bridenstine said.

"Yeah, but you can still launch, can't you?" It was a question that sounded a lot like an order.

Bridenstine stood his ground. "Sir, the only thing we need to be worried about are Bob and Doug. We need to delay."

"Why don't you wait five or ten minutes?" Trump said, suggesting they see if the weather would pass.

That also was an impossibility, Bridenstine said. Missions to the space station had an instantaneous launch window: in order to catch up to the space station orbiting overhead, they had to launch at a precise second, or not at all. But explaining orbital dynamics to Trump was pointless.

He clearly was not happy, but Trump didn't complain further. Shaking his head, he got up and walked out, saying only, "Back to the plane."

As Bridenstine was walking over to give Trump the news, Musk's adrenaline, as he said later, was running at "100 percent." When the launch was scrubbed, the hormone just drained from his body. "It went to zero," he said. "I collapsed and slept the longest I had in probably a year."

AFTER A THREE-DAY delay, everything appeared back on track. The weather on May 30, 2020, the Saturday of Memorial Day weekend, was much improved. Hurley and Behnken made their way back inside the spacecraft. Trump and Pence returned to the Kennedy Space Center. Trump was planning to use the event as part of his reelection campaign. If the rocket lifted off successfully, he'd use the backdrop of the Kennedy Space Center for a speech that would take credit for America's space enterprise—even though the Commercial Crew program began under Obama—and serve it up as a metaphor for his vow to Make America Great Again.

With about fifteen minutes left in the countdown, Trump said he wanted to speak to Hurley and Behnken. There was no way. They had no time and needed to stay focused. It wasn't going to happen.

As the countdown clock ticked, Bridenstine was ensconced in an office at the Kennedy Space Center, monitoring the ground controllers over the communication network. All was going well until, again, it wasn't. With just a few minutes left in the countdown, one of the sensors at the pad showed an unusually cold temperature reading, an indication that the superchilled liquid oxygen might be leaking. This was not good. Bridenstine sought out Pence and asked for his advice: "Should we go tell the president?"

Yes, Pence said, and he once again accompanied Bridenstine to give Trump bad news.

"Sir, we have a challenge with a temperature sensor," Bridenstine said. "And if we don't get it resolved, we're going to have to stop the launch."

Trump, already annoyed at being told he couldn't talk to the astronauts, was not pleased. He dismissed the news as if it were irrelevant. "It's going to go," he said.

Bridenstine tried to spin the situation into the best-case scenario—this all could be a faulty sensor and it could snap back to the correct reading. But the evident displeasure on the president's face turned his stomach in knots, as if the first human spaceflight in nearly a decade wasn't enough to worry about. Just then, one of Bridenstine's aides texted him. The sensor was now within the acceptable limits. They were go for launch.

Minutes later, the Falcon 9 lifted off from Launch Pad 39A. "Go NASA! Go SpaceX! Godspeed Bob and Doug!" shouted a commentator on the NASA broadcast.

Inside the capsule, they were all business, test pilots and astronauts, monitoring the health of the vehicle as it tore through the atmosphere, through the sound barrier, through the moment of maximum dynamic pressure, when forces exerted peak stress. They went higher and higher, faster and faster, until the sky turned black, the second stage separated from the first, and Dragon separated from the second. They were in orbit. SpaceX had joined a club whose only other members were the three most powerful nations on Earth.

It was a smooth, comfortable ride for Hurley and Behnken. After they completed their checkouts and confirmed everything looked good with the ground teams, and after they broadcast a video from space, giving a tour of the capsule's relatively roomy interior, they waved goodnight to their kids. Then, exhausted, both fell asleep immediately, their heads bobbing weightlessly as the spacecraft flew on its own toward the space station.

Soon, however, Hurley woke up. He was freezing. Behnken was hanging out in a T-shirt, as if he were at the beach, but Hurley could feel his hairs standing on end. He was always like that in space, always cold, always wearing long sleeves, like an office worker complaining to a coworker about the thermostat. He furtively snuck over to the touchscreens and raised the temperature, being careful not to wake Behnken up. Then he fell right back asleep.

Twice during the flight, they took over manual control of the space-craft, flying it with the touchscreens that had caused so much conster-nation. But those were only tests to see how Dragon might perform in the event that the autonomous system failed. Everything was working as expected, and for some nineteen hours the spacecraft flew itself all the way to the station. Hurley and Behnken could have dozed some more, or sat back and watched movies as if they were on a cross-country flight. There was little for them to do but monitor the mission from their screens, as if watching it on television. Even when they arrived—thirteen minutes ahead of schedule—and Dragon began the delicate choreography of docking, the former fighter jet pilots were mere bystanders as the spacecraft inched itself toward the station and gently parked itself.

TWO MONTHS LATER, the trip back to Earth was perhaps more treach-erous than the one up. On its reentry through the atmosphere, Drag-on's heat shield would have to withstand temperatures reaching 3,500 degrees Fahrenheit. Flames would engulf the entire spacecraft, leaving scorch marks across the marble-white capsule. Then the parachutes would deploy: first the drogues, the smaller chutes used to initially slow Dragon's descent; and then the four mains, which had given SpaceX so much trouble.

If it all went well and the capsule hit the water softly, Hurley and Behnken were convinced that it would bob up and down with the waves so much that they would get seasick. "I'm expecting a little bit of vomiting maybe to happen in the end," Behnken said before the flight. "So when we get to do that in the water together—it's kind of a weird thing to say—but I'm looking for that kind of celebratory event." For generations, voyagers christened their ships by smashing a bottle of champagne across the hull. Hurley and Behnken would christen their spacecraft with a bit of celebratory vomiting.

In all, the spacecraft would slow from 17,500 miles per hour to zero, shedding an extraordinary amount of energy. The air surround-ing the Dragon would turn into a fiery plasma of charged particles as electrons separated from their molecules, and as the plasma built up around the capsule the astronauts would lose the ability to communicate with the ground.

"Dragon, SpaceX, we show two minutes until predicted comm blackout. We will see you on the other side at 18:42," said Michael Heiman, the SpaceX engineer on the communication loops, using the coordinated universal time.

"Dragon copies. 18:42. We'll talk to you then," Hurley said, sounding as calm as if he were out for a leisurely bike ride.

The blackout period was scheduled to last about six minutes. Six minutes that felt like six hours. In SpaceX's mission control, barely anyone spoke. Barely anyone moved. Row upon row of computer screens were suddenly devoid of data. In addition to the loss of communications, the plasma cut off all telemetry as well. Musk sat still in the front row, in between Koenigsmann and Shotwell, saying nothing.

Fire roared around the capsule, illuminating the interior with an eerie glow. Sitting in the capsule, Behnken felt like Dragon "had come alive," as he would later recall. "You can hear the rumble outside the vehicle." Cut off from the ground, Dragon was firing its thrusters autonomously, keeping the capsule stable and pointed in the right direction. The roar of the plasma and the continuous thruster firings sounded, Behnken would say, "like an animal coming through the atmosphere."

At five minutes, a flicker of data began to come through on Heiman's computer. The connection was poor, but it was enough to check if the communications had come back online.

"Dragon, SpaceX, comm check," he said.

Musk turned his head, scanning the pair of screens in front of him, his elbows on the table, fingers intertwined. But there was only silence. Thirty seconds passed—an eternity.

"Dragon, SpaceX, comm check," Heiman tried again.

One second passed. Two. Then the sound of Hurley's voice, straining against extraordinary physical pressure: "Dragon has you loud and clear. We're about 3.9 Gs."

Shotwell clasped her hands together and tilted back in her chair, closing her eyes. Then she collapsed forward as if in prayer.

"Copy. We've got you five by five as well, Doug," the flight director said, using a term that meant the communication was clear. "Looking good, and you can expect an automated chute deployment."

"Copy," Hurley responded. "Automated chute deployment."

Inside the capsule, Behnken and Hurley could feel the parachutes deploying, a jolting force that clattered their teeth and felt "like getting hit in the back of the chair with a baseball bat—just a crack," as Behnken would say.

SpaceX had done rigorous testing of the new Mark 3 chutes, but it was something of a sprint. And so the sight of them fully inflated allowed Koenigsmann to finally exhale. "We were worried about the chutes," he later recalled. "We did a lot of testing, but it was done rather quickly and that was a concern."

Splashdown came in the Gulf of Mexico, off the Florida coast, in seas so calm that neither astronaut had to even think of reaching for the barf bag tucked into the pants leg of their space suit. It was as if all they had to do was return their seats to their upright positions and stow their tray tables.

"On behalf of the SpaceX and NASA teams, welcome back to planet Earth," Heiman said, doing his best impression of a commercial airline pilot. "And thanks for flying SpaceX."

MUSK WASN'T PLANNING on flying to Houston to greet Hurley and Behnken after their NASA plane landed just outside Ellington Field near the Johnson Space Center. But he had been so stressed about the fiery reentry through the atmosphere that he wanted to see them in person and confirm that they were indeed alive.

The sight of the astronauts, weary but walking fine despite feeling the force of gravity for the first time in two months, was a huge relief. "My entire adrenaline just dumped," he said during the ceremony welcoming them home. "Like, thank God. I'm not very religious, but I prayed for this one."

He had founded SpaceX eighteen years earlier with the goal of flying people. Human spaceflight was so embedded in the company's ethos that Musk had insisted that its first capsule, even though it was designed for cargo and not humans, have a window as a symbol of his intentions. Now SpaceX had successfully flown two astronauts, setting the stage for bigger ambitions to come.

"I do think what this heralds, really, is fundamentally a new era in spaceflight, a new era in space exploration," he said. "We're going to go to the moon, have a base on the moon, and send people to Mars and make life multi-planetary."

Bridenstine was just as relieved, and took the opportunity to review the recent turbulent history with SpaceX and make amends. "We had some significant challenges. We might have had some disagreements on parachutes, and we needed to change the titanium because of its reactivity with nitrogen tetroxide," he said, looking at Musk. "And I sent a tweet, Elon, and I know you remember this. I said, 'It's time to deliver.' And I tweeted it at Elon Musk. And I want to tell you, Elon, you responded magnificently. And you have in fact delivered."

Bridenstine was pumped up, thrilled at the success of the mission and what it portended, and so carried away by a moment of triumph that he couldn't help but slip into a stump speech. He was at heart a politician and he knew how to capitalize on the moment. After congratulating Hurley and Behnken and thanking them for their service, he quickly pivoted to the White House's next goal—the moon. The restoration of human spaceflight had happened on his watch, even if the Commercial Crew program began under President Obama. But the Artemis moon program was a Trump initiative. And now that Hurley and Behnken were back home safe, it was the top priority.

"Today we're flying into low Earth orbit, and in a few short years we want to be flying to the moon," he said. "Not just go once or twice, but we want to go sustainably, with a purpose. We're going to learn how to live and work on another world for long periods of time."

By then, Hurley and Behnken had left the ceremony. They were exhausted and needed to go through medical checkups and get reunited with their families. But later, when they heard that the NASA administrator had turned their mission into a rally about returning to the moon, they felt like they were being used as part of a political campaign. It was, as Hurley said, "completely inappropriate." They didn't need a ticker-tape parade, but Behnken and Hurley should have at least had their day before the NASA administrator started talking about his boss's next space goal.

"You're totally glossing over the fact of the accomplishments of the last decade," Hurley said, "and what the agency has done and SpaceX has done to get us back into launching from the United States. And you're already jumping ahead for something that's years away?"

Bridenstine couldn't help himself. NASA had, as he said, real "momentum," and he needed to keep it rolling. But whether that was enough to get it back to the moon for the first time in some fifty years remained to be seen. The International Space Station, from which Hurley and Behnken had just returned, was in low Earth orbit, 240 miles away, about the distance between New York City and Washington, D.C.; the moon, a far more ambitous destination, was 240,000 miles away. The Artemis program was already facing enormous challenges, and the pressure from the White House was mounting.

PART III

...

Beyond

2020–2025

CHAPTER 14

Super Hardcore

Jeff Bezos had vowed to spend $1 billion of his own money on Blue Origin. He'd personally pitched NASA's leadership—twice—on the company's Blue Moon spacecraft. He'd said "space is the most important work I'm doing," and to prove it, he'd expanded the company's facilities and workforce. Still, Blue Origin had little to show for its efforts. Not a rocket to orbit. Not a major government contract leading to executing on an actual space program. Nothing that would begin to put it in the same league as SpaceX.

On April 30, 2020, however, twenty years after Bezos founded the company, it appeared this was on the verge of changing. Blue Origin led a team of companies that won the lion's share of the nearly $1 billion NASA was awarding to build the lunar lander spacecraft that would ferry astronauts to the moon for the first time since the Apollo era. These were only the initial contracts, a way for NASA to place several bets and see which ones would pay off before making the final awards a year later. Still, it was a significant victory that put Blue Origin at the center of one of NASA's landmark human exploration campaigns. The company walked away with $579 million. Dynetics, a subsidiary of Leidos, the defense contractor, finished second, with $253 million.

SpaceX, much to the delight of Blue Origin's leadership, was a distant third with just $135 million.

With the White House under President Donald Trump pushing the return to the moon, Blue Origin had gone all-in on the contract, convening an all-star assembly of the biggest and most powerful aerospace companies in the world: Lockheed Martin, Northrop Grumman, and Draper, which had provided the guidance and navigation on the Apollo moon missions. With Blue Origin serving as the prime contractor, Bezos publicly trumpeted the companies' efforts as a "national team for a national priority." If SpaceX had dominated the rocket launch market, flying its Falcon 9s at an increasingly high rate and reusing them over and over, while restoring human spaceflight missions to the International Space Station, Blue Origin sought to differentiate itself by staking its claim on the moon. It was, after all, Bezos's lifelong passion, the muse that had inspired his "vision speech" in Washington the previous year.

"This is the kind of thing that's so ambitious it has to be done with partners," he said at the time. "This is the only way to go back to the moon fast. We're not going back to the moon to visit. We're going back to the moon to stay."

With the win, it felt like Blue Origin had emerged from a long quiescence, awakening to finally give SpaceX the competition so many had longed for. Bezos had said in 2016 that Blue Origin was going to compete for every contract SpaceX went after. Now it had.

IN THE MONTHS leading up to the announcement, Blue Origin had kept up its lobbying campaign, continuing to woo White House officials and trying to convince Vice President Mike Pence that it could help make his 2024 moon landing timeline a reality. After Bob Smith's disastrous meeting with Marc Short—Pence's chief of staff, who'd told him "You have a *Washington Post* problem"—Bezos had gone out of his way to praise Pence and his Huntsville moon speech. Then, on January 25, 2020, he hosted Trump's daughter, Ivanka; her husband, Jared Kushner; and Trump adviser Kellyanne Conway at a party at his

$23 million mansion in Washington, D.C., which he'd purchased shortly before Trump's victory in the 2016 election. But tensions lingered between the Trump White House and Bezos, particularly over a highly anticipated Pentagon cloud-computing contract that was worth $10 billion over ten years. Amazon Web Services, which had a similar contract with the Central Intelligence Agency, was considered the favorite. But in 2019, the Department of Defense awarded the contract to Microsoft. Amazon protested the decision in unusually harsh terms, alleging that Trump had directly interfered with the procurement as retribution for Bezos's ownership of the *Post*.

"There was significant political interference," AWS's chief executive, Andy Jassy, said at the time. "When you have a sitting president who is willing to share openly his disdain for a company and the leader of that company, it makes it really difficult for government agencies, including the DoD, to make an objective decision without fear of reprisal."

The *Post*'s unrelenting coverage of the Trump administration was also making life more complicated for Blue Origin. On January 31, 2020, the *Post* wrote yet another editorial critical of Trump's space policies. This time, the headline read, "NASA Keeps Falling Victim to Presidential Whims." The editorial board, which was separate from the news department, criticized Pence for doing "his best John F. Kennedy impression to do the almost impossible: return humans to the lunar surface within five years 'by any means necessary.' The mandate seemed unreasonably ambitious; for one thing, those necessary means aren't nearly as available to NASA today as they were during the Cold War, when all aspects of government were committed to whatever-the-cost victory in the space race."

While Pence had proposed landing the first woman on the moon, the piece argued that space exploration could be done more efficiently by robots and "doesn't actually require humans." On one point, the *Post* was in agreement with the White House, arguing that "private companies increasingly interested in low-orbit adventuring should be entrusted with as much as they're able to carry out, to save NASA money and to ensure that exploratory work continues even as the whims of politicians shift." The piece disclosed—as it always did, and no less awkwardly—that "one of those companies is owned by Amazon chief executive Jeff Bezos, who also owns the *Post.*"

Blue Origin moved to separate itself from the editorial by crafting a response that it hoped would ingratiate itself with the White House and demonstrate that while both the news organization and the space company were owned by Bezos, they had very different views about Trump's space policy. Conspiracists in the White House may have imagined that Bezos would simply pull strings and order the *Post* to publish a counter-editorial. But there was a sacrosanct line that Bezos would not cross. As Martin Baron, the *Post*'s executive editor, wrote in his memoir, Bezos "had never ordered up an editorial, quashed one, or even reviewed one before publication (and, from what I've been told, hasn't done so since)." (Later, when Bezos killed the *Post*'s endorsement of Kamala Harris in the days leading up to the 2024 election, Baron would have harsh words for his former boss. "This is cowardice, with democracy as its casualty," he would write on social media. Trump "will see this as an invitation to further intimidate owner @jeffbezos. Disturbing and spineless at an institution famed for courage.")

Bezos didn't interfere with the *Post*'s space editorial. Instead, as if it were just any other company, Blue Origin spent some $50,000 to purchase a full-page ad on February 18 on page A5 of the *Post* to make its case, hoping the White House would take notice. Weighing in at more than eight hundred words, the ad, under Bob Smith's name, was a point-by-point cataloguing of the importance of space in everyday life, a distillation of Bezos's passion for space and the company he was working to build. It did not hold back in its criticism of the paper's editorial, charging it was representative of "uninformed critiques that come from many corners and have helped stymie well-intentioned prior efforts to move our nation forward into space. It fails to recognize the massive shifts in the space industry that allow us to make greater strides and the emerging threats that require us to re-double our efforts."

Echoing Pence's Huntsville speech, the ad evoked the threat of China, saying its "space dominance objectives and the rise of hypersonic weapons heighten the need for resilient space assets." And it favorably compared the leadership of the White House to Kennedy's call for a moon landing during the Cold War space race. "That kind of leadership is on display across the political spectrum today," the ad gushed. "It is visible in the President's Budget Request and in the administration's re-establishment of the National Space Council."

Then came a doozy of a last line that put Bezos squarely in line with the Trump administration and in direct conflict with the editorial board of his own news organization: "This inevitable expansion will not be stopped by those that waver and merely critique but will be forged by those across government and industry who are unapologetic and who are unafraid to build and to dream."

As Blue Origin had hoped, the White House did notice the ad. Pence was thrilled and excitedly called Bridenstine. "Did you see this?" he said. Bridenstine hadn't, but then read it with glee and took to Twitter to praise Blue Origin. "Thanks @blueorigin for responding to the @washingtonpost Editorial Board's ill-informed criticism of NASA's Moon to Mars efforts. @POTUS Trump & @VP Pence have laid out a vision to use the Moon as a proving ground for missions to Mars, which enjoys bipartisan support in Congress."

If Blue Origin once had a *Washington Post* problem, it had been blunted. But at the same time, NASA was realistic about the company's capabilities. Blue Origin had yet to even make it to Earth orbit, let alone fly anything to the moon. Landing people on the lunar surface in five years seemed like a long shot, if not an impossibility.

That was one of the main reasons why NASA had awarded SpaceX a contract as well, keeping its proposal alive for the final round. While Blue Origin and Dynetics were developing landers that resembled the ones used during Apollo, SpaceX's Starship was entirely different. It was massive and massively complicated, a fully reusable system that would require multiple refuelings in Earth orbit before going to the moon, something that had never been done before. Starship was a risky bet, but one NASA was willing to make, especially given SpaceX's stellar track record of upending expectations. Some in NASA may have been skeptical about SpaceX's ability to launch Starship, but they weren't going to count out the company that, unlike Blue Origin, had proven to be one of the space agency's most reliable partners.

"It's obviously a very different solution set than any of the others," Bridenstine said at the time. "But it also could be absolutely game-changing. So we don't want to discount it. We want to move forward. If they can have success, we want to enjoy that success with them."

THAT SUCCESS, IF it was to come, would be born from the ashes of failure. While SpaceX had briefly flown its Starhopper prototype twice, the next phase of its campaign seemed to be designed to see how many spacecraft it could blow up in the least amount of time and in the most spectacular fashion. The company would end up destroying at least ten prototypes, a series of metal-contorting collapses and Earth-rattling explosions that sent gigantic fireballs and clouds of gas vapors into the sky, as if the company was competing for destruction style points. Watching from afar, skeptics—including some inside NASA—might have concluded that the company didn't know what it was doing, and that Starship was little more than a stainless steel monument to Musk's ego. The repeated failures also gave Blue Origin confidence that the "tin can" SpaceX was intent on flying would never get off the ground.

But the early tests were reminiscent of SpaceX's attempts years earlier to perform another feat no one thought possible: landing the first stage of its Falcon 9 rocket. By 2020, the landings had become routine. But while they looked effortless, they came only after years of failure—boosters crashing and exploding with such frequency that SpaceX compiled them into a blooper reel fit for ESPN titled "How NOT to Land an Orbital Rocket Booster." In hindsight, the early failures looked comical and simplistic—a lack of hydraulic fluid here, an engine sensor failure there—so that SpaceX engineers could look back and laugh at themselves. "Well, technically it did land," they wrote over footage of one spectacular failure, "just not in one piece."

After Starhopper, the first Starship prototype to be tested was called Mark 1, a rudimentary cone of steel that wasn't pretty and wasn't supposed to be. It was supposed to show how well it would be able to withstand the pressure required to fully fuel the spacecraft ahead of flight. Some engineers on the team did not have a good feeling for how the test was going to go. The welds were bad, they thought. The whole thing could burst; but they wanted to proceed with the test anyway.

It burst. On November 20, 2019, SpaceX filled the cone with liquid nitrogen and began bringing it to maximum pressure. When it exploded, the dome that had been serving as the top of the prototype shot hundreds of feet into the air, clocking about as much airtime as Starhopper had, while the top of the test vehicle crumpled as it if were a soda can and a cloud of gas burst into a white column.

Musk continued to show up at Boca Chica with frequency, often on weekends and with little notice. That sent employees scrambling not only to get back to work, but also to find Fiji water, which he seemed to prefer. (Later, his assistant told them that any water was fine. They saw him drinking Fiji water because the bottles fit well into the cupholders of his private jet.) Musk wanted to ramp up production of Starship quickly, staying up all night one Saturday shortly after the Mark 1 failure to help his team design how the production site would be laid out. Musk decreed that he wanted the concrete for the manufacturing site's foundation to be poured that very day.

His team tried to explain that it would be tough to find a concrete company to come on a Sunday. Musk reacted as if he didn't know what weekends were. He was here. They should be too, he insisted.

His team called the concrete company. There was, of course, no answer. For good measure, Musk's assistant called as well, but by then some other issue was occupying Musk's attention, granting them a reprieve until Monday. Musk was in a rush and didn't want to bother building permanent structures. The floors would be concrete, but much of the manufacturing would happen outside, in open tents and hangars. There was no time to hire architects. No time to file for building permits or hire construction crews.

Once, on a visit to Boca Chica, when I asked him what the area would look like in the future, Musk conceded, "It will definitely get fancier than it currently is. The reason it's not fancier is just because it would have taken too long to build the buildings. Since it was going to take so long to build the buildings, we just built it"—Starship—"outside. My new thing is management by rhyming. If the schedule is long, it's wrong; if it's tight, it's right."

TO MEET MUSK'S demands, SpaceX needed to staff up to work shifts around the clock, and in February 2020 it held a job fair at the site. The company was inundated with applicants from a community beset by unemployment and desperate for work. It made for an odd juxtaposition: SpaceX, the hard-charging technology start-up, was situated in the middle of one of the poorest regions of the country—outside of Brownsville, where a third of the residents lived below the poverty line and dreamed not of going to space but of escaping the depressed South Texas economy. The area was full of boarded-up houses, shuttered businesses, and schools surrounded by fences topped with barbed wire.

SpaceX was welcomed by many local officials as a Walt Disney–like corporate messiah poised to spark an economic revival. The state set aside millions to help the company build its facilities at Boca Chica and largely bought into SpaceX's vision of transforming the area into a commercial spaceport that would be sending people to the moon and Mars.

"You know the term 'visionary'; they're the ones who make the world go 'round," Eddie Trevino, the Cameron County judge, told me when I visited him in 2019.

The job fair was packed, especially since Musk had tweeted about it to his millions of followers, saying it was to help "staffing up 4 production shifts for 24/7 operations, but engineers, supervisors & support personnel are certainly needed too. A super hardcore work ethic, talent for building things, common sense & trustworthiness are required, the rest we can train." Local news covered the event. Musk was there in person, wearing a T-shirt that read OCCUPY MARS.

With Musk dialed in, production picked up to the "super hardcore" levels he demanded. Soon SpaceX's engineers were pumping out prototype after prototype, so that when one burst or exploded, the next would be ready to roll off the production line. With so many, SpaceX switched to calling them Serial Number. The first, SN-1, burst during a pressure test that crumbled the cylinder test vehicle and sent it leaping into the air. "So . . . how was your night?" Musk wrote on Twitter, posting a video of the failure. "It's fine, we'll just buff it out." He added: "Where's the Flextape when you need it?!" The next month, SN-2 survived its pressure test. But a few weeks after that, in April 2020, SN-3 collapsed, this time because of a problem with the test itself. SN-4 survived its pressure test and on May 29 successfully lit its engine—a big milestone, even if the prototype later exploded after what appeared to be a fuel leak.

Another test. Another fireball. As spectacular as it was, the explosion went largely unnoticed at NASA, coming the day before SpaceX successfully launched Behnken and Hurley to the International Space Station. When Bridenstine had publicly chastised Musk for spending too much attention on Starship and not enough on Dragon, Musk had argued that SpaceX could focus on both simultaneously. It appeared Musk was right. Despite Bridenstine's public castigation and the subsequent meeting at SpaceX headquarters, the pace of production at Boca Chica did not slow down. If anything, it sped up.

On a Saturday less than a week after Hurley and Behnken successfully lifted off—but before SpaceX had flown them home safely—SpaceX employees received an email from Musk. "We need to accelerate Starship progress," he wrote, adding that it needed to happen "dramatically and immediately." He urged employees working at its headquarters in Hawthorne, just outside of Los Angeles, to consider volunteering to work in Texas. "For those considering moving, we will always offer a dedicated SpaceX aircraft to shuttle people," he wrote, adding: "Please consider the top SpaceX priority (apart from anything that could reduce Dragon return risk) to be Starship."

In the early days of Boca Chica's development, SpaceX employees wasted almost an entire day traveling there from Hawthorne, or from its facilities in Cape Canaveral. There were no direct flights, so they had to make a layover, often in Houston or Dallas. Not content to lose so much time, SpaceX started using its own planes to transport workers, eventually holding at least two flights a day. Housing was also a problem. SpaceX workers crammed the hotels in Brownsville, some forty minutes from the production facility in Boca Chica, and hired shuttle buses to move employees between shifts.

Musk also ordered a fleet of Teslas to be delivered to the Brownsville airport, which employees could reserve.

All sorts of employees made the sojourn to South Texas: vice presidents, engineers, lawyers, accountants, people from human resources, even interns. The workers most in demand, though, were welders, who were needed to manufacture the stainless steel Starship. In his hair-on-fire company-wide missives, Musk urged anyone with any welding skills to move to Boca Chica. As production picked up, SpaceX started erecting actual buildings, but not always following county regulations. More than once, the Cameron County fire marshal descended on the site, angry that SpaceX had not obtained the proper permits.

Late one Friday afternoon in late 2020, he even threatened to arrest Shyam Patel, the site's manager, saying, "If he spends the weekend in the tank, he'll apply for permit." The fire marshal, Juan Martinez, told me it went down this way: "I let them know that any violation against our fire code is a class B misdemeanor, which is straight to jail. I was just giving them information that was pertinent for them. I do take fire violations and life safety issues very seriously." Either way, Patel had left on a flight back to Hawthorne, but another SpaceX employee was able to hold off Martinez until Monday. A team of engineers then spent the weekend creating a spreadsheet of all the permits the company would need, and a 120-day schedule for how they would obtain them. That appeased the fire marshal, who never did arrest Patel. Martinez also told me that SpaceX started doing a much better job meeting code and obtaining permits. More recently, he said, "we've had zero issues." (SpaceX declined to comment on this.)

To accommodate the growing population, SpaceX started bringing Airstreams so that employees could live and work on-site. It built out a tiki bar, meal services, and other accommodations, though parking was always a challenge, especially with the growing number of gawkers descending on the site. SpaceX security put out cones to direct traffic, but Musk hated the color. The orange was cloying, and, some employees believed, somehow triggered his Asperger's. Whatever the reason, Musk wanted white cones, so the security teams used those instead, especially when they knew the boss was going to be around.

SPACEX'S URGENCY WAS not just about making Starship a viable candidate for NASA's Artemis program. There was an election coming up, and there were no assurances that, if Biden won, the new administration would keep funding the project. Starship had become an extraordinarily expensive proposition, and if Musk wanted to keep up the torrid pace of manufacturing and testing—and then, eventually, flying—SpaceX would need additional revenue.

In the middle of 2019, the company embarked on another fundraising round, this time bringing in $536 million, and more than $1 billion for the year. The value of the company rose to more than $30 billion, making SpaceX more valuable than Tesla, Musk's electric car company. Much of what was driving the growth was another major and risky endeavor that Musk hoped would become a new line of business: Starlink, which Musk had said was essentially "like rebuilding the Internet in space." Others, he knew, had tried and failed. In fact, as he said at the time, no one had "successfully gone into operation without going bankrupt."

It was not going to be easy for SpaceX, either. The company was licensed by the Federal Communications Commission to put up more than eleven thousand satellites, many times more than were currently operating in orbit. If SpaceX were successful, it would transform the company from a rocket company to an Internet provider. Musk had said that SpaceX's rocket business generated about $2 billion a year and could possibly grow to $3 billion. But revenue from Starlink could be "more like $30 billion a year," he said. "Total Internet connectivity revenue in the world is about $1 trillion, and we think maybe we can access about 3 percent," he said.

That would require an enormous effort to build and launch the satellites at an incredibly rapid rate. Already there were signs the company was not moving fast enough—or at least not fast enough for Musk. In 2018, during a visit to the company's satellite manufacturing center in Redmond, Washington, Musk fired the Starlink leadership team—including Rajeev Badyal, SpaceX's vice president of satellites—and brought in new management from SpaceX's headquarters, with the goal of launching the first batch of satellites by the middle of 2019.

In May 2019, SpaceX did, sending up sixty satellites on a Falcon 9 rocket. "This was one of the hardest engineering projects I've ever seen done, and it's been executed really well," Musk said. Still, he was stressed. "I think it is important to acknowledge that there is a lot of new technology here," he said. "So it's possible that some of these satellites may not work. In fact, there's a small possibility that all of the satellites will not work." He added, "I do believe we will be successful, but it is far from a sure thing."

Plus, he had competition—at least potentially. Amazon was also pursuing its own Internet satellite constellation, called Project Kuiper, and would eventually earn approval to launch more than 3,200 of its own satellites. There were an estimated four billion people on Earth without access to broadband, as well as airlines and shipping companies that wanted better Internet service for operations in remote areas. Then, of course, there was the military. At a time when computing power and data were as important as bombs and ammunition, having reliable Internet in war zones would be vital, especially at a time when drones and other autonomous weapons were changing the nature of combat. The Pentagon cheered on the private sector pursuing these new technologies, even if it didn't fully think about the consequences of placing such power into the hands of private enterprise.

The market for satellite-based Internet was simply huge. Amazon planned to invest $10 billion into the project, and Bezos had put some of his most trusted leaders in charge of it. One of the first things they did was hire Rajeev Badyal, the Starlink manager who'd been fired by Musk.

Once again, Musk and Bezos would find themselves in direct competition. Once again, SpaceX would have a massive head start. But Amazon had a worldwide customer base, years of experience, and a track record of successfully breaking into new areas—from books to music to cloud computing and artificial intelligence.

Musk did not seem concerned. "It's always good to have competition," he said. Still, on Twitter he posted "@Jeff Bezos copy" followed by a cat emoji.

BEZOS WAS THE target of the taunt. But Musk could also have been referring to China, which was increasingly looking to rival not only the United States but its commercial space enterprise. In a quest to find a SpaceX or Blue Origin of its own, China had begun allowing private investment to flow into its space sector. As a result, several new companies sprouted up, designing new reusable rockets as well as satellites that were smaller, more affordable, and packed far more computing power.

By 2019, Chinese space companies were second only to the United States in terms of the number of start-ups receiving investment, the amounts invested, and the total number of deals. And the numbers were rising. That year, investors poured $314 million into twenty-two ventures, up from $288 million into ten companies in 2018.

It was clear who their idol was. "We wanted to be the Chinese SpaceX, the Chinese Elon Musk," Yang Feng, the CEO of satellite start-up Spacety, told an interviewer.

Like Amazon and SpaceX, China would pursue an Internet constellation of its own, one that would flood low Earth orbit with more than ten thousand satellites beaming the Internet to ground stations in remote areas all over the globe. The effort, as a report from the Center for Strategic and International Studies noted, "enjoys significant support from the Chinese Communist Party." China infamously censored information within its own borders, blocking access to websites and restricting freedom of the press. If China were to become a popular Internet provider in countries in Asia, South America, and Africa "users could be blocked from accessing entire sections of the internet or expose personal and identifying data to network operators," the report said. "This could further exacerbate the suppression or censorship of information in countries where there is already a widespread information vacuum or civil unrest."

Thousands of satellites would require lots of rockets, touching off a hot new market in China. By one count, there were sixteen rocket start-ups in the country by 2019, in various stages of development. Notably, most of them—nine—were pursuing what Bezos and Musk believed to be the holy grail of rocketry: the ability to land and reuse the boosters so that they could be flown over and over.

NONE QUITE MATCHED Starship, however. What had started out as a Musk vision quest, a blurry mirage in the distance that would some-day, somehow, get humanity to Mars, was coming into focus. Musk staked SpaceX's future on its success. The colossus of a rocket was eating up a huge amount of the company's resources and—perhaps most importantly—Musk's attention. In August 2020, a few days after the Dragon capsule carrying Behnken and Hurley splashed down safely in the Gulf of Mexico, the SpaceX employees in Boca Chica prepared for another test—this time of SN-5, a full-scale prototype of the spacecraft.

Its design still seemed to mimic the architecture of agriculture more than aerospace. "No, This Isn't a Flying Grain Silo. It's SpaceX's Starship Prototype," was CNN's headline.

However seemingly unfit to leave terra firma as it may have been, the silo did fly. It was only a short, precarious hop to five hundred feet, while unsettlingly large chunks of debris were blown across the launch site, and the lone engine caught fire during the descent. And yet there was something undeniably magnificent about the rocket—how it lifted off slowly, tilting ever so slightly as it steered itself a few hundred yards toward the landing pad and then, at its apogee, hovered for a split second before descending. After so many failures and fireballs, Starship was alive. At last, Musk was pleased. "Progress is accelerating," he tweeted.

Even Trump noticed. Although NASA had nothing to do with the test, he took credit for SpaceX's success. Trump retweeted

NASAspaceflight.com, a group of SpaceX and space enthusiasts who had been diligently chronicling Starship's progression, perhaps confusing the account with NASA's official Twitter handle. "NASA was Closed & Dead until I got it going again," Trump wrote. "Now it is the most vibrant place of its kind on the Planet . . . And we have Space Force to go along with it. We have accomplished more than any Administration in first $3^1/_2$ years. Sorry, but it all doesn't happen with Sleepy Joe!"

The presidential election was just three months away, and Trump was again trying to use space as a campaign issue against Joe Biden. While it was undeniable that his administration had placed a priority on space in a way not seen since Kennedy, the MAGA-ization of exploration had begun to rub many at NASA and in the space community the wrong way. After SpaceX launched Hurley and Behnken to the space station, Trump had turned the event into a de facto campaign rally. "When I came into office three and a half years ago, NASA had lost its way," he said. "And the excitement, energy, and ambition, as almost everyone in this room knows, was gone. There was grass growing through the cracks of your concrete runways. Not a pretty sight. Not a pretty sight at all."

The speech and the launch were quickly repurposed into a campaign ad that claimed Trump had "Made Space Great Again." It showed a nostalgic panoply from Apollo—Walter Cronkite wiping his forehead in shock and Neil Armstrong taking his first steps on the moon—alongside the astronauts of the moment, Hurley and Behnken, in training and waving goodbye to their wives and children.

Karen Nyberg, Hurley's wife and a former NASA astronaut, who was no fan of the president, objected strongly. "I find it disturbing," she wrote on Twitter, "that a video of me and my son is being used in political propaganda without my knowledge or consent. That is wrong." The ad soon came down.

WHILE SPACEX HAD been publicly showing off its Starship testing campaign—fireballs and all—Blue Origin was keeping a lower profile. Despite the size of the "national team," the companies worked well together in what they called a "badgeless environment." Outsiders could not tell who worked for Blue Origin, or Lockheed, or Northrop. They were one squad, with a common goal: win the final lunar lander contract and beat SpaceX.

But as COVID raged throughout 2020 and into 2021, Bezos dedicated himself to Amazon, which had become a vital resource during the pandemic. As Blue Origin was starting its push on the Blue Moon proposal, Bezos wrote a note to Amazon employees, saying, "My own time and thinking is now wholly focused on COVID-19 and on how Amazon can best play its role. I want you to know Amazon will continue to do its part, and we won't stop looking for new opportunities to help." To keep up with demand—sales of cold and cough medicine jumped 862 percent—Amazon hired tens of thousands of new employees. And Bezos, who had largely concerned himself with the future of the company and long-term projects, was now focused on daily operations once again.

At Blue, where he had been working on Wednesdays and occasionally Saturdays, he became largely absent. "He just went deep into Amazon," one former employee recalled. "We didn't see him for a year."

CHAPTER 15

A Chinese Flag in the Lunar Soil

He was focused on the flight controls and the computer readings, not looking out the window. But as the ascent stage of the lunar module lifted off, Buzz Aldrin couldn't help but peek outside at the moon one more time to take in its "magnificent desolation," as he later described it. There, not too far away, was the American flag he and Neil Armstrong had erected less than twenty-four hours earlier, its vibrant red, white, and blue a splash of color against the dead gray of the lunar surface and the inky dark vastness of space beyond.

More than a symbol of national prowess, the flag—outfitted with a horizontal pole to make it seem like it was flying in the airless void—was a statement of victory in the space race against the Soviet Union. The photograph of Aldrin saluting it in his white space suit would become one of the most iconic of the mission, an image that, like Marines raising the flag over Iwo Jima or Martin Luther King Jr.'s speech at the Lincoln Memorial, would become embedded in the national consciousness, reprinted in textbooks for generations of students.

But the sight outside Aldrin's window that July day in 1969 was anything but inspiring. As the spacecraft lifted off, the thrust of its engine kicked up the lunar dust into a tempest and rocked the flag. Armstrong "was studying the attitude indicator," Aldrin recalled in his memoir, "but I looked up long enough to see the flag fall over."

Astronauts on the next five Apollo missions all planted flags. But they did not fare much better. Even if they survived the liftoff of the ascent vehicles, decades of exposure to the harsh lunar environment likely turned them to tatters. On the moon, temperatures could reach 212 degrees Fahrenheit during the day, and nights could last two weeks, at as much as 238 degrees below zero. "Even more damaging is the intense ultraviolet radiation from pure, unfiltered sunlight," the planetary scientist Paul Spudis wrote in *Air & Space Magazine*. "Even on Earth, the colors of a cloth flag flown in bright sunlight for many years will eventually fade and need to be replaced. So, it is likely that these symbols of American achievement have been rendered blank, bleached white."

Others have surmised that over time the flags likely disintegrated. Either way, the flags that started out as icons of humanity's greatest achievement also came to represent an abdication of the great American space quest started by President Kennedy—and an unfulfilled pledge: "We leave as we came," Eugene Cernan, the last person to walk on the moon, said as he departed the lunar surface, "and, God willing, as we shall return, with peace and hope for all mankind."

NEARLY FIFTY YEARS later, on December 1, 2020, a Chinese spacecraft, Chang'e-5, touched down near the Ocean of Storms on the lunar surface. Over a two-day prospecting expedition, the unpiloted spacecraft used a drill and a scooper to gather more than four pounds of lunar rocks and soil, known as regolith, from a volcanic plain. Then a return capsule lifted off with the material, bound for Earth. No country had brought home samples since the Soviet Union in 1976.

The successful robotic mission was another milestone in the country's step-by-step approach to moon exploration. China was, at that point, the only country to land on the moon in the twenty-first century. Chang'e-5 was the third time it had done so, following a landing on the far side—a historic first—in early 2019. Now it was aiming to set up a permanent presence on the same real estate that the United States had been eyeing for its Artemis program: the lunar South Pole. "A new wave of lunar exploration has been emerging in the world, with participants aiming to make sustainable missions to deepen knowledge of the moon and exploit resources there," Zhou Yanfei, the deputy general designer of China's human spaceflight program, said during a conference in late 2020.

For a country that hadn't flown a person to space until 2003—some forty-two years after the Soviet Union and the United States—China's space program had made dramatic improvements. It was planning to send a rover to Mars and assemble a space station of its own in low Earth orbit. With three modules, the station, called Tiangong, or "Heavenly Palace," would be continuously inhabited, like the International Space Station.

But perhaps nothing better demonstrated China's ambitions in space than a little-noticed gesture that came at the end of the Chang'e-5 lunar sample return mission. On December 3, 2020, the spacecraft deployed a Chinese flag on the lunar surface. It was small, weighing less than an ounce, but it was bright red with yellow stars—a source of immense pride for China's Communist Party. And unlike the American flags planted during Apollo, it was designed to last. Chinese scientists and engineers spent more than a year working to develop a new composite material "that could stand the harsh environment and be dyed in China's vivid national colors," state media reported. "The Chinese national flag shines an even brighter red from the moon, and from now on it will be a grand reminder for stargazers from all over the world of the excitement and inspiration we felt from Apollo missions more than half a century ago."

ANOTHER THING CHINA had in its favor: its authoritarian government. Putting aside its controversial record on human rights, the Communist Party was good for the country's space exploration goals. Without having to worry about the messiness of democratic rule—of changing politics, administrations, and priorities—China could set its course and stay on it. In the United States, different presidential administrations had pointed NASA toward many destinations: to the moon, then to Mars and an asteroid, then the moon again. No deep space human exploration program had survived from one presidential administration to the next since the Apollo era. NASA was unable to gain much momentum in any one direction before a new administration arrived and spun them around. It happened with such frequency that NASA's leaders started referring to it as Lucy and the football—after the *Peanuts* cartoon, in which Lucy Van Pelt takes the football away just as Charlie Brown goes to kick it, over and over again.

Jim Bridenstine knew the history—and the fact that once his tenure as NASA administrator came to an end, Artemis might very well be axed by the next administration. Ensuring that it survived—on top of finding a way to make the near-impossible 2024 lunar landing happen—would require support from Democrats. So in the summer of 2019, Bridenstine launched a political-style campaign, trying to bring on board elected officials from both sides of the aisle. He crisscrossed the country, glad-handing his former congressional colleagues, posing for photos, touting Artemis and its promise to land "the first woman" on the lunar surface. In all, he met with more than thirty members of Congress or their staffs, pushing for the billions in additional funding needed to make the program a reality. Rocket fuel may make the rockets go, but in the world of spaceflight an equally important maxim was "No bucks, no Buck Rogers."

Prying money out of Congress for space exploration was no easy task. While members loved to say they supported NASA—just like they favored lower crime and better education—getting them to actually increase the agency's budget was another story, especially when they suspected the 2024 timeline was dictated by politics, not exploration goals. No matter how hard Bridenstine pressed, Eddie Bernice Johnson, the chair of the House Committee on Science, Space, and Technology, wasn't buying it. "Rhetoric about American leadership in space and advancing the role of women in spaceflight is all well and good," the Democrat from Texas said. "But it is not a substitute for a well-planned, well-managed, well-funded and well-executed exploration program.

To date, Congress has not been given a credible basis for believing the president's moon program satisfies any of those criteria."

Like any good politician, Bridenstine was undeterred. He even ventured into what many of his fellow Trump political appointees would consider enemy territory, visiting Nancy Pelosi in her California district. Pelosi and Trump had been trading insults for months; he had dubbed her "Crazy Nancy," and she referred to him as an impostor who was in "over his head." If Bridenstine could squeeze an endorsement for Artemis out of her, it would be a political coup. Bridenstine's staff reached out to hers about conducting a town hall–style meeting at NASA's Ames Research Center in Mountain View, California. Pelosi's staff suggested they do it on Women's Equality Day, August 26, 2019.

Before he left for Silicon Valley, however, Bridenstine ran the idea by Pence while they were backstage during an event at the National Air and Space Museum to commemorate the fiftieth anniversary of the Apollo 11 moon landing.

"Sir, I'm heading out to do a town hall with Speaker Pelosi about Artemis," Bridenstine told him. "If you tell me I should not do this, I won't."

Pence told him to go: "We need the resources, and she could be helpful."

Bridenstine was nervous about how it would go. As a member of the House, he'd been no ally of Pelosi. But at the event, Pelosi was cordial and her introduction of Bridenstine appeared heartfelt. "Mr. Administrator, you bring us honor by being here," Pelosi told Bridenstine while she was sitting on stage. "As far as having a woman step foot on the moon, our hopes are riding on you."

Bridenstine talked about how diverse NASA's astronaut corps had become. "This time when we go to the moon, we go with, in fact, all of America, under the Artemis program," he said. "I'll tell you why this is so important to me as the NASA administrator. I have an 11-year-old daughter, and I want her to see herself as having all of the opportunities that I saw myself as having when I was growing up." He added: "Maybe, just maybe, the first person who ever walks on Mars will be a woman."

For a while the event transcended politics. Pelosi and Bridenstine sat strapped in side by side in Ames's vertical motion simulator, pretending they were landing on the moon. Bridenstine had gotten what he came for: her tacit endorsement.

Pence was thrilled. And he expressed it publicly on Twitter to put Pelosi on the record in a rare moment of comity: "Great to see @SpeakerPelosi join us in supporting Artemis, which will land the first American woman on the Moon by 2024! Thank you @NASA for all of your hard work."

BIPARTISAN CONGRESSIONAL SUPPORT was vital, but Bridenstine and his staff had also been working on another strategy to make the Artemis program bulletproof. In May 2020, Bridenstine and Michael Gold, a veteran of the space industry whom Bridenstine had recruited to lead NASA's legal and international efforts, announced what were called the Artemis Accords, bilateral agreements that would lay the legal framework for how nations would operate on the moon.

During Apollo, the United States went to the moon by itself. Now it would do so with a broad international coalition, with NASA as the leader. The accords would dictate norms of behavior—an extension of the values of a free and democratic society in an attempt to ensure that the settlement of the moon would follow American ideals. Signatories would vow to be transparent about their activities on the lunar surface, to explain where they would operate and what they would do. They'd promise to share any scientific discoveries with the world and be required to care for ailing or injured astronauts of other nations.

The accords would build upon the tenets of the Outer Space Treaty of 1967 to "promote peaceful purposes" that would allow nations "to participate safely in outer space," Bridenstine told me at the time. The effort was meant to put pressure on China, which many feared would not adhere to such standards as it forged its own path to the moon. There was another motivation for standing up the accords: They would make the Artemis program appealing to any incoming administration. Kill it, and they'd be nullifying any space exploration agreements that the United States, through NASA and the State Department, had signed with other countries.

Bridenstine was well aware that if Trump lost the 2020 election, his successor might keep alive the streak of jerking NASA in yet another direction. "When it comes to deep space exploration programs, we often say, 'Failure wasn't an option, it was a certainty,'" Gold recalled. "Part of the push with the accords was to make it not only more international, but to make it more attractive politically, especially with Democrats. If Artemis failed, it would be crippling. We wouldn't have a future of going beyond low Earth orbit."

The model Bridenstine and Gold strove for in the moon campaign was not Apollo, but rather the International Space Station. The continuously inhabited, football-field-sized laboratory had endured for two decades, over multiple administrations, in large part because it involved a broad international alliance. In all, fifteen countries participated, represented by the space agencies from the United States, Russia, Europe, Japan, and Canada. In late 2019, about a year before the presidential election, Bridenstine dispatched Gold to Paris to meet with the heads of the European Space Agency, float the idea of the Artemis Accords, and persuade countries to sign on. It did not go well. The Europeans had witnessed time and again the fickleness of the American space program, and had been burned by its ever-changing goals. They said, effectively: *In a year, there might be a new president, who will throw away your plans. If we support you, we'll just look silly. Why should we believe anything you're saying?*

Then they invoked a bit of American pop culture that Gold didn't realize they were aware of: Lucy and the football. The Europeans did not want to end up as Charlie Brown, muttering "Good grief."

Gold knew the history as well as anyone. "You're right," he told them. "You shouldn't listen to what I'm saying. But you should listen to what our Democratic leadership is saying." He reminded them that Pelosi had endorsed the program, as well as other Democrats, including Senator Maria Cantwell of Washington.

The first countries that signed included U.S. allies, countries that had long been partners on the ISS—the United Kingdom, Canada, Japan, Italy—and others, such as the United Arab Emirates, with relatively new but up-and-coming space programs. The first tranche of signatories, nine in all, was only the beginning, Bridenstine announced in October 2020. Ultimately, the United States would assemble "the biggest, most diverse coalition of nations ever in the exploration of the moon and beyond."

THE LAST PART of the strategy to make Artemis an enduring program came after Biden defeated Trump in the 2020 election, at the eighth and final meeting of the National Space Council during Pence's tenure. At the Kennedy Space Center on December 8, 2020, Pence took a victory lap, celebrating with Trumpian, chest-thumping flourish the restoration of human spaceflight from American shores; the creation of the Artemis program and the Space Force; and "taking steps to unleash America's commercial space companies." Not everyone was impressed. "Listening to the National Space Council meeting, you might get the mistaken impression that Trump invented space policy," tweeted one longtime space wonk.

Pence and Bridenstine introduced what they said was the cadre of Artemis astronauts, the group that would train to return to the lunar surface. It was a bit of theater designed to mimic the naming of the Mercury 7, the early pioneers who led America's space program at the dawn of the Space Age. In all, there were eighteen astronauts, whom Pence introduced one by one, as if in a beauty pageant. "It really is amazing to think that the next man and the first woman on the moon are among the names that we just read, and they may be standing in the room with us now," Pence gushed. The subtext was clear: Pence wanted to put faces to the program. "If you kill the Artemis program," he was saying, "you'll take the moon away from these American heroes."

But for all the flag-waving theatrics, there was some substance to the meeting—a pointed discussion of China. "China poses the greatest national security threat to the United States, and that includes China's actions in space, where China is pursuing weapons capable of destroying our satellites," warned John Ratcliffe, the director of National Intelligence. And Pence reiterated that the United States was in a new space race with China. "As the world witnessed, China recently landed an unmanned craft on the moon," he said, "and, for the first time, robotically raised the red flag of Communist China on that magnificent desolation."

The Artemis program had real momentum, support from Democrats, and the beginnings of an international coalition. Though NASA's budget was still a tiny fraction of total federal spending, the Trump administration had secured significant budget increases for the agency. During Trump's first year in office, NASA's budget was about $19 billion. By his last year, the president's budget request had jumped to more than $25 billion, as it started spending money on a spacecraft designed to land humans on the moon for the first time in fifty years.

What NASA didn't have was a clear public statement of support for Artemis from the incoming Biden administration. The president-elect had spoken barely a word about space during the campaign. During his decades in government, he had never been known as a space enthuiast. Would Biden keep Artemis? The Space Force? Or was the United States going to play another round of Lucy and the football? It was a mystery. And as Bridenstine left NASA to return to his home state of Oklahoma on January 20, 2021, he wondered if he had done enough to ensure Artemis would survive.

Since the end of the Apollo era, presidents had looked at Mars as the next giant leap, the next great adventure. Mars, as *The New York Times*'s John Noble Wilford once wrote, "tugs at the human imagination like no other planet—with a force mightier than gravity." In 1969, a task group appointed by President Nixon concluded in a report that NASA would be able "to carry out a successful program to land a man on Mars within 15 years." President Obama in 2010 also looked to Mars, and directed NASA to send humans by the 2030s, giving rise to what he called "the Mars generation." And, even though his own adminstration created the Artemis program, Trump had wondered why Mars wasn't the destination, and was annoyed when told NASA couldn't get there while he was president.

Bridenstine and much of the space community felt strongly that the United States needed to establish a presence on the moon before going to Mars—a far more difficult feat. But would Biden or a subsequent president feel the urge to give in to Mars's tug and change course?

IF ARTEMIS WAS a little-known program, the Space Force had become a punch line, derided by late-night comedians and the inspiration of a Netflix lampoon starring Steve Carell. "Reporters asked Trump who should lead this Space Force," Jimmy Fallon said in his show's opening monologue one night. "And he said, That's easy: Buzz Lightyear."

"Hello, citizens of Earth," said Stephen Colbert on his show. "And if this is a rerun two years from now, hello to all our fighting boys in the asteroid belt. Go give the Astro-Kaiser hell. Because tonight there's big news about: *Spaaaaaace Foooooorce!*" Space Force became a rallying cry at Trump's rallies, and his campaign sold Space Force merchandise. To many people, the service seemed like a farce—storm troopers with laser guns, skirmishes with aliens—just another stunt from Trump to boost his own ego.

On February 2, 2021, just a couple of weeks after Biden's inauguration, Jen Psaki, the new White House press secretary, was asked about the Space Force. She treated the question dismissively, as a joke. "Wow. Space Force! It's the plane of today," she said, referring to an earlier inane question about the color scheme of Air Force One.

In reality, the Space Force and its $15 billion budget was fast becoming an integral part of the Pentagon, forging an identity as the first new military branch since the creation of the Air Force in 1947—a point Bloomberg reporter Josh Wingrove pressed Psaki on.

"It's not," he said. "It's an entire branch."

"It's an interesting question," Psaki said. "I am happy to check with our Space Force point of contact. I'm not sure who that is. I will find out and see if we have any update on that."

Yes, the Space Force was born during the Trump administration, but its roots far preceded him, and the idea of space as a warfighting domain had long been in the mainstream of national security consciousness. So much of the Pentagon and intelligence community's apparatus was loitering in orbit, a fleet of satellites that had become integral to modern warfare. In recent years, however, China and Russia had proven that the satellites were sitting ducks, within reach of missile attacks or more subtle forms of interference, from lasers and jammers to cyberattacks. Intelligence analysts had even noticed that Russian spacecraft were getting uncomfortably close to some U.S. satellites, perhaps casing them the way burglars would a bank.

People who took the time to understand the Space Force—and the threats it was tasked with deterring—understood its importance. "Creating a Space Force is arguably an excellent idea, one for which Trump may deservedly go down in history, along with all the other things he will be remembered for," my former colleague David Montgomery wrote in an excellent piece about the Space Force in *The Washington Post Magazine*. "No, really. I'm tempted to laugh at myself as I type these sentences because I, too, greeted news of the Space Force with incredulous guffaws. . . . What I missed at the time, though—and what everyone else mocking Space Force doesn't seem to appreciate—is the sheer range of problems that could ensue if other countries are able to establish extraterrestrial military supremacy."

Republicans were eager to pounce on Psaki's comments, using them to paint her and the White House as ill-informed and anti-military. "It's concerning to see the Biden administration's press secretary blatantly diminish an entire branch of our military as the punchline of a joke, which I'm sure China would find funny," Congressman Mike Rogers, a Republican from Alabama and one of the first champions of the Space Force, said at the time. "The Space Force was passed with near unanimous support in Congress, the same type of 'unity' President Biden is supposedly working towards. Jen Psaki needs to immediately apologize to the men and women of the Space Force for this disgraceful comment."

Psaki did not apologize, but her office put out a tweet that evening that confirmed the administration's commitment to the importance of space as a warfighting domain. The next day, Psaki was asked by Kristin Fisher, of Fox News, about Rogers's request that she apologize. "I did send a tweet last night," Psaki said. "You may not all be on Twitter. Maybe they're not on Twitter. That said, we invite the members of the Space Force here to provide an update to all of you on all of the important work they're doing, and we certainly look forward to seeing continued updates from their team."

It was a defensive, unsatisfying answer that seemed to put the burden on the Pentagon to justify the existence of a branch of the military that had been operating for more than a year. Fisher didn't let it go: "But big picture here: Does the Space Force have the full support of the Biden administration, or is the president at some point perhaps going to get rid of it or in some way diminish it?"

"They absolutely have the full support of the Biden administration, and we are not revisiting the decision to establish the Space Force," Psaki said, finally giving an expansive answer that had been guided by the White House national security team. "The desire for the Department of Defense to focus greater attention and resources on the growing security challenges in space has long been a bipartisan issue informed by numerous independent commissions and studies conducted across multiple administrations. And thousands of men and women proudly serve in the Space Force. As you know, it was established by Congress, and any other steps would actually have to be taken by Congress, not by the administration."

Fisher—who happened to be the daughter of two NASA astronauts—was not done. She asked what the Biden administration planned to do with Artemis. Psaki didn't know the answer. "I am personally interested in space," she said. "I think it's a fascinating area of study. But I have not spoken with our team about this particular program, so let me see if we can get you a more informed overview of that."

In her office on the ninth floor of NASA headquarters, Bhavya Lal, the Biden-appointed interim chief of staff at NASA, was like an eager student in a classroom with her arm raised and waving frantically, desperate to be called on. She knew the answer to the question.

With the exception of a few dissenters, the transition team the Biden administration had sent to NASA was overwhelmingly in favor of keeping the Artemis program. The 2024 timeline was, in all likelihood, unachievable and always had been—a political stunt from the start. But the idea of returning astronauts to the moon, of including a woman among them, and doing it with an international coalition to stand as a bulwark against China, was something many felt the Biden administration should embrace. Even with the bitter aftertaste that would come from swallowing a Trump-era program.

In addition to her appointment at NASA, Lal also served on Biden's national security transition team. She directed staffers in both realms to work up memos "emphasizing how Artemis isn't just a science issue or a discovery issue. It's a national security issue," Michael Gold recalled. The team wanted to emphasize "how allowing China to beat America to the moon would have dramatic geopolitical and national security repercussions." The memos went directly to Jake Sullivan, the new national security adviser, along with something else: an invitation to NASA for an in-person briefing. Lal wanted Sullivan and the National Security Council to understand how important the new moon race had become.

When they met, Lal told Sullivan and his team that China was planning to build an international space coalition of its own. The race with China was not like the one between the United States and the Soviet Union during the Cold War. It wasn't about planting a flag or demonstrating power and prowess by leaving footprints in moon dust. Now even more was at stake. If China's taikonauts got there first, it would be a major diplomatic and political coup that could help dictate how future settlements would be established and operated.

Officially, China pushed back on the idea that it would make any sovereignty claims on the moon, which would have been a violation of the 1967 Outer Space Treaty. It continued to maintain that it would go peacefully, in pursuit of science and exploration, while building a community of international partners. But its plans were so shrouded in secrecy, it was hard to determine exactly what its motivations were. Lal and others feared China would lay claim to the South Pole of the moon as it had in the South China Sea, where it constructed artificial islands and patrolled the open waters as a form of intimidation.

For years, China had ignored a ruling by an international tribunal at The Hague that rejected its territorial claims in the South China Sea. It continued to fortify its outposts and conduct military operations. Tensions between the United States and China had been mounting. Early in the Trump administration, the Pentagon accused China of unlawfully seizing a U.S. Navy underwater drone that was collecting oceanic data from the region. In 2018, the Pentagon charged that China had performed "an unsafe and unprofessional maneuver" at sea when one of its warships, the *Lanzhou*, passed within forty-five yards of the USS *Decatur*, an American destroyer. What if, instead of making claims to the Gaven Reefs or other maritime shoals, China were to seize Shackleton Crater on the moon? What if it controlled the water there?

Lal told Sullivan that while the moon was big, there weren't that many spots where the water was located. NASA was only studying about a dozen possible locations, all relatively flat, well-lit, and with access to the permanently shadowed regions where the water lay. Access to them would be key for any nation interested in exploring the lunar surface. Whatever country got there first would be making a statement about its technological and economic capabilities, which would attract allies and possibly even help shift the balance of power on Earth.

Sullivan listened intently. He was sold. Artemis had to continue.

PSAKI WAS BACK in the White House briefing room on February 4, 2021, two days after Fisher's question about Artemis. She had an answer.

"Kristin, who is back today again, asked a great question about the Artemis program, which I dug into, and I'm very excited about it now to tell my daughter all about it.

"To date, only 12 humans have walked on the moon," she went on. "That was half a century ago. The Artemis program, a waypoint to Mars, provides exactly the opportunity to add numbers to that, of course." She explained that lunar exploration had bipartisan support in Congress, and then made clear that the new Biden administration would "support this effort and endeavor."

Bridenstine, watching from his home in Tulsa, was thrilled Artemis would live. He had ensured there would be no Lucy and the football. His successor would have a series of challenges to make a moon landing a reality, but Bridenstine could leave NASA knowing it was on a good path.

The much-delayed SLS moon rocket, being built by Boeing as the primary contractor, was finally going through a major test of its systems and engines, one of the last steps before its first flight. The Orion spacecraft was ready to fly. And NASA was getting ready to award the final contract for the spacecraft that would land astronauts on the moon. Just as important, NASA's budgets were increasing, if incrementally. Democrats were getting on board. And now the White House had publicly endorsed the program. NASA was as close to the moon as it had been in fifty years.

CHAPTER 16

Can't Get It Up
(to Orbit)

W inning the NASA contract to build the spacecraft that would land astronauts on the moon was going to be a coup for Bezos. Finally, he'd be on par with Musk and SpaceX. Finally, Blue Origin would be part of a significant NASA project, one that was personally important for Bezos at a time when his life was entering a new chapter. He was preparing to step down as CEO of Amazon, a move that would allow him to focus more attention on his space venture. He was also considering the idea of flying on New Shepard's first-ever human flight, fulfilling his lifelong dream of going to space. It was a good time for other reasons. His personal life was settling as well, following a period of tumult. After his divorce, Bezos had gone to war with the *National Enquirer* after the tabloid threatened to publish a story about his affair with a woman named Lauren Sánchez, complete with intimate photos. (Bezos countered with a stunning and brilliant blog post that said, "Rather than capitulate to extortion and blackmail, I've decided to publish exactly what they sent me, despite the personal cost and embarrassment they threaten.")

On April 15, 2021, NASA gave all three finalists in the competition for what NASA called the Human Landing System—Blue, SpaceX, and Dynetics—a heads-up that it would announce the winners the next day. The space agency had previously indicated that it wanted to choose two, both to foster competition and to have a backup in case one of the companies failed. Bezos and the leaders of Blue had every reason to be confident. They had won the bulk of the initial contract award, a haul of more than half a billion dollars. They had assembled the National Team, harnessing not just a generation of aerospace expertise from Lockheed Martin, Northrop Grumman, and Draper but their longstanding relationships with NASA. To help lead the program, Blue had hired John Couluris, a SpaceX veteran who oversaw the company's first flight to the International Space Station and whose father worked on the lunar module during the Apollo era. It all added up to what felt like an unsurmountable edge. One of the engineers at Lockheed Martin had even put the design of the Blue Moon lander on his wedding cake.

The next day, NASA convened a call to reveal its decision. Kathy Lueders, the head of the human spaceflight division, joined some ten minutes late, then got to the point in a calm monotone. Budgets were tight, so there would be only one winner, and it was not Blue Origin.

Lueders promised that there would be more opportunities to work with NASA in the future, and encouraged Blue Origin to keep at it. "We want you involved in the program," she said. "Other companies will be able to on-ramp later. This doesn't close the door."

Bob Smith and the rest of Blue's leadership were stunned. "There was this silence," Couluris later recalled. "Like you couldn't believe it."

LOSING THE CONTRACT was simply inconceivable—especially when Blue learned that the single winner was SpaceX. This wasn't just a loss; it was an embarrassment. Bezos had personally pitched the Blue Moon spacecraft in his vision speech, making it a central part not only of his company's effort to return astronauts to the moon, but of his legacy. It was a rare loss in a career that had been synonymous with triumph.

"It was grueling," a former top executive told me. "It was really painful for many reasons. We were the underdog when it came to launch. SpaceX had the Falcon 9; we only had New Glenn, which wasn't launching. And so when we kept losing to SpaceX on launch, we could rationalize that we were going to catch up by going to the moon: We're the lunar people. Elon doesn't want to go to the moon. He wants to go to Mars. Jeff even had his vision speech about it, laying out the vision for the moon. And then SpaceX came in and won, and it really, really bothered him.

"He wants Blue Origin to be what he's remembered for, not Amazon. He wants it to be his legacy. Jeff was going to be the one who opened the moon for humanity, and now SpaceX is doing that."

Almost immediately, Bob Smith entered a state of denial and anger, combining the first two stages of grief. "The customer is wrong," he told his team repeatedly. NASA had made a mistake. It was as simple as that. He vowed to get the decision reversed. The company had options under the laws overseeing federal procurement. It could file a protest with the Government Accountability Office. It could sue. It could lobby Congress to put pressure on NASA and come up with more money to fund a second lander. None of these was a preferred route. But Blue felt like it had no other choice. This decision simply could not stand.

Their anger, however, seemed to blind them from several truths. SpaceX's bid came in at $2.9 billion; Blue Origin's was more than twice that, at $6 billion. In a document detailing her decision, Lueders made it clear that Blue's price was a big issue, saying she did "not have enough funding available to even attempt to negotiate a price from Blue Origin that could potentially enable a contract award."

While NASA's budgets had increased, it was not enough to cover the billions in development costs for two spacecraft. And the agency had wanted the companies to invest their own money in the project. This was part of a new way of doing business with the commercial space sector. Private industry would have to put some skin in the game. As a benefit, they would be able to leverage the government's investment to build systems that could also open new lines of business. SpaceX had done this with its Falcon 9 rocket, which not only flew missions for NASA but also launched commercial satellites. Blue Origin was also looking for commercial business and had one of the world's wealthiest benefactors as its founder. And yet its costly bid would stick the government with much of the bill for developing its lander.

Then there was the relationship that NASA and SpaceX had forged over the years. Yes, it had been tense at times. Musk's employees often rankled their government counterparts, whom they saw as old and stodgy. And yes, NASA officials sometimes feared employees at SpaceX had the dangerous combination of ignorance and arrogance. But they had now been working side by side for more than a decade. SpaceX had proven it could successfully fly cargo and supplies to the International Space Station. Then it had flown astronauts there. All those missions, the years of working together in the trenches of mission control through triumph and failure, had bonded them. Blue Origin had none of that history.

The afternoon of the contract announcement, Blue issued a terse statement that hinted at its incomprehension and anger: "The National Team doesn't have very much information yet," it said. "We are looking to learn more about the selection."

While SpaceX celebrated its victory, Blue Origin behaved like a losing political candidate calling for a recount of the votes. It conceded nothing.

ABOUT AN HOUR after the announcement, John Couluris, Blue's senior vice president for Lunar Permanance, pulled together the entire National Team, about 850 employees from all four companies, and tried to console them over Zoom. He played "Tubthumping" by Chumbawamba, hoping the lyrics—"I get knocked down, but I get up again"—would resonate.

"Look, you're going to win some, you're going to lose some," he said. "We've got to recognize that. Although we don't understand the decision, we'll wait for the selection board to see how NASA judged it. We'll figure out what we did right, what we did wrong. We'll figure out next steps. But again, we get knocked down—that's okay. Lick your wounds and get up again."

Late that night, after a long and exhausting day, Brent Sherwood,

Blue's senior vice president of Advanced Development Programs, tried to console his staff as well. "After a crushing loss," he began in an email, "it's hard to make a team feel better. Major competitions are really hard."

"We were all astonished by today's news," he continued. "Given our tremendous team, our solid design, our many risk-reduction tests, our great customer relationship . . . and our vision for lunar development, it seems preposterous that we didn't win. Furthermore, after NASA's insistence on a strategy of choosing two providers. . . . Their abrupt reversal in choosing only one seems equally preposterous."

Then, without mentioning SpaceX, he wrote: "Finally, their choice of winner seems . . . preposterous. I share your astonishment, disappointment, disorientation, confusion and even anger.

"We're not just giving up on HLS. Critical meetings started today, and continue over the weekend and into next week, to determine next actions. We are studying NASA's decision logic and considering potential steps about this procurement outcome." He added, "You are in the right place: the frontier business unit of a major independent space company, backed by the most successful businessman in history, whose passion project we are privileged to contribute to."

The company had embarked, he reminded them, on a wide array of "astounding systems and products," though few of them had been revealed publicly and many were only in the idea phase. They included a spacecraft called Blue Ring that would act as a sort of service vehicle for space, flying across different orbits between Earth and the moon; new propulsion technologies, including nuclear power; a reusable upper stage for the company's New Glenn rocket that would make it fully reusable, like Starship. The list included a commercial space station called Orbital Reef, intended to replace the aging International Space Station, as well as other "concepts ranging from space suits to rovers to moving asteroids to lunar elevators."

The email revealed the depth and breadth of Bezos's ambitions: to be the "Everything Company" for space, an Amazon-like behemoth that would build the infrastructure to allow space to truly open up. But maybe that was the problem. Maybe without Bezos's day-to-day involvement, Blue was stretched too thin, focused on too many things at once, while not delivering on any of them.

THAT WEEKEND, BEZOS, Smith, and the rest of Blue's leadership team continued to stew. They thought NASA had rushed the award, not waiting for Bill Nelson—Biden's pick to succeed Bridenstine as NASA administrator—to be confirmed. Surely he should have had a say in such a consequential decision.

On Sunday morning, two days after the loss, Bezos urged his team to look well into the future. "How would we go about this if NASA did not exist?" he said. It was a provocative question that revealed the depth of his passion for space, and it forced his team to reimagine everything they had done. Immediately, they started brainstorming a new Blue Moon lander design and the systems that would go with it.

But at the same time the engineers were going back to the drawing board, Smith was assembling the company's attorneys. He and Bezos were still livid about the loss and were convinced NASA had made an unjust decision. Blue was going to challenge it.

On April 26, ten days after NASA's announcement, Blue Origin filed a protest with the GAO that argued the agency had erred. NASA, it complained, had "executed a flawed acquisition for the Human Landing System program and moved the goal posts at the last minute. In NASA's own words, it has made a 'high risk' selection. Their decision eliminates opportunities for competition, significantly narrows the supply base, and not only delays, but also endangers America's return to the moon."

In an interview with *The New York Times*, Smith reiterated what he had been telling his team about NASA making a mistake. Sounding like a schoolteacher condemning an unruly student to detention, he said: "It's really atypical for NASA to make these kinds of errors. They're generally quite good at acquisition, especially its flagship missions like returning American astronauts to the surface of the moon. We felt that these errors needed to be addressed and remedied."

Those comments and the protest did not endear the company to

NASA. "Realizing now that it gambled and lost, Blue Origin seeks to use GAO's procurement oversight function to improperly compel NASA to suffer the consequences of Blue Origin's ill-conceived choices," NASA's lawyers wrote in a brief.

The protest also irritated Musk, who took Blue Origin's criticism of NASA as criticism of SpaceX. His company's victory wasn't some sort of fluke. SpaceX had been making real progress with Starship, flying the prototypes in Texas one after the other. Blue had done almost nothing by comparison. Its relatively tiny New Shepard rocket only flew to an altitude of just some sixty-five miles. SpaceX first reached orbit in 2008. Blue Origin still hadn't flown a rocket to orbit. How was it supposed to suddenly fly humans to the moon? "Oh, stop teasing, Jeff," Musk had tweeted in reaction to Jeff's vision speech in 2019, in which he'd unveiled Blue Moon.

When news of the protest broke, Musk lampooned Blue, writing on Twitter that the company "can't get it up (to orbit) lol." He took the image of the Blue's lander and, over the bulbous fuel tank, crossed out the "Moon" in "Blue Moon" and replaced it with "Balls."

That evening, I reached out to Musk. When I suggested that Bezos had truly wanted to win the contract, he became more sincere and less juvenile. "The BO bid was just way too high," Musk wrote me. "Double that of SpaceX and SpaceX has much more hardware progress. I think he needs to run BO full-time for it to be successful. Frankly, I hope he does."

The sentiment echoed what many in the space industry had been saying for years. How could Blue be successful with Bezos working there only on Wednesdays and some Saturdays? It didn't seem to have Amazon's drive or success. While relying on Bezos's personal fortune, it didn't need to fight for resources and had gotten complacent. At least that's how Musk and Gwynne Shotwell saw it.

"They're two years older than us, and they have yet to reach orbit," Shotwell once said at a conference. "They have a billion dollars of free money every year from [Bezos]. . . . I think engineers think better when they're pushed hardest to do great things in a very short period of time, with very few resources. Not when you have 20 years. I don't think there's a motivation or a drive there."

A couple of days after Blue filed the protest, *Saturday Night Live* announced that Musk would host an upcoming show. I tweeted that with the dust-up between Musk and Bezos over the contract, the show's writers "now have a whole lot of material to work with. The opening monologue could be . . . something."

Elon replied, in typical cheeky fashion: "Romeo + Juliet fish tank scene"—a reference to a scene in the 1996 film version in which Leonardo DiCaprio and Claire Danes meet for the first time.

Then he added, "Did my heart love till now? forswear it, sight!

For I ne'er saw true beauty till this night."

BIZARRE MOVIE REFERENCES and impulsive tweets about blue balls may have been Musk's way of letting off steam and goading his supposed rival into becoming an actual one. But it was also a prelude to the sort of behavior that was becoming more common—and coarse. Musk was beginning to spew noxious comments and conspiracy theories that would be embraced by some on the far right while being denounced as anti-Semitic, misogynist, anti-LGBTQ, and anti-immigrant by virtually everyone else. And it only intensified after he purchased Twitter in 2022 for $44 billion.

The White House would condemn Musk for promoting what it called a "hideous" conspiracy about Jewish people fomenting hatred against whites. Advertisers, such as Disney, fled Twitter, which Musk renamed X. The goodwill that Musk had built up over years as a champion of renewable energy, electric cars, and space exploration began to evaporate, tweet by pernicious tweet. Instead of focusing on those endeavors, he started to insert himself into politics, visiting the border with Mexico, urging Congress not to fund Ukraine as it fended off the Russian invasion, and criticizing the government's response to the COVID pandemic, tweeting at one point that "my pronouns are Prosecute/Fauci."

He was also railing against regulators, including the Federal Aviation Administration, which issued SpaceX launch licenses and was charged with protecting people and property on the ground. In December 2020, with Musk pushing his engineers to work ever faster, SpaceX planned to launch a Starship prototype. The FAA had not signed off on the flight, but SpaceX launched anyway—a short hop, several miles high, that ended in a crash landing and fireball. No one was hurt, and the failure was not unexpected. But the FAA was concerned that the conditions would "exceed the maximum public risk allowed by federal safety regulations." The FAA grounded SpaceX and ordered an investigation, including "a comprehensive review of the company's safety culture, operational decision-making and process discipline."

If people at SpaceX knew they were in violation of their license, no one had said so, Hans Koenigsmann, then SpaceX's vice president of flight reliability, told me: "There was confusion on everyone's side." Afterward, however, Koenigsmann felt it was clear SpaceX did not have permission to launch and shouldn't have. He wrote a report saying so, which did not go over well with Musk. "He was very sensitive about it," Koenigsmann said. "That's basically where Elon and I disagreed at the end."

Furious about the FAA's actions, Musk took to Twitter, blasting the agency as an overly burdensome bureaucracy that was impeding SpaceX's progress. "Unlike its aircraft division, which is fine, the FAA space division has a fundamentally broken regulatory structure," he wrote. "Their rules are meant for a handful of expendable launches per year from a few government facilities. Under those rules, humanity will never get to Mars."

The tweet was "not helpful," a source at the FAA told me at the time. Musk and his tens of millions of followers on social media could bring enormous pressure on regulators. But the FAA felt it had to stand firm. If it had approved the launch and people ended up getting hurt, the person told me, "then we're in a situation where we're second-guessed— Did you do everything you could? And were you influenced by Elon and his fan club?"

Koeingsmann, one of SpaceX's earliest employees and often its public face, would find his responsibilities diminished. "You did an awesome job over many years, but eventually everybody's time comes to retire," Musk wrote him in an email a few months later. "Yours is now."

AMAZON, WHICH WAS planning to launch its own constellation of Internet satellites to rival SpaceX's Starlink, tried to use Musk's behavior against him as the companies sparred before the Federal Communications Commission. In one unusually combative filing, Amazon went after Musk personally, saying he was a dangerous rogue who openly and repeatedly defied lawful regulations. SpaceX's violation of its launch license was part of a pattern of reckless behavior, the company wrote.

"Try to hold a Musk-led company to flight rules? You're 'fundamentally broken,'" Amazon's lawyers wrote. "Try to hold a Musk-led company to health and safety rules? You're 'unelected & ignorant.' Try to hold a Musk-led company to U.S. securities laws? You'll be called many names, some too crude to repeat. In the words of the *Wall Street Journal*, Elon Musk wages a 'War on Regulators,' the public servants charged with uniformly applying the same rules to all. As the *Journal* reported, Federal agencies say [Musk is] breaking the rules and endangering people. . . . Rather than engaging in a give-and-take with government authorities, Mr. Musk's default response includes making public, sometimes crude, remarks via Twitter disparaging them.

"Whether it is launching satellites with unlicensed antennas, launching rockets without approval, building an unapproved launch tower, or re-opening a factory in violation of a shelter-in-place order, the conduct of SpaceX and other Musk-led companies makes their view plain: rules are for other people, and those who insist upon or even simply request compliance are deserving of derision and ad hominem attacks."

As controversial and off-putting as Musk was becoming to many, there was simply no denying SpaceX's dominance in the rocket launch market. It was making a statement in Boca Chica, where it would fly the rudimentary Starship prototypes to an altitude of about six miles, to perfect the art not just of liftoff, but of landing. Like the Falcon 9 rocket, Starship Super Heavy booster would fall through the atmosphere and land "propulsively," meaning by reigniting its engines to slow it down. On its descent, the Starship spacecraft would first fall horizontally, like a splayed skydiver belly-flopping through the air. Then, shortly before hitting the ground, it would right itself into a vertical position, while firing its engines for a soft touchdown.

The acrobatic landing attempts were thrilling to watch, not just because the first few landing attempts of ship prototypes ended in fiery crashes fit for Hollywood, but because SpaceX really was pushing the envelope, attempting something truly original. In the space world, the events were must-see TV. Every attempt was broadcast live by SpaceX, and nobody knew whether it would end in fireball or triumph. The December 2020 prototype flight was in violation of FAA rules, and it culminated in a fireball that spewed pieces of rocket across the launch site—but by SpaceX standards it was meaningful progress.

The engines "worked great," Musk wrote on Twitter, adding: "the crater in the right spot was epic!!"

A FEW WEEKS after that crash, the FAA cleared SpaceX to launch its next prototype, known as SN-9, which also crashed. The next attempt, in March 2021, landed precariously, with an apparent leak that resulted in an explosion a few minutes later.

Finally, in May, a mere two weeks after winning the lunar lander contract, SpaceX stuck the landing with its SN-15 prototype. SpaceX had served notice: Blue Origin could hire all the lawyers it wanted to fight in court, but SpaceX was going to make its case on the launch pad.

Shortly before the successful test, I interviewed Musk in SpaceX's hangar at the Kennedy Space Center, where the company refurbished its used Falcon 9 boosters to get them ready to fly again. On this day, there were five of them, stacked side-by-side, all of them sooty from their previous flights. As usual, Musk was late. The interview was supposed to start at 4:00 P.M., but it kept getting pushed back—to 6:00, then to 8:00. I was getting nervous that he would cancel. At 5:49 the next morning, SpaceX was scheduled to launch its third human spaceflight mission in less than a year, a contingent of two NASA astronauts, including Megan McArthur, Bob Behnken's wife, as well as one each from Japan and France.

Finally, Musk met me at 11:00 P.M.—a ridiculous hour, given the imminent launch, but then again, Musk appeared never to sleep. He was in a good mood and took a moment to tour the hangar, marveling at the rockets, studying them as if he were in a museum. He was more than happy to chat about Starship's progress. After landing the SN-15 prototype, he said, one of the next big hurdles would be to integrate the Starship spacecraft on top of the Super Heavy booster and attempt to launch it to orbit.

"I think we're close to getting to orbit with Starship," he told me. "The overarching goal of Starship is to be able to transport enough tonnage to the moon and to Mars, to have a self-sustaining base on the moon and ultimately a self-sustaining city on Mars." He said that Starship's booster could be ready to fly again as soon as an hour after landing, a prospect he called "insane."

"It's difficult to explain exactly how profound Starship is from the design perspective," he said. "There's nothing like it. No one has ever proposed such a thing."

It was typical Musk. Boastful, wildly optimistic, setting impossible timelines to meet near-impossible goals, pushing the edges of physics to will some science fiction version of the future into reality. Maybe it would happen. Maybe not. But what struck me wasn't so much his comments but what was going on around us inside the hangar. It was nearly midnight on a Thursday, and the place was humming with SpaceX employees working on rockets. Before the interview, we had been warned that the work would not stop for us. We'd have to deal with the noise and the tumult. The only thing that would stop work was the launch of four humans into space; many others were plowing ahead on other projects.

SpaceX is often compared to a Silicon Valley company, and in some respects it is. Its headquarters is full of brilliant programmers and software designers writing the code that enables rockets to land with precision and spacecraft to autonomously dock with the space station. But here in the hangar, SpaceX was more Detroit than San Francisco—workers in T-shirts, jeans, and hard hats bending metal. Country music playing. Pickup trucks in the parking lot. An oversized American flag hanging from the rafters. At the stroke of midnight, security guards started clearing us out: All the workers not assigned to the launch had to leave, their shift cut short. Otherwise, they would have kept on going, keeping up the breakneck 24-hour operations that at SpaceX were the norm.

At Boca Chica, the workload was just as intense, perhaps even more so. That summer, Musk was driving as hard as he ever had, like some World War II general pushing for a production surge. He had ordered a rush to get the booster and spacecraft stages of Starship mated and flying together, an effort that would require an influx of people and resources. Upset with the pace of progress, he scolded one of his lead Boca Chica site engineers. "This is not a volunteer organization," he said in July. "We are not selling Girl Scout cookies. Get them here now."

In an email to employees that same night, after 1:00 A.M., he demanded: "Anyone who is not working on obviously critical path projects at SpaceX should shift immediately to work on the first Starship orbit. Please fly, drive, or get here by any means possible."

Were Blue Origin's employees working in shifts around the clock? I doubted it. I had once toured its sprawling factory at Cape Canaveral in August 2021. It was an impressive piece of real estate that stretched for hundreds of yards. But I couldn't help notice how clean the place was. How quiet. Blue was chasing SpaceX, in a so-far fruitless effort. Would they ever catch up? Watching what was happening all around me, it didn't seem possible.

THE SN-15 LANDING was a show of force, a message to NASA that it had made the right choice with SpaceX. But Blue didn't back down. In addition to its legal protest, the company had taken to lobbying Congress to compel NASA to pick a second company to build a lunar lander. The team met for an hour every day at minimum, in sessions that resembled a presidential campaign's war room. What intel were they getting? Who was reaching out to whom? "It felt like an all-out battle, and we never gave up," one lobbyist recalled.

Senator Maria Cantwell, the Democrat from Blue's home state of Washington, was happy to help. During Bill Nelson's confirmation hearing, she lamented the fact that NASA had already awarded the contract, saying it had taken her by surprise. NASA would need two providers competing against each other and serving as backup in case one failed, she argued. "I think there needs to be redundancy, and it has to be clear in this process that it can't be redundancy later," Cantwell said. "It has to be redundancy now."

Nelson agreed with her. "Competition is always good," he said.

Privately, he and his deputy, Pam Melroy, a retired Air Force colonel and astronaut who had flown on three Space Shuttle missions, were not pleased with the way the award had been handled. They would have wanted two companies competing against each other for the ultimate prize of landing NASA astronauts on the moon. Melroy was left feeling "pretty mystified," she later told me. "I actually would have done it very differently."

But their hands were tied. Blue's protest meant the fate of the procurement was up to the GAO. All work on it had to stop. Nelson wasn't able to speak with Blue or SpaceX while the proceeding was litigated.

Cantwell, however, filed an amendment to an unrelated bill that would direct NASA to fund "not fewer than 2 entities" in the program and authorize $10 billion for the effort. It was not clear how NASA would actually do that, however. The contract had already been awarded. There were no do-overs in the federal acquisition system. And the $10 billion was merely "authorized," a near meaningless term, since the way Congress actually spent money was through a separate process known as appropriation.

Still, SpaceX saw Cantwell's amendment as parochial interference and an unprovoked attack that needed to be thwarted immediately. SpaceX dispatched its own lobbyists, who distributed a one-page paper titled "Oppose the Cantwell NASA Amendment." It warned that it "undermines the federal government procurement process, rewards Jeff Bezos with a $10 billion sole-source hand-out and will throw NASA's Artemis program into years of litigation." That, in turn, would hand "space leadership to China."

Blue Origin competed and lost, SpaceX said, "because its bid was technically inferior, too expensive by a wide margin and did not meet the terms of the solicitation." If adopted, the amendment would set a bad precedent of rewarding a company that had won in early rounds of various competitions only to lose in the end. "Blue Origin has received more than $778 million from NASA, the Air Force and the Space Force since 2011, *and it has not produced a single rocket or spacecraft* capable of reaching orbit. In fact, the government has chosen not to proceed with Blue Origin after every major development contract."

For years, Bezos and Blue Origin had endured Musk's taunts. Typically, they said nothing. Musk, Bezos and his team thought, was increasingly becoming a Trump-like figure, an unhinged belligerent who would say and do anything. Best to let him spiral out of control and self-immolate. But the fight over the lunar lander had gone too far, and Bezos was ready to punch the bully in the nose.

Blue distributed a paper of its own on Capitol Hill seeking to tell "the truth" about the Cantwell amendment. It started with a jab: "What is Elon Musk afraid of . . . a little competition?"

"Elon Musk repeatedly talks about the value of competition, but when it comes to NASA's Human Landing System program, he wants it all to himself," it went on. Next to each of SpaceX's allegations about giving Bezos and Blue a corporate bailout, stifling competition, and slowing the Artemis program, it countered with: "Lie . . . Lie . . . Lie." Instead, the amendment "would ensure competition—two providers ensure greater safety and mission success, promote competition and control costs."

Blue also kept the pressure on NASA. On July 26, 2021, three months after the contract award, Bezos sent a letter to Nelson that repeated many of the same arguments in the company's protest. Blue had won the first round of the contracts, Bezos wrote, and since then the company had developed a spacecraft "that could achieve a human landing in 2024." SpaceX's Starship, by contrast, was a far riskier bet because it would be refueled in Earth orbit, requiring a fleet of tanker spacecraft. That approach "locks every trip to the moon into 10+ Super Heavy/Starship launches just to get a single lander to the surface," he wrote.

The letter, which was made public, ended with an extraordinary offer: If NASA would award Blue Origin a lunar lander contract, Bezos would contribute $2 billion to cover the additional costs. He would also pay for a mission to low Earth orbit to test out how the lander would perform, a flight that would cost tens of millions of dollars. "I am honored to offer these contributions and am grateful to be in a financial position to be able to do so," he wrote.

If anything, however, the offer annoyed NASA. While the general public might have thought that refusing a $2 billion gift was madness, the agency's leaders knew that under the rules of federal procurement there was no way they could accept it. Blue had already made its bid. The procurement was over. If Bezos wanted to contribute an additional $2 billion, he should have made it part of the company's original proposal; instead of bidding $6 billion, it could have bid $4 billion. But even then, Blue's bid would have been $1 billion more than SpaceX's.

Bezos's letter never got a response.

AT THE END of July, the GAO roundly denied Blue Origin's protest. In rejecting its arguments, the agency ruled that "NASA did not violate procurement law or regulation when it decided to make only one award." NASA's "evaluation of all three proposals was reasonable, and consistent with applicable procurement law, regulation and the announcement's terms."

With the denial, NASA's leaders were eager to finally put the issue behind them and get back to work with SpaceX. Landing humans by 2024 was increasingly being revealed as a fantasy, but in embracing Artemis, the Biden administration also sought to embrace the sense of urgency their predecessors had infused into the program.

Only Blue wasn't done fighting. It was planning to escalate its fight to federal court, a move that would once again tie up the procurement by months. When I tweeted that a source had told me about Blue's plan to file a federal lawsuit, Musk responded by tweeting an image of the Blue Origin lander, this time with the large, round fuel tank deflated. It was an unflattering photo that looked unmistakably like a sagging, withered testicle.

"Somehow this wasn't convincing," Musk wrote.

He kept tweeting. In response to one of Blue's infographics, he wrote, "The sad thing is that even if Santa Claus suddenly made their hardware real for free, the first thing you'd want to do is cancel it." And: "If lobbying & lawyers could get u to orbit, Bezos would be on Pluto."

Blue would lose the federal lawsuit as well. In a ruling on November 4, 2021, some seven months after NASA issued the contract to SpaceX, Judge Richard Hertling dismantled Blue's arguments and made it clear the company had no shot at overturning the award. Using italics for emphasis, the ruling mocked Blue as a sore loser: "Blue Origin is in the position of every disappointed bidder: *Oh. That's what the agency wanted and liked best? If we had known, we would have instead submitted a proposal that resembled the successful offer, but we could have offered a better price and snazzier features and options.*" Finally, Bezos capitulated. He wrote on Twitter that it was "not the decision we wanted, but we respect the court's judgement [*sic*], and wish full success for NASA and SpaceX on the contract."

NELSON, WHO BY then had been confirmed as the head of NASA, was relieved the issue had finally been resolved. Returning astronauts to the moon by 2024 was always going to be tough. Now it was plainly impossible, and one of the reasons was that work on the lander had had to stop while Blue pursued its legal actions.

"We've lost nearly seven months in litigation," Nelson told reporters in November, a few days after the judge's ruling. "And that likely has pushed the first human landing to no earlier than 2025."

Jake Sullivan, the national security adviser, had told Nelson in a meeting that President Biden was following Artemis's progress and had made it clear that he "does not want to be second to China on the moon." While Artemis was on hold, China had continued to move steadily. In May 2021, it successfully landed a spacecraft on Mars and became only the second country, after the United States, to operate a rover on the Red Planet. At the time, Nelson had congratulated China. But at a congressional hearing a few days later, he held up a photo taken by China's Zhurong rover as evidence of how serious the country was about deep space exploration. "They're going to be landing humans on the moon," he told Congress. "That should tell us something about our need to get off our duff and get our Human Landing System program going vigorously."

In June, China reached yet another milestone, launching the first crew to inhabit its new space station. In a broadcast from aboard Tiangong, the Chinese astronauts, all members of the People's Liberation Army, gave military salutes to President Xi Jinping while positioned in front of Communist Party flags.

That same month, the China National Space Administration made a startling announcement: It would partner with Russia to build what it called the International Lunar Research Station, a moon base that would be inhabited and operational by the 2030s, a rival to NASA's Artemis. China started inviting countries to join its effort "to cooperate and contribute more for the peaceful exploration and use of [the] moon in the interests of all humankind, adhering to the principle of equality, openness and integrity." Eventually other countries, such as Venezuela, Belarus, Pakistan, South Africa, and Egypt, would sign on as partners, as China sought to build a coalition of fifty nations.

Like Pence's 2024 date, Nelson's 2025 was aspirational. Many at NASA believed that 2028 or even 2029 was more realistic, given the fact that the space agency had yet to fly its SLS moon rocket and had only just settled the turmoil over the lunar lander contract. It was a distinct possibility that China could get to the moon first.

Three days after Nelson's comments in November, Ye Peijian, the head of China's lunar exploration program, gave a rare glimpse into the country's plans. "As long as technological research for crewed moon landings continues, as long as the country is determined," he told an interviewer, "a Chinese crewed moon landing is entirely possible by 2030."

CHAPTER 17

The Gremlins of Unknown Unknowns

n early June 2021, in the middle of his ugly fight with NASA over the lunar lander, Jeff Bezos finally had some good news to share: He was going to space.

The Blue Origin capsule he'd travel in would just scrape the edge of space, floating outside the atmosphere for only a few minutes before falling back to the ground. But he would see Earth from above and the vast darkness of the cosmos beyond. It would be a daring mission—and risky. Blue Origin had never flown humans before, but it was planning to begin a space tourism business, ferrying paying customers out of the atmosphere and back. Bezos wanted to make a statement that the company had built a vehicle that was robust and safe, one he felt comfortable flying in himself. His brother, Mark, would join him, as would the top bidder of an online auction. The date of the flight was set for July 20, 2021, the anniversary of the Apollo 11 moon landing.

The flight would come just a few weeks after Bezos stepped down as CEO of Amazon and mark a new chapter in his life, one in which he would dedicate himself to other pursuits, Blue Origin chief among them. Musk may have won the lunar lander contract, but he had yet to go to space himself. Bezos would also edge out Richard Branson, the billionaire British entrepreneur, whose Virgin Galactic had been referring to itself as the "world's first commercial spaceline" as it also worked to fly tourists on suborbital space trips.

Branson, who was seventy-one, had also been impatiently waiting to leave Earth. It would cap a career of daredevil stunts that had made him a "one-man publicity circus" for his myriad businesses. In 1987, his attempt to cross the Atlantic Ocean in a hot-air balloon ended when he had to bail out by jumping into the sea. Later, he crossed the Pacific by balloon, but ended up so far off course that instead of touching down in Southern California, he crash-landed on a frozen lake in Canada. He had also attempted to set a powerboat speed record across the English Channel in seas so choppy it "was like being strapped to the blade of a vast pneumatic drill," as he recalled in his memoir.

Going to space would be the ultimate adventure, although he and Bezos had different plans for getting there. Unlike a rocket that launched vertically, Virgin Galactic's spacecraft, known as Space-ShipTwo, was tethered to the belly of a mothership and hoisted to some 40,000 feet. The two would separate, and then the pilots would fire SpaceShipTwo's engines, propelling the craft straight upward. The vehicle would remain in space for a few minutes before falling back to Earth and gliding to a runway.

To get to this point, Virgin Galactic had overcome a series of technical challenges, as well as a crash in 2014, when its spaceplane came apart in a harrowing accident that killed one pilot and seriously injured the other. Branson had almost been forced to shut the company down. But in May 2021, it sent SpaceShipTwo to space for the third time, in a flight its CEO called "flawless." Virgin Galactic had said it would do one more test flight, and then, finally, it would be Branson's turn—probably sometime in the fall.

In the meantime, Branson was actively training. "One good thing about COVID is it enabled me to get as fit as I've felt since I was in my twenties," he told me at the time. "It's great to be able to really work on getting your body fit for spaceflight, and I'm going to enjoy every minute of it." In a gracious tweet, he seemed to concede that Bezos would beat him to the heavens: "Many congratulations to @JeffBezos & his brother Mark on announcing spaceflight plans. Jeff started building

@blueorigin in 2000, we started building @virgingalactic in 2004 & now both are opening up access to Space—how extraordinary!" But he added, somewhat cryptically: "Watch this space. . . ." as if there would be more to the story.

There was. A few weeks later, while I was on a plane to Seattle to interview Bob Smith at Blue Origin's headquarters, I got a text from Branson's press person. "Are you free?" she wrote. "Urgent update." Branson was going to move his voyage up. He'd be on Virgin Galactic's next test flight, scheduled for July 11, allowing him to beat Bezos by nine days.

When my plane landed, I scurried to find a quiet place in the airport, by an empty baggage carousel, to interview Branson. "I've been itching to go, and they said they wanted somebody to properly test the astronaut experience," he told me. "And I was damned if I was going to let anyone take that seat."

"Oh, come on," I said. "You moved up your flight so you could be first."

No, no, he said, chortling, as if the suggestion had never occurred to him. He was just "incredibly excited"; the change was "honestly not" intended to best Bezos. "I completely understand why the press would write that," he said. "It's just an incredible, wonderful coincidence that we're going up in the same month."

Spoken in his moneyed British accent, it sounded sincere, even charming, as if in his world, on his private island in the Caribbean, such "incredible, wonderful coincidences" fell from the sky like coconuts from palm trees. Branson would never publicly denigrate a rival. He'd never stoop to vulgar schoolyard taunts about blue balls or not being able to "get it up." He was no Musk. Still, as I hung up, I couldn't help but think Branson was a brawler. Decades earlier, he had founded Virgin Records and signed the Sex Pistols when no one else wanted them, building a successful label that eventually included the Rolling Stones. With Virgin Atlantic, he had famously taken on British Airways. And now, in the most polite way possible, he was tormenting Bezos.

• • •

BLUE ORIGIN'S TEAM was livid. In our interview the next day, Smith told me he wished Branson and Virgin Galactic well and said, "We really sincerely hope they have a great safe flight." But he couldn't help but take a dig at his rivals, intimating that by moving up Branson's flight, Virgin was prioritizing schedule over safety, one of the cardinal sins of human spaceflight, especially for a company that had already had a fatal accident. "Launch fever" lay at the heart of what caused both the Space Shuttle Challenger and Columbia disasters, when NASA had allowed lapses to slip by unheeded because of the desire to meet some arbitrary deadline.

"We hope they haven't rushed anything. That would be a concern," Smith said. "Schedule pressure is a real thing in getting launch fever. And hopefully that's not manifesting itself here."

He also charged that Branson wouldn't even make it to space—not technically. The Air Force and Federal Aviation Administration awarded astronaut wings for pilots who had soared above fifty miles, the barrier Virgin Galactic was aiming to break. But another boundary stood at sixty-two miles, or 100 kilometers: the Kármán line, named for Theodore von Kármán, one of the founders of NASA's Jet Propulsion Laboratory. According to the Fédération Aéronautique Internationale, the World Air Sports Federation, a record-keeping organization, the Kármán line was where space began. Blue Origin agreed. Its passengers would fly past sixty-two miles; Virgin Galactic's would not.

"We are going to the internationally recognized boundary of space," Smith said. "We've committed to that from day one. All of our flights go beyond that." He added that Blue's astronaut customers would have "no asterisks" beside their names.

Then Smith showed me a chart that Blue Origin had prepared, comparing New Shepard to Virgin's SpaceShipTwo. Blue's customers would fly in a rocket, not a "high altitude airplane," it read. The windows in the New Shepard capsule would be "the largest windows in space"; SpaceShipTwo's would be "airplane-sized." New Shepard had an escape system in case of an emergency; SpaceShipTwo did not. Blue's impact on the ozone layer was "minimal," while Virgin's was "high" because of its type of propellant. Finally, New Shepard had flown to space fifteen times; SpaceShipTwo had done it just three times.

It was a defensive, puerile gesture that made Blue seem insecure, and I was surprised the company had spent the time to develop it. I suspected Smith and his staff knew it would not go over well publicly because they showed it to me only on an off-the-record basis. But then, two days before Branson's flight, they posted the chart on Twitter.

Indeed, the reaction was harsh. "Space Tourism Rivalry Gets Extremely Petty Ahead of Branson's Spaceflight," read the headline in *The Verge*. Nicola Pecile, one of Virgin Galactic's pilots, lashed out on Twitter. "Hey BO, we at VG are truly big supporters of your program too. But this pissing contest about the Kármán line is so childish that it is getting really embarrassing to watch," he wrote. Internally, many of Blue Origin's employees were embarrassed. While many of them hated SpaceX and Musk, they didn't view Virgin Galactic and Branson the same way. It was much more of a friendly competition, and many of the companies' employees were friends.

Branson largely ignored Blue's tweet. But when he was asked about the rivalry with Bezos during an interview on CNBC, he couldn't help but reply: "Jeff who?"

MUSK, MEANWHILE, MADE a show of support. Two days before the Virgin Galactic flight, he tweeted at Branson that he would "see you there to wish you the best."

"Thanks for being so typically supportive and such a good friend, Elon," Branson replied. "Great to be opening up space for all—safe travels and see you at Spaceport America!"

Musk showed up in the middle of the night at the house where Branson was staying, hours before Branson was to report to the launch site. Excited and unable to sleep, Branson went down to the kitchen at about 2:30 A.M. to find Musk there, comfortable and barefoot. "We made a pot of tea and sat outside under the stars and caught up," Branson recalled. Later that morning, he posted a photo of the two of them. "Big day ahead," he wrote. "Great to start the day with a friend. Feeling good, feeling excited, feeling ready."

It was a welcome show of comity. The bickering between the billionaire space barons was threatening to turn human spaceflight—a daring act of discovery and exploration undertaken at the time by only some 550 people—into an unseemly battle of egos between some of the wealthiest and most privileged men in the world. "If space travel lost its novelty in the early 1970s, it might now be in the process of losing its dignity," the historian Jeff Shesol wrote in an op-ed in *The New York Times.*

But while Branson appeared to be taking the high road, the pageantry surrounding his flight made it clear that it was going to be nothing like a typical NASA launch. The launch site was in the New Mexico desert at Spaceport America, a building that looked like a flying saucer that could, at any moment, lift off and make its own foray into space. A pair of tents hosted Branson's guests—one for friends and VIPs, and one for the media—and each was outfitted with lavish breakfast buffets. Instead of a stoic *three, two, one* countdown, there was a party-like atmosphere along the tarmac. There was even a musical guest, Khalid, who debuted a new song as if he were appearing on *Saturday Night Live.* The company's live broadcast of the flight was anchored by Stephen Colbert. Musk took it all in with his young son perched on his shoulders.

The flight itself was anticlimactic. There was none of the thunder of a rocket launch. No fire shooting out of the booster. Instead, Virgin Galactic's mothership took off from a runway, like any plane taking off at a commercial airport. When it got to about 40,000 feet, the mothership released the spaceplane, which fired its engine. On the ground, we could see the faint flash of light and the silent contrail it left as it tore straight up through the cloudless sky, a glimmer of sun reflecting off the silvery ship until it seemed to disappear. Off went Branson and the rest of the crew, all Virgin Galactic employees: Sirisha Bandla, vice president of government affairs; Colin Bennett, the lead operations engineer; and Beth Moses, Virgin's chief astronaut instructor.

The pilots flew SpaceShipTwo to an altitude of 53.41 miles before gliding back to the runway, touching down just fourteen minutes after release. In all, Branson was in space for only a few minutes. Still, he talked like a man who had touched the face of God. In several rounds of interviews that went on for hours, far eclipsing his time in weightlessness, he rhapsodized about the experience. "Just magical," he said. Always on-brand, he didn't miss a chance to plug his company: "Having flown to space, I can see more clearly how Virgin Galactic is the spaceline for Earth."

Which is what Bezos wanted for Blue Origin too, of course. But after Branson's flight, he offered an olive branch: "@richardbranson and crew congratulations on the flight," he wrote on Instagram. "Can't wait to join the club!"

IN THE YEARS leading up to his own spaceflight, Bezos had gone through a metamorphosis. Long gone was the skinny-necked nerd in the ill-fitting Dad jeans and baggy sweaters whose hair was staging a mutiny. Now he was bald and buff, with bulging biceps, a reinvention resulting from hours in the gym. A lot of people attributed the change to his desire to go to space. In reality, he could have boarded the capsule as the same old schlub, and it would have been fine. Flying on New Shepard did not require defined pectorals—only decent bladder control.

Being able to endure some ninety minutes without access to a bathroom was one of the conditions Blue Origin required of its passengers. Liftoff could encounter all manner of delays, from bad weather to technical issues, that would require urinary forbearance. The astronaut Blue's capsule was named for, Alan Shepard, had his own ability to hold it pushed to the limit on the flight that would make him the first American in space. As he waited out an unexpected hours-long delay on the launch pad, the orange juice and coffee Shepard had downed with his steak and eggs that morning conspired against him. "Man, I gotta pee," he finally pleaded over the radio to mission control. A quick run to the restroom was out of the question. Do it in the suit, they told him. So

Shepherd did. On subsequent missions, NASA would issue adult diapers to astronauts for just this sort of predicament. But Shepard had only his undergarments; thankfully, he was alone on the flight.

The ninety-minute hold-it policy wasn't the only requirement Blue Origin had imposed. Its passengers would also have to be able to climb seven flights of stairs to reach the capsule (a requirement that was later obviated by an elevator). They would need to be able to endure three times the force of gravity for two minutes on ascent, and 5.5 Gs on the way down. In other words, if you could endure a fairly intense roller-coaster ride, you'd be fine on New Shepard. Finally, to fit in the seats, participants would need to be between five feet and six feet four inches tall and weigh between 110 and 223 pounds.

If flying to space was the ultimate adventure, it was not regulated as such. The Federal Aviation Administration concerned itself only with protecting people and property on the ground. The flying passengers, not so much. If they were crazy enough to strap themselves into a rocket built by a quixotic billionaire, so be it. Like leaping out of a perfectly good airplane or attaching a bungee cord to their ankles before jumping off a bridge, space tourists only had to sign an "informed consent" waiver acknowledging the enormous risks and decent possibility that their "experience of a lifetime" could end in their demise. Over the years, some members of Congress had squawked about the dangers of such loose requirements and called for stricter regulation. But for now, going to space was something anyone could do.

If they could afford it. Initially, Virgin Galactic charged some $250,000 per seat; when they went on-sale after Branson's flight, it charged $450,000. Blue Origin executives thought the value of a ticket to space could be far higher, and they set up an auction for the open seat on their inaugural flight. Some 7,600 bidders from 159 countries made an offer, giving Blue an invaluable database of potential customers and a sense of what they were willing to pay. Before the results were revealed, several employees put their guesses into a hat. None came close to predicting the winning bid—an astounding $28 million. It had come from a cryptocurrency entrepreneur named Justin Sun; after he pulled out, the seat went to Oliver Daemen, an eighteen-year-old from the Netherlands, whose wealthy father ran an investment group.

Though the auction was a novel approach, private citizens going to space was not new. Between 2001 and 2009, Russia had flown several paying customers to the International Space Station on its Soyuz spacecraft in trips that reportedly cost between $20 and $35 million. And well before that, NASA had planned on regularly flying private citizens on the Space Shuttle. Its original designers believed the ship would travel as often as sixty times a year, so regularly that NASA would need to supplement the ranks of its professional astronaut corps with regular people. That touched off debate within the agency about whether it was prudent, and later, skirmishes over whom to send.

NASA ultimately decided on a teacher, choosing Christa McAuliffe from an applicant pool of more than eleven thousand educators. The effort, known officially as the Spaceflight Participant Program, was expected to continue on subsequent flights with a journalist and then, probably, an artist. By the time of McAuliffe's mission on Challenger in 1986, the list of journalists had been winnowed down to forty finalists. Walter Cronkite, the CBS news anchor who had provided the play-by-play of the Apollo 11 moon landing, was considered the favorite.

He never flew, of course. NASA canceled the program after Challenger exploded, killing McAuliffe and the six others also on board. Space travel, NASA decided, was far too new and too dangerous to be opened up to the masses.

"Anyone who has lived with large rocket engines understands that their awesome power is produced by machinery churning away at very high temperatures, pressures and velocities. A thin and fragile barrier separates combustion from explosion," Michael Collins, the command module pilot during Apollo 11, wrote in *The Washington Post* two days after the disaster. He wondered "if Christa or her family really grasped the seriousness of riding that gigantic pile of machinery on Launch Pad 39, despite the realistic briefings NASA conducted. . . ."

"We tend to pooh-pooh danger, and if you go into the VIP stands before a launch, there is a carefree, holiday atmosphere, like being at the company picnic," Collins continued. "Ride one of the beasts and you get a different perspective."

BLUE ORIGIN'S ENGINEERS took this to heart. They'd be flying the company's founder and benefactor. If Bezos died, there was a good chance Blue Origin itself would cease to exist. When I visited Smith at Blue's headquarters a few weeks before Bezos's flight, he assured me they were not going to be cavalier with the boss on board. "We didn't take any shortcuts," he said.

Getting the vehicle ready to transport humans to space and back had been a long and slow process that embodied the slow-but-steady spirit of the company's mascot, the tortoise. In 2015, after Blue first reached space and then landed the New Shepard booster so that it could be reused, Bezos predicted the company would be flying people within two years. Instead, it had taken six. Over that time, it had flown the vehicle to space fifteen times in test flights without anyone on board, "gradually stepping up and expanding the envelope, pushing the vehicle to its design limits," as Gary Lai, the senior director of the New Shepard design team told me in an interview at the time.

The company had extensively tested the capsule's emergency escape system on the ground and twice during flight. If anything went wrong with the rocket, the spacecraft, which had its own engine, would jettison itself away, much like SpaceX's Dragon capsule could. On one of the abort system tests, Blue even made sure there was a parachute failure to see how the spacecraft would land under two chutes instead of three. It worked fine.

But the company's engineers were well aware of potential blind spots. "As an engineer, you can never dispel the gremlins of unknown unknowns," Lai told me. "There are always going to be things that you wonder, *Well, what if I forgot about this?*" Such thoughts had kept him and his team up at night. "Our standard is to fly anybody for a space tourism mission, and for them not to actually take any substantive risk," Lai told me. One way to frame the question was "whether it's safe enough to put our own children on it. And I would be confident putting my own kids on this vehicle."

Bezos was certainly comfortable with the risk. "He, more than anyone else, has been through all the technical details and has been there through all the major decision points," said Ariane Cornell, the director of astronaut and orbital sales. "So, I'm not surprised at all that he said, 'Hey, let's go.' I mean, it's really been his dream, like a lot of us, to fly to space since he was a little kid."

To join Bezos, his brother, and Daemen, Blue invited Wally Funk, an eighty-two-year-old pilot and member of the Mercury 13—a group of women who had been privately trained for NASA's astronaut program in the early 1960s but never got the chance to fly.

Now, she would.

THE CREW DROVE to the launch pad in Rivians, the electric vehicle start-up that, under Bezos, Amazon had invested in heavily. It was a public endorsement for one of Tesla's main rivals and seen as a slight to Musk and SpaceX, which ferried its astronauts to the launch pad in Teslas.

The morning of his spaceflight, Bezos wore his cowboy hat, which was maybe not the best choice. True, Bezos had spent enough time in Texas to be able to legitimately claim the state. He'd spent every summer as a child on his grandfather's ranch; he'd worked on heavy equipment, ridden horses, and even castrated cattle. By rights, he could wear the hat. But combined with the tight, shiny, one-piece space suit that looked like it belonged on a NASCAR driver (and reportedly required a last-minute alteration to the crotch by his tailor, who flew into Texas to make the adjustments, according to Bloomberg's Brad Stone), and the fact that he was about to fly on a rocket that looked unmistakably like an erect penis, it opened Bezos, and his flight, to ridicule. On his late-night program, Colbert showed a picture of the New Shepard rocket and said, in mock alarm, "You can't show that on CBS! It's a family show." Producers then blurred out the rocket as if it were porn.

Bezos's and Branson's flights were mocked as inconsequential joyrides by billionaire "space cowboys" who had nothing better to do with their money, reducing to folly what was supposed to be a transcendent experience. Ever since humans began traveling into space, they have come back swooning, proclaiming that nothing can prepare a person for the view. The rapture of seeing Earth from above even has a name— "the overview effect"—an experience that has turned astronauts into evangelists, preaching the gospel of the cosmos.

"Orbiting Earth in the spaceship, I saw how beautiful our planet is," Yuri Gagarin said in 1961, as the Cold War space race was reaching a fever pitch. "People, let us preserve and increase this beauty, not destroy it." In the decades since, hundreds have come back singing a similar refrain. Even the stolid, reticent Neil Armstrong found his muse out in the beyond. "It suddenly struck me that that tiny pea, pretty and blue, was the Earth," he said. "I put up my thumb and shut one eye, and my thumb blotted out the planet Earth. I didn't feel like a giant. I felt very, very small."

The first time former NASA astronaut Mike Massimino saw Earth from space, he felt he should avert his gaze, like the scene below him was a secret that was supposed to remain invisible. "It just seemed so beautiful," he said. "How can something so beautiful be tolerated by human eyes?"

After his flight, Bezos summed up the experience this way: "OH. MY. GOD." At the post-flight press conference, his explanation was detailed but succinct. "The zero-G piece may have been one of the biggest surprises because it felt so normal," he said. "It felt almost like we were as humans evolved to be in that environment, which I know is impossible, but it felt so serene and peaceful. . . . The most profound piece of it, for me, was looking out at the Earth and looking at the Earth's atmosphere. Every astronaut, everybody who's been up in space, they say this: that it changes them, and they look at it and they're kind of amazed and awestruck by the Earth and its beauty, but also by its fragility. And I can vouch for that."

He was right. Blue's customers would come back similarly transformed. After his flight, William Shatner, the actor who played *Star*

Trek's Captain James T. Kirk, grew teary as he thanked Bezos for the experience. "What you have given me is the most profound experience I can imagine," he said. "I'm so filled with emotion about what just happened. It's extraordinary. I hope I never recover from this. I hope I maintain what I feel now. I don't want to lose it."

Another Blue Origin customer, George Nield, a former FAA official, was also overwhelmed, choking up as he told me about when he flew to space in 2022. The Earth below was so bright, he said, it shimmered in high definition. But the real surprise was the view above the atmosphere: the pitch darkness of space was unlike anything he had ever seen. "It's not like, oh, it's dark outside," he told me. "No, this is *bright* black. It was just jarring—and a very beautiful and emotional experience."

Bezos's *oh my god* description perhaps lacked the poetry of Massimino feeling like he should look away, or Gagarin's call for peace, or Nield's description of a black so deep and luminescent it was bright, but it was sincere. Bezos was clearly moved.

Sitting in the capsule in the moments before liftoff, Bezos told his fellow crewmates, "Guys, if you're willing, let me invite you—when we get up there, there's going to be all kinds of adrenaline, all kinds of excitement, novelty. Take a minute, take a few seconds, to look out and calmly think about what we're doing. It is an adventure, it is fun—but it's also important."

AFTER HIS FLIGHT, Bezos said he would dedicate himself to building the infrastructure for routine space travel. His voyage was "the first step of something big," he said. "I know what that feels like. I did it almost three decades ago with Amazon. Big things start small. But you can tell when you're on to something. And this is important: We're going to build a road to space so that our kids and their kids can build the future, and we need to do that to solve the problems here on Earth."

Blue Origin had been building according to its motto: "Step by step, ferociously." It was an ethos that also captured another of Bezos's projects—the 10,000 Year Clock. Constructed inside a mountain on his property in West Texas, not far from where Bezos had launched to space, the clock was a monument to long-term thinking. It would tick once a year; the "century hand" would advance every one hundred years; and the cuckoo would chirp on the millennium. Taking that long view, Bezos believed that Blue Origin, far more than Amazon, would define his legacy. Bezos was getting older, however. At fifty-seven, his own clock was ticking. Now the emphasis had to be on ferocity.

Aside from Bezos's successful spaceflight mission, the company had been struggling in virtually every other area. If it was building a road, it was filled with wrong turns, dead ends, and potholes. The loss of the moon contract to SpaceX was a symptom of much bigger problems. If Blue Origin was ever going to compete with SpaceX, or even be relevant in the space industry, Bezos would have to fix them himself.

CHAPTER 18

Toxic, Limping, Abysmal

Jeff Bezos's employees had been trying to warn him about how bad things had gotten at Blue Origin. In 2019, a mid-level engineer wrote an exhaustive twenty-three-page memo to Bezos and his leadership team, detailing the "toxic" company's "systemic" woes and charging it was "limping towards constantly shifting finishing lines." Instead of operating like SpaceX, a start-up hungry to upend an industry long dominated by military contractors, Blue was instead "shaping a future akin to the next Boeing or Lockheed Martin. A company so beholden to the next contract, that it lacks its own vision, and does not substantially move the capability and inspiration of spaceflight forward," according to the memo, which I obtained at the time.

Shortly after Bezos's flight, a top executive wrote another memo, albeit anonymously, putting the blame squarely on Smith: "He treats his entire leadership team like puppets and his personal insecurity as a leader prevents any delegation of authority." The head of human resources had even hired an outside consultant to interview executives in an attempt to get to the root of the problem and help Smith guide the company. "There was unanimous agreement among his leadership team that Bob is a micro-manager, which is detrimental to his performance as a CEO," the memo read. "When Bob heard the debrief report, he left the meeting and refused to meet with the consultant or his leadership team on this subject again."

Musk, the anonymous executive noted, also had serious issues as a manager—but he was building rockets that flew and bringing in revenue. At Blue, employees were frustrated with the glacial pace and leaving in droves. Nearly every program at the company was facing a delay, including the BE-4 engine that would be used to power Blue's New Glenn rocket.

Bezos had announced a deal to sell the engines to the United Launch Alliance, the joint venture of Boeing and Lockheed Martin, in 2014. Now, seven years later, it still hadn't delivered a single engine and progress "is abysmal," the executive wrote. That was a problem because ULA planned to use the engines for its next-generation rocket designed to fly national security satellites for the Pentagon and intelligence agencies. No engine meant no rocket, and no rocket meant the Space Force couldn't stay ahead of its adversaries.

Bezos once said that New Glenn would be ready to fly by 2020. Instead, in February 2021, Blue put out a press release that made it seem as if its customers were the problem, and said the first flight would come in late 2022. That was fiction, the executive wrote in the memo: "New Glenn will slip well beyond the announced delay of first flight in 2022."

The anonymous memo concluded with a warning: "I am certain of this—Blue will never, never, ever achieve acceptable company execution and success with Bob Smith as the CEO."

Even if Bezos never read the two memos, there were plenty of other warning signs, none more public than an essay written in October 2021 by Alexandra Abrams, the company's former head of employee communications, along with twenty unnamed Blue employees. It was posted by the whistleblower site *Lioness,* which publishes stories of workplace misconduct. The essay alleged that the company's "culture sits on a foundation that ignores the plight of our concerns, and silences those who seek to correct wrongs." One former executive, the post alleged, "frequently treated women in a condescending and demeaning manner, calling them baby girl, baby doll or sweetheart and inquiring about their dating lives."

The *Lioness* post broke the dam, and soon many others came forward to speak and pass along the memos they had written. In an article the next day, I wrote that Blue in 2019 had fired a longtime top executive for inappropriate behavior. One employee, who was not among the signatories to Abrams's essay, told me that the executive had once said about her in a meeting: "I apologize for her being so emotional. It must be her time of the month." Blue Origin pushed back against the allegations, saying that the company "has no tolerance for discrimination or harassment of any kind. We provide numerous avenues for employees, including a 24/7 anonymous hotline, and will promptly investigate any new claims of misconduct."

The turmoil had caused an exodus of talent. In a short period, the company lost its chief operating officer, comptroller, and senior engineers on New Glenn, New Shepard, and BE-4. The list went on and on. People I had known for years were leaving, lamenting that the dream they had signed on to achieve was foundering. Where, they wondered aloud, was Bezos? And why wasn't he doing something about Smith?

After I prepared a follow-up story on the problems at Blue, I gave its communications team a heads-up that it was going to publish the next morning. The company had provided me with a statement saying that it took "all claims seriously and we have no tolerance for discrimination or harassment of any kind. Where we substantiate allegations of misconduct under our anti-harassment, anti-discrimination and anti-retaliation policy we take the appropriate action—up to and including termination of employment."

A few hours later, Bezos tweeted out a statement of his own, a preemptive strike, one that I couldn't help but think was aimed directly at me. He posted an image from a *Barron's* magazine cover story from 1999, right when the Internet bubble was bursting. The headline read "Amazon Dot Bomb" and the story began: "The idea that Amazon CEO Jeff Bezos has pioneered a new business paradigm is silly."

Above the image, Bezos wrote on Twitter: "Listen and be open, but don't let anyone tell you who you are. This was just one of the many stories telling us all the ways we were going to fail. Today, Amazon is one of the world's most successful companies and has revolutionized two entirely different industries."

It was an unfair comparison. My story didn't say Blue Origin was going to fail. It simply pointed out that some employees, particularly women, didn't feel like it was an equitable place to work, and that there were serious leadership challenges that had held it back, allowing SpaceX to charge ahead.

Unable to sleep that night, I reached for my phone to look at Bezos's tweet again. This time I noticed that Musk had responded with a tweet of his own.

There were no words, just an emoji of a second-place medal.

IF ANYTHING, MUSK wished he had something resembling a rival. That would keep him and his team hungry and sharp. In an interview with the *Financial Times* at the end of 2021, Musk said that while Bezos had a "reasonably good engineering aptitude," he "does not seem to be willing to spend mental energy getting into the details of engineering. The devil's in the details." Musk admitted that "I'm trying to goad him into spending more time at Blue Origin, so they make more progress. As a friend of mine says, he should spend more time at Blue Origin and less time in the hot tub."

By then, SpaceX had flown multiple astronaut missions to the International Space Station, and in September 2021 had even upstaged Bezos's flight with a tourism mission of its own that was far more daring. Instead of an up-and-down, suborbital mission, SpaceX flew four private citizens to orbit around Earth for three days, the first time a crew composed entirely of amateurs had accomplished such a feat. Commissioned by Jared Isaacman—the billionaire founder of Shift4 Payments, a tech company—the flight, dubbed Inspiration4, reached an altitude of 363 miles, higher than the ISS and even the Hubble telescope.

The flight raised $250 million for St. Jude Children's Research Hospital, and joining Isaacman was Hayley Arceneaux, a childhood cancer survivor who worked as a physician assistant at St. Jude; Sian Proctor, an artist and space enthusiast; and Chris Sembroski, then an engineer at Lockheed Martin. Proctor won a contest to fly on the mission; Sembroski flew after a friend, who had also won a seat, had to back out.

If spaceflight was going to open up to the masses, ordinary people not only had to fly, but they had to push boundaries. "Technology doesn't automatically improve," Musk once told me. As evidence of that, he pointed out that we had been to the moon more than fifty years ago but had never returned. "That's nuts," he said. Technology only improved with hard work by people willing to take risks. And Inspiration4 was truly risky. If its crew were harmed or killed, it would have been a setback to the cause of civilians in space to rival the Challenger disaster. "If that went wrong, they'd say, 'We told you it shouldn't be done this way. Let's just keep it to government astronauts forever,'" Isaacman told me. For him, the risk was worth it. "If we're going to go to the moon again, and we're going to go to Mars and beyond, we've got to get a little outside our comfort zone and take the next step in that direction," he said.

AFTER THE INSPIRATION4 flight lifted off, Bezos was gracious: "Congratulations to @ElonMusk and the @SpaceX team on their successful Inspiration4 launch last night," he wrote on Twitter. "Another step towards a future where space is accessible to all of us."

Musk took the high road as well: "Thank you," he replied.

Since losing the lunar lander project to Musk, Bezos had been pushing his team to act more like SpaceX—that is, to be original. His challenge to his staff two days after the project was announced—to imagine "What if NASA didn't exist?"—was a thought exercise that touched off a soul-searching discussion. Not just about the best way to get to the moon on their own, but about Blue itself. What did it stand for? What did it want to accomplish? If they were no longer chasing a

NASA contract, or nipping unsuccessfully at SpaceX's heels, they were liberated to brainstorm about the future they wanted to create.

For so long, they had played NASA's game, restricted by its rules, or what Blue perceived those rules to be. SpaceX, meanwhile, seemed to be playing in an entirely different arena, somehow dictating the rules as it saw fit. As a SpaceX executive once told me, the company's philosophy was: "Don't try to win the game everyone else is playing. Change the game." And so it had. It had put touchscreens in its Dragon spacecraft when NASA's fighter pilot astronauts demanded sticks. Even though NASA had built a mechanism to dock with the International Space Station, SpaceX essentially told them, *Yours sucks, ours is better,* and insisted on creating its own. SpaceX overcame enormous skepticism from NASA that it was safe to load superchilled, superdense propellent onto spacecraft with astronauts already strapped inside. Perhaps most significant of all, SpaceX had also convinced both NASA and the Pentagon—which initially were wary of flying missions on used Falcon 9 boosters—that using "flight proven" rockets was actually better. "If it was an airplane, would you rather fly on the first flight out of the factory or after it's flown a bunch of missions? There's a lot of merit to flying on something that has proven itself in flight," Musk once told me during an interview in SpaceX's hangar at the Kennedy Space Center, which at that moment held several boosters with the soot marks that verified they worked.

Nothing shifted the paradigm as much as Starship. Standing 397 feet tall, it was an entirely novel behemoth that looked like it belonged on the cover of a science fiction novel. It looked like the future. Blue Moon, by contrast, looked like the spacecraft that transported Neil Armstrong and Buzz Aldrin to the lunar surface in 1969, down to the same four spidery legs. It could have stood in as a replacement for the Apollo Lunar Module on display at the National Air and Space Museum. But with Artemis, NASA didn't want to repeat Apollo. If it was going to embrace this new space age, the space agency had to leverage the new innovative approaches that sprang from industry, even if they didn't fit the mold.

Having lost with a conservative, traditional proposal, Blue would now embrace the SpaceX approach. It would build the lunar lander that it wanted, one designed to enable a permanent presence on the moon.

"We were not going make the same pandering mistake we made the first time, which was to parrot NASA's architecture back to them, because the SpaceX award made it evident that they didn't really care what the architecture was," a former executive told me. "Okay, lesson learned. We're not going to do that again."

Starting on that Sunday morning call, Bezos and his team started sketching out what this would look like. John Couluris, who ran the Lunar Transportation team, recalled, "Jeff kept pushing us: 'Well, can you land on the moon? Can you do more?'" They would start with a lander, for cargo only, not humans, called Mark 1. "And as we started to work on that vehicle, more blossomed out of that. Like, this vehicle can be that first stake in the ground for anywhere on the moon. It can start to ensure permanent presence on the moon."

Over time, Blue Origin developed ideas for a reusable lander that would remain in orbit around the moon, along with another spacecraft used to refuel the lander that would fly back and forth between Earth and the moon. This was nothing like the Apollo lander, or even what Blue had proposed to NASA earlier. This new Blue Moon represented a step forward, requiring new technologies, including the ability to transport and transfer propellants in space, an exceedingly difficult task.

If the company was going to enable "a million people living and working in space," it would need not only new spacecraft, but an entire infrastructure, which engineers began to call the Blue Lunar Architecture. A moon base would require manufacturing, drilling, power, transportation. Blue would have to develop the technologies to allow lunar pioneers to live off the land—that is, to use the regolith, or moon dirt, to build habitats and solar cells, and to extract water and separate out its hydrogen and oxygen. In space talk, this was known as in-situ resource utilization, or ISRU. If SpaceX had dominated the rocket race, Blue would focus on ISRU.

In the middle of 2021, Blue Origin went on a hiring spree, posting a couple dozen openings that seemed straight out of *Star Trek*: a "space resources mining systems lead," a "space resources chemist," a "space resources materials scientist"—all of whom would be part of a team "focused on the development of ISRU technologies related to the manufacturing of useful products from regolith; such products could include, but are not limited to, alloys, photovoltaic cells, building materials or solar reflectors." To give it an edge in this particular niche of space development, Blue made its first acquisition, buying Honeybee Robotics, which had built the drills, sensors, and scoopers on several of NASA's Mars rovers.

Most importantly, Bezos was engaged in a way he hadn't been since the beginning of the COVID pandemic. Inspired by his spaceflight and galvanized by the lunar lander loss to SpaceX, he started giving Blue much more of his attention. Employees who had grown up at Blue when Bezos was distracted now were getting feedback on their work from the boss directly, and learning how intense and demanding the man could be. As they worked on the new Blue Moon system, Bezos pushed his team to continue to iterate and find better solutions, going through proposals "sentence by sentence, word by word," as one former executive recalled.

"Literally two weeks after he came back to Earth, after the first human flight, the discussions that we had with him changed," the person said. In addition to the weekly "Strategy, Architecture and Technology" meetings, in which senior executives would review the various programs the company was undertaking, Bezos also started holding three-hour business review meetings where he would dig into the details of the company in a way he had not done in years, if ever. He would sometimes scold and lecture the senior management team in a "brutal" way, the former executive recalled. "This is stupid," he would say. "How could anyone think this is a good idea?"

The loss to SpaceX left the impression that Bezos looked at Blue as a "mismanaged disappointment" that still was not operating on Amazon's level. "He did form this sort of hypothesis that we were a bunch of bozos playing with his company and we didn't really know what we were doing," the person said. "A lot of meetings would turn into people doing their best to get off the stage with their skin intact."

He had stepped down as CEO of Amazon, but he was not retired.

"He's now the CEO of Blue," as another person told me.

ON NOVEMBER 14, 2021, after months of working on the new moon plan, three of the company's top executives wrote a memo laying out in detail the answer to the question Bezos had asked that Sunday morning after it lost the lunar lander contract: How would they get to the moon without NASA?

Copying the Amazon format, the memo was six pages long, written in the form of a press release, with a Q&A at the end. "Blue Origin Commissions First Moon Base," read the headline. "Permanent Armstrong Station to Revolutionize Science and Develop Lunar Resources and Commercial Business."

Dated in the year 2032, it began: "Blue Origin announced today the successful commissioning of humanity's first permanently off-world base on the Moon. Armstrong Station has begun 24/7 operations for various exploration, science, research, manufacturing, and tourism businesses. Located on Shackleton Ridge at the Moon's South Pole, the base hosts government customers and now commercial customer visits for the first time. The newly christened base is one giant leap toward Blue Origin's vision of enabling millions of people living and working in space to benefit Earth."

It quoted Bezos as saying, "The Moon is a gift to humanity. It is the most accessible planetary surface in the solar system, just four days away with daily launch opportunities and only a slight communications delay. The Moon is where we will learn to manufacture using materials and energy that don't come from Earth."

In this telling of the future, Blue already had robots scouring the lunar surface to harvest resources, and mining operations had begun, "producing four products locally: oxygen, which makes up much of the mass of rocket propellant; metal alloys for structural elements; solar cells to expand the base power capability; and volatile elements from ice for rocket fuel, water, and trace chemicals."

The commissioning of Armstrong Station had been the culmination of a series of ambitious events, the release went on, starting with Blue's first robotic landing on the moon in 2023, followed by the first human landing in 2028, with successive flights in the years after. The next step in the base's development would be increased power generation to allow for an expanded human presence: "The base today supports continuous presence, and the next exciting phase will be expansion based on solar cells made on the Moon."

"Living and working on the Moon shows us the true value of Earth," Bezos was quoted as saying. "Carl Sagan said, 'On it, everyone you love, everyone you know, everyone you ever heard of, every human being who ever was, lived out their lives. . . . The Earth is a very small stage in a vast, cosmic arena.' Today, we humbly add another address. A new home. Not a replacement, but one that provides a unique perspective to how precious the Earth truly is."

THE MEMO WAS, of course, science fiction, a hopeful vision of what the company might one day accomplish. In reality, Blue Origin had yet to even reach Earth orbit. SpaceX, meanwhile, launched thirty-one times in 2021, setting a new record for the company; it also completed its one hundredth booster landing and had one of its Falcon 9 rockets fly its eleventh mission, also a record. By the end of 2021, it had launched nearly two thousand Starlink satellites and was beginning to become a legitimate Internet service provider.

Musk was still unsatisfied. Success begat not rewards and repose but more suffering, much of which Musk inflicted on himself. He was living in Boca Chica, overseeing the development of Starship, and pushing his team to mass produce its Raptor engines and work around the clock, holding meetings that sometimes started at 11:00 P.M. In an email the day after Thanksgiving, he lit a fire under his team with a dire, alarmist note.

"The Raptor production crisis is much worse than it seemed a few weeks ago," he wrote. While he had planned on taking time off for the holidays, that was no longer an option—for him or the rest of the Raptor team. "We need all hands on deck to recover from what is, quite frankly, a disaster."

Despite everything it had accomplished, SpaceX suddenly faced a "genuine risk of bankruptcy," he wrote, if it didn't start pumping out engines so that it could start flying Starship on a regular basis. The desperate, anxious tone was Musk in fight-or-flight mode, manufacturing a crisis to motivate, or scare, his employees to continue to push beyond their limits. The company was dominating every aspect of the launch market—but Musk believed that rest led to complacency, which led to death. And so he continued to push.

"Physics does not care about hurt feelings," he told Walter Isaacson, his biographer. "It cares about whether you got the rocket right."

BY THE END of 2021, Bill Nelson was in a fight of his own. Since his first day as the head of NASA, he had been pushing to get the White House's budget office to give the agency the resources it needed to fund a second lander.

As impressive as SpaceX was, Starship was a risky, unproven rocket, and Nelson disagreed with the decision to award it a solo contract. He wanted another option in case it faltered. More than most, Nelson knew how Washington worked and how to get what he wanted. He had served in the House of Representatives from 1979 to 1991, and then in the Senate from 2001 to 2019. He also had a personal relationship with President Biden, and he enlisted the White House in his effort to reopen the lunar lander endeavor.

In a meeting with Jake Sullivan, the national security adviser, who had worked during the early days of the administration to ensure the Artemis program would continue, Nelson raised his concerns about China and its ambitions to get to the moon first. "He told me face-to-face that the president does not want to be second to China on the moon," Nelson recalled. "And so, I kept reminding the folks in the White House budget office that that was the case."

Assured NASA would have the budget in the coming years, Nelson announced in March 2022 that there would be a competition for a second lander. Having won the first contract, SpaceX would not be eligible, leaving Blue Origin as the clear favorite.

"We think, and so does Congress, that competition leads to better, more reliable outcomes," Nelson said at the time. "It benefits everybody. It benefits NASA. It benefits the American people. I promised competition, so here it is."

Bernie Sanders, the liberal senator from Vermont, tried to block Nelson. He had been on a crusade against the privatization of space in general and Bezos in particular. "If Mr. Bezos wants to go to the moon, good for him," he said on the Senate floor. Citing Bezos's net worth of $186 billion at the time—which had grown significantly during the pandemic, in tandem with Amazon's stock price—he said Bezos should pay for the mission himself, "not U.S. taxpayers." He kept up his campaign in an op-ed, writing in *The Guardian* that NASA "has become little more than an ATM machine to fuel a space race not between the U.S. and other countries, but between the two wealthiest men in America—Elon Musk and Jeff Bezos." He added that "at this moment, if you can believe it, Congress is considering legislation to provide a $10 billion bailout to Jeff Bezos's Blue Origin space company for a contract to build a lunar lander."

That was not quite correct. The measure didn't appropriate $10 billion to Blue Origin. Instead, it authorized that amount for the entire lunar lander program—SpaceX and another provider that would probably be Blue—between 2021 and 2025. And the funding would have to be appropriated on an annual basis during that time.

Ultimately, Sanders's measure to strip funding for the lunar lander program fell short. He got seventeen votes, while seventy-eight senators voted against it.

In more than fifty years, the U.S. government hadn't spent a dime on a spacecraft designed to carry astronauts to the moon. Now NASA would have two.

· · ·

JUST AS IMPORTANT, NASA also had some swagger. Which it hadn't had in years. Once it had struck out boldly, in deed as well as word, from "light this candle" to "one giant leap" to "failure is not an option" and "we shall return" to a period when it seemed to have gone on mute. But in February 2021, NASA landed the Perseverance rover on Mars. As it descended through the thin atmosphere of the Red Planet, it deployed a parachute with a very odd color scheme. It looked almost like a piece of modern art: mysterious reddish-orange and white splotches—one resembling a fan, another an accordion—along with long, slender rectangles and stubby squares—that, like a Rorschach test, could have been open to any interpretation.

The design was actually a binary code, grouping zeros and ones to spell out a secret message: "DARE MIGHTY THINGS." It was a mantra at NASA's Jet Propulsion Laboratory, which designed Perseverance, and came from a speech by Theodore Roosevelt in 1899 titled "The Strenuous Life." "Far better it is to dare mighty things," it read, "to win glorious triumphs, even though checkered by failure, than to take rank with those poor spirits who neither enjoy much nor suffer much, because they live in a gray twilight that knows not victory nor defeat."

NASA was on a winning streak. A regular rotation of flights was sending astronauts to space from U.S. soil. NASA was poised to launch the James Webb Space Telescope, which would look back through time to the beginnings of the universe. Through its partnerships with private industry, NASA was building new space suits and developing commercial space stations to replace the aging International Space Station. A fleet of privately built robotic spacecraft were also beginning to fly in a series of missions to the moon. Maybe some of them would make it; others would not. In this new paradigm, failure was acceptable. NASA wanted to "take shots on goals," as its leaders started saying, knowing some of them would miss.

No place was the renewed energy more evident than at the Johnson Space Center in Houston, home to the agency's astronaut corps. After the Space Shuttle was retired in 2011, the astronauts had been dispatched to Russia to train to fly on the only spacecraft capable of taking them to orbit, the Soyuz. Now they could be assigned to one of three different American spacecraft, two designed for low Earth orbit—SpaceX's Dragon, Boeing's Starliner—as well as NASA's Orion capsule, designed to take them to the moon. Eventually they would train on Starship, and Blue Origin was still pushing to get NASA to fund a second lunar lander as well.

AFTER YEARS OF delays and cost overruns that made it a poster child for government inefficiency and raised questions about America's ability to do great things, even NASA's Space Launch System rocket was finally making progress. Set for late 2022, its debut mission was called Artemis I—the kickoff of NASA's campaign to return to the lunar surface. This was the flight that would send the Orion capsule around the moon and deeper into space than any spacecraft designed for humans had ever flown—1.4 million miles in all. No one would be on board; it was a test to see how the SLS and Orion performed ahead of the crewed missions to come. The plan called for Artemis II to send a quartet of astronauts around the moon, and then for Artemis III to land them on the surface.

Tall and powerful, the SLS, manufactured by lead contractor Boeing, proved to be a complicated and fickle vehicle, prone to hydrogen leaks and faulty sensors that caused a series of delays. But at 1:47 A.M. on November 16, 2022, it lifted off in a stunning display of power that lit up the Florida Space Coast and shot Orion toward the moon.

Leading up to the flight, NASA's leaders talked about the sheer force of the vehicle, the might of its two side-mounted solid-rocket boosters, the power of the four RS-25 engines strapped under the rocket. They indeed put on a pyrotechnic show, fit for the Fourth of July. But another, little-discussed feature of the system proved just as meaningful. Orion was outfitted with sixteen cameras, and on the first day of the mission, already some 57,000 miles from Earth, the spacecraft sent back a high-resolution image of the pale blue dot suspended in partial shade. It was a new image for a new generation that rose above the pettiness of contract disputes and Twitter spats and courtroom feuds. A single frame that muted the discord of feuding billionaires and restored a sense of awe to the act of exploration.

On Day 6, Orion was quickly approaching the moon, beaming back photos of it growing larger and more detailed until the capsule flew just eighty-one miles over the surface and eventually over Armstrong and Aldrin's Tranquility Base. The photos revealed the moon rough's contours, a tapestry of ancient craters pockmarking the landscape, breaking up the hills and valleys and ridges, at once inviting and forbidding.

On Day 13, Orion reached its maximum distance, 268,563 miles from Earth and more than 40,000 miles past the moon. From that vantage point, the moon and Earth were coupled together, alone in a dark and cold universe, as if holding close for warmth.

On Day 20, Orion whizzed around the moon again, this time ducking around the far side, forcing it to lose communication with Earth for about thirty minutes. When it reemerged, the spacecraft's cameras showed the planet as an impossibly thin crescent in the distance. The spacecraft was finally heading home for a splashdown in the Pacific Ocean, off the coast of San Diego, on the fiftieth anniversary of the Apollo 17 moon landing.

On Day 23, as Earth loomed large, Orion turned its cameras back toward the moon, now the size of a pea, snapping a photo of where the spacecraft had been just a few days before. NASA's caption of the photo read, "Until next time."

TAKEN TOGETHER, THE catalogue of images captured by Orion revealed a new moon, significantly different from the one that beckoned during Apollo. That one seemed to be a lonely and desiccated rock. The moon we were returning to with Artemis, we now knew, was wet, with life's key ingredient just below the surface at the poles. That allure would inspire new missions, in a new space age, led not just by NASA but by the growing commercial sector, which was preparing a fleet of robotic spacecraft designed to touch down on the lunar surface. Other countries beyond the United States and China were eyeing the moon as well. In 2023, India became the fourth country to land a robotic spacecraft on the lunar surface when its Chandrayaan-3 spacecraft touched down near the South Pole. Japan became the fifth when its lander touched down, sideways but intact, in early 2024.

This new moon race, however, was not about planting a flag and crossing a finish line, but rather about setting new precedents that would ultimately guide how humanity would form these frontier settlements and what the rules would be. It was "a race about the race," as the aerospace analyst Todd Harrison put it, about how you got there, the partnerships you leveraged, the resources you could access.

Wanting to stay in the lead, NASA would continue to push for a human landing by a new generation of astronauts. The crew picked for Artemis II was different from any that had gone before. It consisted of three Americans: Reid Wiseman as commander; Christina Koch, who would be the first woman to orbit the moon; and Victor Glover, the first African American. They'd be joined by Jeremy Hansen, the first Canadian. Artemis III would fly the first woman to walk on the surface. But the mission would not end there. The landing would be one in a series of steps that would stretch well into the future toward a new moon, carrying the promise to not merely arrive, but to eventually build a permanent presence.

AFTER ORION SPLASHED down in the Pacific Ocean off the California coast on December 11, Nelson was ebullient, saying the mission marked the "beginning of the new beginning."

"It's historic because we are now going back to space, to deep space, with a new generation," he said, calling it a new era of human spaceflight, "one that marks new technology, a whole new breed of astronauts, and a vision of the future." The heat shield, one of the key tests of the mission, worked "beautifully," he said.

NASA's career civil servants were also thrilled with the successful splashdown. But the point of the flight was a specific test. They wouldn't fully declare success until they reviewed the data and inspected the spacecraft's systems, including the heat shield, which had to withstand temperatures that reached 5,000 degrees Fahrenheit as it plunged through the atmosphere. On its return home, Orion would perform what was known as a "skip reentry," where it would plunge into the atmosphere traveling at Mach 32, or thirty-two times the speed of sound. Then it would skip back out of the atmosphere and dive back in, a manuever that would help lower its speed and guide it to a more precise landing location. But the double dip would strain the heat shield, and there was no way to replicate those conditions on Earth.

"Initial indications are very favorable," said Mike Sarafin, NASA's Artemis I mission manager, after the flight. "But there's more ahead of us in terms of exactly understanding what the reentry flight test told us."

NASA didn't have to wait long to see that there had been serious problems. Across the heat shield there were more than a hundred places where it wore away in ways engineers did not expect. Instead of a smooth, even charring, the heat shield was marked by pothole-like divots. In February 2023, NASA's Orion team told Nelson he should come back to the Kennedy Space Center, where the capsule was being inspected after the flight. Nelson was a politician, not an engineer, but even he could tell immediately that NASA had a problem.

"It had these chunks that had come off," he recalled. "Common sense would tell you that there was something wrong."

CHAPTER 19

Corporate Alchemy

Jeff Bezos's secret laboratory was perched at the end of a dead-end street north of Los Angeles in a residential neighborhood of ramblers and bungalows far more modest than the Beverly Hills mansions a few miles to the south. The outside of the squat redbrick building was unremarkable in every way except for one: It was protected like a fortress. With a locked gate, and a tall iron fence surrounding it, it was adorned with signs prohibiting unpermitted vehicles and giving notice that THE AREA IS UNDER 24-HOUR VIDEO SURVEILLANCE.

In an earlier incarnation, the building housed a bakery. But instead of chefs baking bread, there was now a team of a few dozen scientists and engineers—geologists, chemists, metallurgists, electrical and mechanical engineers, roboticists. Unlike their counterparts at Blue Origin who were working on rocket technology, this group was focused instead on ISRU, developing the technologies to allow future astronauts to use the resources found on the moon and elsewhere in space.

The moon may have been a barren wasteland by comparison to Earth, but it did have a precious commodity in abundance: moon dirt, or regolith. Millions of years of asteroids hitting the lunar surface had ground it into a fine grain that was surprisingly rich in the kind of resources—iron, aluminum, and silicon—that would be useful in creating a lunar outpost. Building reusable rockets was one thing, but extracting resources from a celestial body 240,000 miles from Earth was a problem of another magnitude that could take years, even decades, to solve. That was why Blue Origin had gone on its ISRU hiring spree in 2021, with a challenge from Bezos as he was looking to drastically boost his space venture's ambitions and lay a foundation for exploration that would span generations: *Prove that it can be done.*

In February 2023, Blue Origin announced that the scientists in its secret lab had made a significant breakthrough: They had successfully taken regolith and transformed it into small solar cells and transmission wire. The project was called Blue Alchemist, and it demonstrated, the company said, that the energy needs of the first moon base could be met by melting moon dirt. "To make long-term presence on the Moon viable, we need abundant electrical power," the company proclaimed in a press release. "We can make power systems on the Moon directly from materials that exist everywhere on the surface, without special substances brought from Earth. We have pioneered the technology and demonstrated all the steps. Our approach, Blue Alchemist, can scale indefinitely, eliminating power as a constraint anywhere on the Moon."

The process of transforming dirt into solar cells wasn't magic, but science, Vlada Stamenković told me during a rare tour of the former bakery, which Blue Origin now called the Space Resources Center of Excellence. The company used what was known as "molten regolith electrolysis," a complicated process of melting the simulated moon dirt in a reactor at extremely high temperatures—about 3,000 degrees Fahrenheit. Then the scientists would run an electrical current through it to produce iron, silicon, and aluminum, with oxygen as a by-product. With a title fit for *Star Trek*—Senior Director of Space Resources—Stamenković was a fantastic tour guide—a brilliant, fast-talking, fast-walking, passionate scientist, whose idealism was infectious.

Over the course of the tour, I met many of the scientists he had hired as we walked and talked our way through an assembly-line-like setup of brilliant scientists and engineers: a geologist, who explained the composition of the lunar regolith; a robotics engineer, who showed me samples of melted lunar regolith simulant, some of which looked like glassy crystals, others like lava. I got to see the "microwave melting vacuum chamber," which looked like a massive pizza oven from the future, and ultimately would house the reactor and simulate the lunar environment. I met a scientist working to extract and then store the oxygen, which would be a valuable commodity in the airless void of space, and an engineer who was developing the "solar cell recipe" that would produce the most efficient panel in the simplest fashion, and the engineers who were then manufacturing them.

And here they were—glass-covered cells about the size and weight of a wafer, though far prettier with purplish hues, evidence that Blue had done it, at least on a small scale and in the controlled environment of a lab. "We tested, failed, retested, and then succeeded to build the real technology," Stamenković said. "We proved it's not sci-fi but feasible and working."

Before coming to Blue, Stamenković had worked at NASA's famed Jet Propulsion Laboratory, the arm of the space agency that had pulled off some of its most daring feats, from Mars rovers to building spacecraft that have visited every planet in the solar system, the sun, and even interstellar space. He had been lured to Blue, however, "because I wanted to be fast in being able to really take on big challenges, to ask what really has to be done, and then just try to solve it." Blue Alchemist was, he said, "an enabler, and that's what I love about it." It could generate power on the moon, yes, but also on Mars and all over the solar system. "We have rocks everywhere." Including, of course, Earth.

Blue Alchemist, he was saying, could help produce steel without any toxic by-products, therefore helping to combat climate change. Their system could create metal—building materials—but in an environmentally friendly way, he said.

This, he said, was a big part of the promise of Blue Alchemist, a way to help explore the solar system but also to benefit Earth. If what Stamenković and his team were working on in this secretive lab came to fruition, it really could make an impact. NASA thought so too, and had awarded the company a $35 million contract as part of a program that would help companies develop technologies to build landing pads, roads, and habitats on the moon; to create nuclear power for energy and in-space 3D printing for manufacturing; and even to lay high-voltage power lines across the lunar surface.

The contracts would help the commercial space industry "develop technologies that could support long-term exploration on the moon and in space," NASA said at the time. But the total dollar amount came to a mere $150 million. If NASA was really serious about a permanent presence on the moon and in space, it would have to invest a lot more than that. Living off the resources of the moon or other celestial bodies seemed far-fetched, perhaps as elusive as turning lead into gold. But the settlement of space couldn't rely just on big, new rockets. There had to be technologies such as Blue Alchemist to allow humanity to evolve from exploration to expansion.

WHILE STAMENKOVIĆ'S SCIENTISTS were working on their break-through, Blue Origin's engineers, given a second chance by NASA in 2022 to compete for a lunar lander contract, were working on the company's proposal. This time, Blue had learned its lesson. It would bid $3.4 billion, far lower than the $6 billion it bid when losing the first round. Bezos was funding the rest of the development cost himself, a figure said to be at least the amount of what Blue was asking NASA to invest, but likely far more. The new Blue Moon proposal the company submitted also had the novel design born from Bezos asking his team to think about what they would build if NASA did not exist. The answer was a technologically ambitious system that would rely on two reusable spacecraft—a lander that would stay in orbit around the moon in between missions to the surface, and a refueling service vehicle that would fly back and forth between the Earth and moon. The key to the entire system was the ability to store propellant in space, something SpaceX would need for Starship as well. As Bezos told me later, the design was "completely different" than the one in its previous bid. "There's no similarities." He added, "It's absolutely the right system. It's elegant. It sets us up to explore the whole solar system."

Being able to fuel spacecraft in space had been seen as a crucial advancement in space exploration, though it had never been achieved. Orbital refueling would allow spacecraft to remain in space for far longer, flying back and forth between destinations, like the Earth and the moon, or even Mars. Now, with Starship and Blue Moon, two companies were working on it.

Having already won a contract, SpaceX was ineligible to bid, meaning Blue's only competition was Dynetics, which had finished third in the previous competition. Still, Blue's executives were nervous, so much so that in May 2023 when NASA called John Couluris, who oversaw the Blue Moon program, to tell him that Blue Origin had won, the reaction was "euphoria," he recalled.

With the lunar lander contract, and its pursuit of technologies to allow for human settlements, Blue had some real momentum. Smith had been hired six years earlier with the mandate to transform Blue from a research-and-development start-up to a revenue-generating operation that competed for and won government contracts. He had overseen tremendous growth, from 850 employees at the beginning of his tenure to 10,000. Blue Origin had opened new factories, expanded its headquarters, built the New Glenn launch pad from scratch, and was making progress on New Glenn and the engine program. But Smith's tenure as CEO had also been a tumultuous one, and many felt the company was not where it should have been.

Now, he could boast of a major scientific breakthrough, as well as putting Blue in partnership with NASA on the agency's flagship exploration program. But his departure from the company was already in the works, and Bezos was thinking about who would replace him to move the company into its next phase. The company also made another change. In the high-ceilinged lobby of its sprawling Cape Canaveral manufacturing campus, the first New Shepard rocket to reach space and land had been on display for years. Now it was replaced with a prototype of the Blue Moon lander. The symbolism was obvious: The small suborbital New Shepard rocket represented the first stage of the company's growth; the moon ship was its future. And it would be led by someone new.

. . .

AFTER MORE THAN a dozen years at Amazon, during which he oversaw some of its biggest product lines—including Alexa, Kindle, Ring, and more recently the Kuiper Internet satellite program that Amazon was developing to compete with SpaceX's Starlink—Dave Limp wanted to try something different. He had always thrown himself into his work, but by the beginning of 2023, he just didn't feel the same passion for the job. He wanted a change, though he wasn't sure exactly what that would be.

At a dinner with Andy Jassy, who had succeeded Bezos as Amazon's CEO, Limp said he wanted to "wind down" his tenure at the company but promised that he would stay "as long as it takes to find a replacement." Bezos called the next day, trying to convince him to stay. Amazon was a massive company with lots of different opportunities. Surely, they could find something for him. But Limp was adamant.

By April, Limp had decided he wanted to go off on his own and start a company in the energy sector. "And so I was pitching him the idea, and he said, 'Dave, I love the idea. I want to talk about it.'"

But Bezos had an idea of his own. Smith was resigning as Blue's CEO, Bezos said. "I think you'd be great for this job."

Limp didn't think so. His expertise was in putting devices in people's homes, not into orbit. He didn't know much about space or the rocket business. That was Bezos's passion, not his.

"I'm not an aerospace engineer," he told Bezos. "I'm not a rocket scientist. I'm not the right person."

Still, Bezos urged him to "sleep on it." Maybe he'd change his mind.

They were in touch a few days later, but Limp had not changed his mind. Fine, Bezos conceded. But he had a favor to ask: "Why don't you come and visit a couple of the facilities and then at least give me your opinion of what you would do to fix some things."

So, Limp did. At Amazon, he and Bezos had spent an enormous amount of time working together. Bezos, Limp concluded, was one of the world's smartest people when it came to e-commerce. And he could tell that Blue Origin Bezos was very similar to Amazon Bezos: "What is the same is Jeff is the most tactically impatient and strategically patient person—it's a bit of an oxymoron—that I've ever met," he told me. That said, "Jeff knows more and has more fun in and around rockets and rocket engines than he knows about e-commerce. And that is saying something."

Finally, after seeing the facilities, the opportunities that Blue was pursuing, and Bezos's passion and vision for the company, Limp became convinced that Blue could be a great opportunity. Still, the company continued to lag far behind SpaceX. It was considered by many in the industry to be a disappointment.

So there was one question Limp wanted Bezos to answer: "Is this a hobby or a business?"

In retrospect, it seemed an insulting question. Bezos had said repeatedly that Blue was the most important work he was doing. He had poured billions of his own money into it. In its ambition and scope, Blue encapsulated his childhood dreams. Yes, it was a business, a big business. When I asked Bezos if the question came as a shock, he said, "I was a little surprised because to me, it's so obviously a business." When Blue Origin reached the moon, building space stations and mining asteroids, it would be bigger than even Amazon, he believed, with its market value of more than $2 trillion. "I think it's going to be the best business that I've ever been involved in, but it's going to take a while," as he said at a conference.

By the time Limp took over the CEO job in December 2023, it had already been a while. Bezos had founded the company nearly a quarter century before. If Blue was going to operate like a business, it needed to act like one. At Amazon, as Limp knew as well as anyone, customers came first. Bezos had long said that "the most important single thing is to focus obsessively on the customer. Our goal is to be Earth's most customer-centric company." If Blue was going to be successful, it needed to not only have customers but prioritize them. After just a few months at the helm of Blue Origin, Limp made some changes to the company's Leadership Principles—an important document, he wrote in a company-wide email, that affected the company's "culture, behaviors, hiring, promotions, decision-making, fellow employees, customer engagements, and how we grow as a company."

Given Bezos's obsession with customers, a few senior leaders who worked under Smith were amazed that the word "customer" was nowhere to be found in the leadership principles. Under Limp, that would change. "Customer Focus" would now be chief among them, Limp wrote in his email to Blue: "Leaders start with their customers and work backward. They work vigorously to earn and keep customer trust."

That was the first change. The second was adding an attribute— Resourceful—that had been lacking as Bezos continued to fund the company from his own fortune instead of relying on revenue generated from actual customers, such as NASA or the Pentagon. As Limp wrote: "Resourceful: Leaders are frugal and accomplish more with less, recognizing that constraints breed self-sufficiency, invention, and innovation." Limp wanted his employees to "show grit," he wrote.

THOSE WERE WORDS that could have been used to define SpaceX, a steamroller continuing to shatter records. In 2023, it launched nearly 100 Falcon rockets, averaging about a launch every four days, a record, as it built out its Starlink satellite constellation, which by the end of the year totaled more than 5,000 spacecraft. In all, SpaceX hoisted more than 2,500 spacecraft that year. Perhaps more impressively, it lifted more than 2.5 million pounds to orbit. No one else was even close. China, by contrast, launched a total of 67 rockets, carrying 215 spacecraft with a total mass of under 300,000 pounds. SpaceX continued flying military and intelligence satellites for the Pentagon and intelligence community; and cargo, supplies, and astronauts to the International Space Station for NASA.

Musk may have been increasingly courting controversy with rapid-fire posts on his social media platform, X. But SpaceX had completed an improbable odyssey, years in the making, from being the distrusted outsider to becoming one of the government's most important partners. It was also pushing the envelope with its private astronaut missions, including another by Jared Isaacman, the billionaire entrepreneur, who in September 2024 flew with SpaceX again, and this time performed the first civilian spacewalk in a risky mission that recalled the early, swash-buckling days of space exploration.

At the same time, SpaceX's army in South Texas was continuing to develop Starship, pushing it through an intense test campaign that en-capsulated the force SpaceX had become: explosions and fiery failures mixed with seemingly impossible triumphs, all while Musk continued to drive at warp speed.

SpaceX liked to talk about its iterative process—how it tested, failed fast, fixed problems, and tried again in a crawl-walk-run ap-proach. So when Starship failed in spectacular fashion on April 20, 2023—the first flight with the spacecraft stacked on top of the Super Heavy booster—SpaceX cheered it as a successful failure. Three en-gines shut off at liftoff, and a couple more failed during the ascent. Then, as it reached an altitude of twenty-four miles over the Gulf of Mexico, the rocket, as tall as a forty-story building, started tumbling. The onboard flight termination system, essentially an onboard bomb designed to destroy the rocket before it could crash, detonated four minutes after liftoff, destroying the vehicle in a fireball. But in test flights like this, SpaceX said, the payload was the data, meaning the purpose wasn't to carry a satellite to orbit but rather to learn how the vehicle operates in flight.

The rocket wasn't the only thing obliterated. Fully fueled with ten million gallons of liquid oxygen and liquid methane, Starship gener-ated sixteen million pounds of thrust at liftoff, or twice the power of the Saturn V rocket that flew the Apollo astronauts to the moon. Its thirty-three first-stage engines would devour forty thousand gallons of propellant a second. At ignition, it delivered a punishing blow to the launch mount and ground infrastructure. As usual, Musk had been in a rush to get Starship flying and said he'd decided to proceed even though a water deluge system, used to dampen the acoustic blast of the rocket, wasn't going to be ready in time. The pad also did not have a flame diverter to direct the engines' fire and exhaust. That, too, "could turn out to be a mistake," Musk conceded at one point. "It will take us probably several months to rebuild the launchpad if we melt it."

They didn't melt the pad; instead, Starship blew it up. As the rocket's engines ignited, chunks of concrete and metal ejected in every direction as if from a bomb blast, denting storage tanks, pummeling the shoreline with shrapnel, and leaving a crater at the launch site. Large pieces of debris crashed into the water, kicking up splashes sixty feet high. Several seconds after launch, the plume of dust and sand was 850 feet high and over time it traveled as far as six miles before coming down on communities nearby. Researchers studying the blast later concluded that "the pressure that was built up under the pad was equal to a volcano." As was becoming habit, the FAA ordered an investigation. But now environmentalists and critics pounced, saying the rocket was too dangerous to operate near a fragile ecosystem.

Still, SpaceX pressed on, rebuilt the pad, made changes to the rocket and ground infrastructure, and kept flying, learning as they went. With a newly installed water deluge system, the following flights didn't blow up the pad, and SpaceX continued to make progress until it felt it could finally try to land the Starship booster. Unlike the Falcon 9, which touched down on a ship at sea or on landing pads, SpaceX intended to catch Starship with the chopstick-like arms of its launch mount. The performance would make good theater—this colossus falling from the sky under a pillow of fire, toward the embrace of its launch mount— but it was also, Musk believed, a fundamental step toward achieving rapid reusability. Getting to Mars would require a fleet of Starships able to leave like an armada, their boosters launching, landing, refueling, and flying again to propel the ships into deep space.

STARSHIP'S FIFTH TEST flight came on October 13, 2024. Liftoff, an hour after sunrise, was picture-perfect. Given the inherent dangers of launching such a Goliathan projectile, laden with millions of gallons of combustible propellant, the FAA had diverted all air traffic well to the north and south of SpaceX's Starship launch site, like an invisible dam disrupting the flow of a river. In the Gulf of Mexico, the U.S. Coast Guard had been out in force well before dawn, preventing boats from entering a strict keep-out zone that had been marked for days. "Navigational hazards from rocket launching activity may include free falling debris and/or descending vehicles or vehicle components, under various means of control," it wrote in an ominous warning to mariners, who should remain "in a heightened state of awareness." Sheriff's deputies evacuated a nearby village and closed roads, and SpaceX issued warnings about sonic booms that could rattle windows for miles.

Ahead of the flight, SpaceX cautioned that "thousands of distinct vehicle and pad criteria must be met" before it could even consider the chopsticks grab. The decision to call *go for catch* would come down to the flight director. If he noticed anything amiss, SpaceX said, the booster would follow another path toward a splashdown in the Gulf of Mexico, far from land. "We accept no compromises when it comes to ensuring the safety of the public and our team, and the return will only be attempted if conditions are right," the company said.

As the countdown clock ticked down to liftoff on the morning of the launch, SpaceX engineer Kate Tice again sought to temper expectations. "We're only going to attempt to catch if everything on the booster and tower checks out after launch," she said during the company's live broadcast of the mission.

On its descent, the Super Heavy booster, as it was called, would gather speed quickly, topping out at about 2,700 miles per hour, or three and a half times the speed of sound. If something went wrong and it suddenly veered off course, the rocket's onboard emergency detonation system would blow it up, preventing it from crashing into Brownsville, population 190,000, or the nearby towns of South Padre Island and Port Isabel. But even a preemptive explosion would result in a shrapnel-spewing fireball, raining jagged chucks of molten steel into the Gulf of Mexico.

At SpaceX headquarters outside of Los Angeles, throngs of employees chanted the final seconds of the countdown in unison—*five, four, three, two, one . . .* —and watched as the engines began to ignite. The rocket was so big, however, that it took a full five seconds for it to begin to move off the pad, and then did so slowly, inching its way skyward in I-think-I-can slow motion.

Soon enough, the might of Starship's engines was winning the tug-of-war with gravity, spouting a tail of flame a thousand feet long, twice the length of the booster. Chatter picked up among the headsetted engineers.

"Vehicle is pitching downrange."

"Max Q." The rocket was enduring maximum dynamic pressure, or peak stress, as it was picking up speed, while also still fighting gravity before leaving the atmosphere.

"Stage separation."

The Starship spacecraft was now flying independently, on a trajectory that would take it more than halfway across the globe for a splashdown in the Indian Ocean.

"Booster boost-back burn." The booster stage was flipping to reorient itself back toward land and firing its engines, essentially going in reverse, back the way it came.

The question now was: Let it fall harmlessly in the Gulf, or fly it back to the landing site?

"Thirty seconds 'til catch decision."

"Confirming," the flight director responded. "Thirty seconds to make that decision."

The computer readouts in mission control showed the engines were performing well. Power and telemetry were "nominal," a word that in the space parlance means everything is A-OK. Despite enduring the punishing force of liftoff just a few minutes earlier, the tower, with its mechanical arms, was also ready for a catch. The flight director checked with his team, arranged in rows before their computers studying the engine pressure, the avionics, the flow of propellant.

Nominal. Nominal. Nominal.

"Ten seconds 'til catch decision."

If the flight director was going to issue the command, he had to do it now.

One second passed. Then two. Three.

"Flight director is go for booster return."

"Operator go-vote is set true. We are go for catch."

Minutes later, the booster descended softly under a pillow of fire, into the arms of its launch site.

As pandemonium broke out at SpaceX's Hawthorne office, the spacecraft continued its journey—flying across the Atlantic, over southern Africa, and then descending toward its target in the Indian Ocean off the coast of Australia. As it plummeted horizontally through the atmosphere, plasma built up around the heat shield. Temperatures reached 2,600 degrees Fahrenheit, but the tiles held, giving hope that one day Starship would carry people, first to Earth orbit and eventually the moon and Mars that would recall another era of progress and dreams, when a stoic astronaut named Neil Armstrong steered his spacecraft to the lunar surface and proclaimed, "The Eagle has landed."

In the darkness over the Indian Ocean, a camera that SpaceX had positioned on a buoy at the spacecraft's intended landing site captured the final moments of Starship's descent. The ship flipped itself to vertical and, firing its engines to slow itself to a hover, made a soft touchdown in a dark, rippling sea so magnificently desolate that for a moment it seemed like the surface of some distant planet.

The callout came from mission control: "Starship has landed."

THREE MONTHS LATER, in January 2025, Bezos was showing me his own mission control room—row after row of flight controllers and engineers in front of their computer monitors. "This is where it all happens," he said. Wearing a tight-fitting black Blue Origin polo and jeans, he was excited but anxious. After years of delays, Blue Origin's New Glenn rocket was standing on its launch pad, ready to fly. In his first year at the company, Dave Limp had pushed hard to get New Glenn to orbit. It seemed the new CEO spent more time at the company's engine manufacturing facility in Huntsville, Alabama, and its facilities at Cape Canaveral in Florida, than at its headquarters outside of Seattle. Blue was still far from perfect, but it was starting to flow, catching a rhythm, with actual hardware beginning to come off the assembly line with regularity. Under Limp's leadership, the company was now cranking out one BE-4 engine a week, he told me, a dramatic increase in production. Workers were being pushed in ways they hadn't been before, working twelve-hour shifts or longer. And, leading up to the launch, engineers were on call at all times—sleep, personal life, everything else be damned, several employees, exhausted but exhilarated, told me.

"I wish I could say there's a magic wand with one simple solution here, but it is a lot of things," Limp told me while sitting with Bezos in Blue's Cape Canaveral manufacturing site, overlooking the factory floor, shortly before New Glenn's maiden flight attempt. When I had visited the factory in 2021, it was clean and quiet. Now it hummed. Below us, engineers were working on the next two rocket boosters, Limp said, pointing them out, as well as eight second stages. Productivity was leading to progress. "One of the amazing things about once you start getting to rate is that it has a flywheel effect," he said. "You've built tooling to make it easier. You've made the design easier, and so I think we've done a lot of that."

Bezos, seated next to him, jumped in. "We are driving a more decisive company. It's happening. I can see it. You know, the main reason I left the CEO-ship at Amazon was so I could come focus full-time on Blue Origin. Dave is driving extraordinary urgency. Dave's right, of course, there are a hundred factors in improving the rate of manufacturing, but culturally, you just see, I think, a shift in terms of decisiveness. You can talk about a problem for a long time, or you can fix the problem."

But how do you infuse that into the company? I asked him.

"One step at a time," he said. "By having a physical presence and by talking to people and saying, 'Okay, why did we take the day to make the decision that could have been made in five minutes.'" Limp helped bring "cultural attributes of urgency. And, you know, I tend to bring energy to things," he said, letting out one of his cackles for emphasis.

Blue Origin had been working on flying New Glenn for more than a decade. Now it was time to fly. With its ability to hoist some fifty tons to low Earth orbit, New Glenn was classified as a "heavy-lift rocket." It was powered by seven first-stage engines burning methane and liquid oxygen. If it worked, great. If it didn't, "we'll pick ourselves up and do it again," Bezos said. Pointing to the rockets in production below him on the factory floor, he said, "We'll be ready to fly again." New Glenn represented "a giant step" for Blue, Bezos told me. All the company's ambitions—flying satellites for commercial customers and NASA, getting its Blue Moon lander to the lunar surface—depended on a successful test flight. The previous June, Blue Origin had been chosen by the Pentagon to fly national security missions along with SpaceX and the United Launch Alliance, allowing it to compete for contracts worth many billions of dollars. It was a long-awaited coup for the company, but first it had to prove to the U.S. Space Force that it could reach orbit successfully. "They all are counting on us to deliver New Glenn and make it an operable vehicle," Bezos said. "So, yeah, it's a giant moment for sure."

The industry needed Blue to be successful if for no other reason than to finally give SpaceX some competition. SpaceX had gotten so big and successful that it had outpaced everyone, leaving the government with few other options. When the Pentagon announced the winner for the first batch of its next launch contracts in late 2024, SpaceX won them all, beating out the United Launch Alliance for a haul worth $733 million. Starlink was also sopping up billions from the Pentagon and National Reconnaissance Office, while Amazon's Kuiper satellite network was still nowhere near operational. Given the continued troubles with Boeing's Starliner capsule, which had flown NASA astronauts Suni Williams and Barry "Butch" Wilmore to the space station, the space agency would be forced to rely on SpaceX to fly them home in March 2025 after they had been stuck in space for nine months.

A successful New Glenn launch would represent hope that the nation's spacefaring ambitions didn't have to be placed largely into the hands of Musk, whose behavior was increasingly making officials nervous despite SpaceX's unprecedented success. But the initial flights of rockets often experienced setbacks. SpaceX's Falcon 1 didn't reach orbit until the fourth try. Blue tried to manage expectations, but there was a lot of pressure to get New Glenn in the air.

In a statement before the flight, Jarrett Jones, a Blue Origin senior vice president, said that the company had "prepared rigorously for it.

But no amount of ground testing or mission simulations are a replacement for flying this rocket."

Privately, at least some of the company's engineers put it another way, using an acronym that embodied the company's new go-for-it culture: "FiFi." *Fuck it. Fly it.*

NEW GLENN LIFTED off on January 16 at 2:03 A.M., an hour designated by the Federal Aviation Administration so that it wouldn't disrupt airline traffic. Right on cue, all seven engines ignited, lighting up the sky like a Phoenix rising. Bezos wore a headset allowing him to listen in to his engineers calling out one successful milestone after another.

"All seven engines at full thrust."

"Vehicle has cleared the tower."

"Engine chamber pressures look good."

"Data quality looks good."

"Passing Mach One. New Glenn is now supersonic."

At three minutes and ten seconds, right on schedule, the main engines cut off. Two seconds later, the first and second stages separated. Eleven seconds after that, the two second-stage engines ignited, propelling the stage to orbit. The booster, which was designed to be reusable, meanwhile headed back to Earth. The company was going to try to catch it on a remotely operated ship in the Atlantic Ocean, several hundred miles off the Florida coast. A successful landing was, as the company repeatedly said, a secondary objective to reaching orbit. "It's probably a little crazy to attempt to land the booster on the first flight," Bezos had told me. No one had ever landed a rocket on the first try before, so Blue had painted the words SO YOU'RE SAYING THERE'S A CHANCE on the side of the rocket, a cheeky nod to the comedy *Dumb and Dumber.*

At seven minutes and seventeen seconds, three of the booster's engines relit to slow it down as it reentered the atmosphere. But as it continued to fall, the telemetry from the rocket cut off. The booster was lost. It didn't matter. The company would try again on the next flight. Shortly after liftoff, Limp congratulated the team in a company-wide email that made it clear the pace of progress had to continue. Blue had finally reached orbit, but it still had a lot to prove.

"We are already looking ahead to our next launch in the spring," he wrote. "It goes without saying building booster and second stages are critical to our success. Let's double down and show everyone how quickly we can get back to flight. For now, celebrate this moment and the success we achieved in reaching orbit."

The company had envisioned a next-generation rocket, called New Armstrong. When I asked Bezos about it, he demurred, saying he wasn't ready to talk about that. "But we are working on a vehicle that will come after New Glenn and lift more mass," he said. When I asked him if it would be fully reusable, like Starship, he said he had created a competition within the company, pitting two teams—one building a "low-cost, high-rate manufactured expendable upper stage," the other a "highly operable reusable second stage"—against each other to see which would win. "This is the very beginning," he added. "This is the beginning of a new golden age of space."

For now, he was happy to celebrate the successful flight. Bezos eventually emerged from the control room pumping his fists in the air as he greeted his guests, who had been drinking pink champagne from mini-bottles. Later, Bezos would celebrate with his fiancée, Lauren Sánchez, eating a late breakfast—pancakes, eggs over medium, and a sausage patty—at the nearby Country Cookin' Diner, a greasy spoon in a run-down strip mall next to a Pizza Hut.

Even Musk was impressed. "Congratulations on reaching orbit on the first attempt! @jeffbezos" he wrote on X, his social media platform.

"Thank you," Bezos responded, adding an emoji of gratitude.

Musk and SpaceX were preparing for the next Starship test flight, scheduled for later that day. Now it was Bezos's turn to demonstrate goodwill on Musk's X. "Good luck today @elonmusk and the whole SpaceX team!!" he wrote.

"Much appreciated!!" Musk responded.

He followed up twenty minutes later, posting with a meme from the movie *Step Brothers*, a 2008 comedy in which Will Ferrell and John C.

Reilly play a pair of immature adults living at home whose parents marry. "Did we just become best friends?" Ferrell's character asked Reilly's character in Musk's meme.

FRIENDS MUSK AND Bezos were not, but they were at least now rivals again, circling in similar orbits that, for the moment, had President Trump at their center. When Bezos went to have dinner with Trump at Mar-a-Lago in December 2024 shortly after Trump had won the presidency for the second time, Musk joined them as dinner was underway. Later, Musk and Bezos chatted amicably at a black-tie inauguration ball on the eve of Trump returning to the White House.

Bezos was aware, of course, that Trump had little interest in the moon. It was, to Trump's New York real estate mind, the low-rent district. Mars was Fifth Avenue. Mars was, of course, Musk's goal as well. And his influence over the president was clear. During his inaugural address in January 2025, Trump had declared that the United States would "pursue our manifest destiny into the stars, launching American astronauts to plant the Stars and Stripes on the planet Mars."

Standing behind him in the Capitol Rotunda, amid the cabinet secretaries and billionaire businessmen—including Bezos—Musk was quick to join the standing ovation, pumping his fists and giving two thumbs up. Musk would soon install himself in the White House as a top Trump adviser, who oversaw the Department of Government Efficiency's blitz through the federal bureaucracy as it sought to cut billions from the federal budget. Given Musk's business interests, which over twenty years had taken in at least $38 billion of taxpayer money in the form of contracts, loans, subsidies, and tax credits, according to an analysis by *The Washington Post,* the position was rife with conflicts of interest. One of the most glaring, however, was one that many in the space community would cheer. Musk's slashing of government programs could lead to the canceling of NASA's SLS rocket, which was billions of dollars over budget, cost $2 billion a launch, and relied on older technology that wasn't reusable. Even long-time supporters, such as Scott Pace, the head of the National Space Council during Trump's first term, said NASA needed "an off-ramp for reliance" on SLS. If SLS was no longer NASA's main heavy-lift rocket, that would directly benefit SpaceX with its Starship rocket. And now that New Glenn had flown, it could help Blue Origin as well.

Bezos had been quick to appease the Trump White House during the first term. After then–Vice President Pence's chief of staff told Blue Origin executives that they had a *Washington Post* problem, the company took out an ad in Bezos's paper, refuting its editorial board's criticism of the administration's space policy. Now, it seemed to many, Bezos was willing to go even further to ingratiate himself. The *Post* opinion pages would write in support of personal liberties and free markets, Bezos declared, earning praise from Trump: "Jeff Bezos is trying to do a real job with *The Washington Post,* and that wasn't happening before." Not only did Amazon pay $1 million on Trump's inauguration, it spent $40 million for the rights to Melania Trump's documentary, the most it had ever spent on a documentary and more than three times the offer of the next-highest bidder. Bezos had once said Trump "erodes our democracy." But after Trump won a second term, Bezos said he was optimistic this time around. Speaking about the man who had once called him "Jeff Bozo," he said Trump is "calmer than he was the first time—more confident, more settled."

There was only one way to account for the change in Bezos's tone, according to Marty Baron, who had praised Bezos's stewardship of *The Washington Post* while he was executive editor: "There is no doubt in my mind that he is doing this out of fear of the consequences for his other business interests, Amazon (the source of his wealth) and Blue Origin (which represents his lifelong passion for space exploration)," Baron wrote in a statement to media organizations.

In a blistering essay published in *The Atlantic,* he charged that Bezos had now faltered badly and had served as Trump's showpiece at the inauguration. Comparing Bezos to the fearless Katharine Graham, who published the Pentagon Papers and the Watergate coverage as publisher of *The Post*, Baron wrote: "Now we know that Bezos is no Katharine Graham. It has been sad and unnerving to watch Bezos fall so terribly short of her standard as he confronts the return of Donald Trump to the White House."

Trump seemed to be having an immediate effect on Blue Origin. For years, it had been focused on the moon, not Mars. But if Musk and Trump were now going to make Mars the destination, Blue Origin would compete to join that effort. "I think we should do both," Bezos told me during our interview at Blue Origin's Cape Canaveral factory. "But I wouldn't want to see us start and stop. We should continue, and we should do the moon, and we should do Mars." The Blue Moon lander, he said, could be converted into a vehicle for Mars as well. "Very little has to be done to achieve that," he said. In other words, Blue Moon could become Blue Mars.

If he was optimistic about Trump's second administration, he was also optimistic about the future of Blue Origin and space exploration. Before the New Glenn flight, when I asked him about what the next ten years held for the company, Bezos told me that it would have to get better at flying New Glenn at "a very high rate." "We will have landed people on the moon," he said. The company would also have begun doing some of the work to allow for a long-term presence there: "in-situ development, producing oxygen, producing hydrogen, maybe producing some early solar cells."

"There are going to be many successful companies in this industry," he said. "SpaceX is going to be successful. Blue Origin is going to be successful. There's somebody founding a company right now that doesn't even exist yet. And that's also going to be very successful in space. That's what's coming." Now, it seemed that success—and keeping pace with Musk—would come not just through engineering but politics as well.

In the months to follow, SpaceX would continue to sprint forward. Starship, already the largest rocket in the world, would grow taller and able to hold even more propellant, making it even more powerful. There would be setbacks, as well, more explosions, more shrapnel raining down from the sky, more FAA investigations. But Musk was already looking ahead to the next Starship, Version 3, which would be ready within a year, he told me in a message in early 2025. It was a design that he said "really converges to greatness."

While he had been gracious toward Bezos for reaching orbit, Blue Origin still had a long way to go before catching SpaceX. "I think probably Blue Origin needs a higher sense of urgency and that's up to Jeff," Musk told me during an interview in May 2025 at SpaceX's headquarters, which had been relocated to Boca Chica in South Texas and was now dubbed Starbase. "Jeff Bezos is an extremely competent, smart human being. And so the more time he applies to Blue Origin, the better the company will be."

I asked him what he thought of Bezos's attempts to curry favor with Trump.

"The president can't improve the sense of urgency of Blue Origin," he said. While the company has "talented people," Musk said, the only person who could drive change was Bezos. At SpaceX, Musk had forged a culture by what he called his "maniacal sense of urgency."

"I'm just wired that way and that's the kind of mindset that I've tried to instill in the people at SpaceX," he said. "You have got to drive hard, and not everyone is cut out for that. Like, people want to have the chill vibes and SpaceX is ultra-hardcore. But if we're not ultra-hardcore, how are we going to get to Mars?"

The SpaceX team was operating at "Caesar 13th legion level," he said referring to Legio XIII Gemina, the unit Julius Caesar commanded across the Rubicon in 49 B.C. "Not saying I'm Caesar," he said, breaking into a mischievous cackle with an arched eyebrow and a Dr. Evil gesture from the film *Austin Powers*. "But maybe a little?"

A modern-day Caesar. It was an interesting analogy. Musk had become the wealthiest person in the world, who, if not an emperor, certainly lorded over an empire. His enormous power, wielded with an indiscriminate cudgel, was prompting investigations from Democrats in Congress, obsessive-compulsive coverage from the Washington press corps, backlash from voters and consumers who were mustering into what he perceived as a growing army of enemies eager to see him fall.

If so, he was armed and prepared to keep fighting.

"The 13th legion," he said, "they're going to win."

SO, THE MOON? Mars?

In some ways, the destination didn't matter as much as the means of ascent. NASA had spent years developing its SLS rocket and Orion spacecraft, which had flown together once. But the space agency was now investing in new reusable vehicles capable of being refueled in orbit. For the first fifty years of space exploration, the United States had been focused on mastering the art of getting *to* space. Now in the new space age, the challenge was developing the technology that would allow movement *through* space—building an enduring presence on the moon, on Mars, and the points between. Musk and SpaceX had always traveled at light speed, but now Bezos and Blue Origin had discovered their own warp drive. They were pushing each other, accelerating like a pair of spacecraft, each in their own orbit, until the liberating force of inertia took over and carried them to a point in the cosmos so remote and dark, it was brilliant.

Plant the Flag

A crew of American astronauts with the first woman among them did not land on the moon in 2024. That goal, as many knew when then–Vice President Mike Pence first declared it, was never feasible. After the Artemis I mission sent the Orion spacecraft around the moon without anyone on board and splashed down in December 2022, NASA had spent about two years studying why the heat shield ended up looking like a lunatic had taken a pickaxe to it. In late 2024, the space agency finally announced that the damage was caused by the way the spacecraft returned to Earth. Instead of diving right into the atmosphere, it performed a skip reentry, dipping into the atmosphere to slow itself down, then popping back out like a stone skipping across a pond, before plunging back in. But as the spacecraft flew in and out of the atmosphere, the heat generated gases inside the heat shield that became trapped. With nowhere to escape, they ended up cracking the heat shield in ways NASA did not predict.

Space Agency leaders made it clear that had there been a crew on board the Artemis I flight, "they would have been safe." But NASA's inspector general, a government watchdog, warned that a faulty heat shield could lead to a repeat of the Columbia disaster. "Should the same issue occur on future Artemis missions," said the IG's report, "it could lead to the loss the of the vehicle or crew."

As a result, NASA announced that it would have to modify the re-entry trajectory for Artemis II, when four astronauts would fly around the moon. That meant yet another delay—with a human landing on the surface (Artemis III) coming no sooner than the middle of 2027, NASA administrator Bill Nelson said. Despite the setback, the revised schedule "will be well ahead of the Chinese government's announced intention" of sending people to the moon by 2030, he promised. But whether or not the United States returned astronauts to the lunar surface before China was no longer his problem. President Trump would soon pick a new administrator; Nelson's time at the space agency was coming to an end.

Despite the delays, the Artemis program had made some progress. A series of uncrewed robotic landers were starting to fly to the moon in scouting missions that would pave the way for human missions. The first of those landings came in February 2024, when a fourteen-foot-tall spacecraft named Odysseus touched down on the lunar surface. One of the spacecraft's spidery legs broke when it hit the surface harder than expected, however, and photos showed the Odysseus listing side-ways at a 30-degree angle, its partially severed leg dangling. But the ship was intact and generating power, a coup for Intuitive Machines, the company that designed and operated it, which could claim the first touchdown of an American spacecraft on the moon since the last of the Apollo missions in 1972; also, it was the first time a commercial venture had reached the lunar surface.

The spacecraft carried a suite of scientific instruments, including six for NASA. It also carried an American flag that had been destined for the moon half a century earlier but instead was catalogued, ware-housed, and largely forgotten. The flag had been designed to be affixed to the left shoulder of a space suit for one of the last three Apollo missions—Apollo 18, 19, or 20. When those were canceled, the flag was sealed in an airtight bag, dated March 2, 1970, and placed in a stor-age room in Building 7 at NASA's Johnson Space Center in Houston.

As Intuitive Machines geared up for its flight, Gary Spexarth, the company's lunar lander production manager, reached out to Mark Schaefbauer, his former colleague at Johnson Space Center, who was now working at a company that made a special kind of cloth used in space suits. Intuitive Machines wanted to install a bit of the cloth on the outside of the spacecraft and adorn it with NASA and Intuitive Machines logos. As they spoke, Schaefbauer suddenly remembered the Apollo-era flag stored in the plastic bag. What if they attached that as well? "How cool would it be for an American flag that was printed in 1970, and made for the moon, to be finally able to get up to the moon?" he said.

Odysseus's landing might have been rough, but as the dust settled, an onboard camera beamed back an image of the spacecraft with the red, white, and blue clearly visible against the grainy, gray surface of the moon.

FOUR MONTHS LATER, on June 1, another spacecraft touched down. China's Chang'e-6 landed at Aitken Basin, the four-billion-year-old impact crater near the South Pole on the moon's far side, marking China's second landing on the part of the moon that perpetually points away from Earth. The spacecraft carried a robotic arm with a scooper and drill that retrieved 4.2 pounds of rocks and regolith that China flew back to Earth. Scientists would be able to study a part of the moon that promised to yield significant insight into the formation of the solar system.

Chang'e-6 also carried a surprise. In 2020, China had deployed a flag on the moon made of polymer fiber that was designed to survive in the harsh environment of space. Now, the country's space agency wanted to take the technology a step further. This time, it developed a flag out of basalt, a type of volcanic rock abundant on the lunar surface, as well as on Earth. China's scientists worked to melt the basalt in a furnace, turning it to lava at some 2,700 degrees Fahrenheit. Then they drew out ultra-thin threads of fiber one-third the diameter of a human hair, which they spun into cloth. It took years, but they were finally able to perfect the technique, and eventually had a small Chinese flag, about twelve by eight inches, that would be able to withstand the moon's extreme environment.

The far-side, sample-return mission was heralded as another success for China's space ambitions. NASA and congressional leaders were in awe, but concerned about what it might portend. China now had not one, but two flags on the lunar surface, including one conspicuously planted in the region of the South Pole. Congressman Brian Babin, a Republican from Texas and the chairman of the House Science Committee, said his fear was that "when U.S. astronauts return to the moon, they will find a No Trespassing sign written in Mandarin."

AS TRUMP BEGAN his second term, his administration vowed it would not come in second place. "We're going to win the modern space race," Mike Waltz, Trump's first pick for national security adviser, pledged. Jared Isaacman, the billionaire private astronaut who became Trump's intitial choice for NASA administrator, promised that "Americans will walk on the Moon and Mars," and that the United States would "never settle for second place."

Of course, it wasn't that simple. This wasn't John F. Kennedy proclaiming the United States would put a man on the moon before the decade was out "because it is hard." The contours of the new space race were dynamic and shifting. It was a race between nations and companies and billionaires; of ideals and values; of military superiority; of prestige and science, but also of dominion and the lucrative economies that could be generated by the resources of space. There was not just one destination but many—the increasingly congested real estate in low Earth orbit; the South Pole of the moon, the goal of the first Trump term; and now Mars, the Musk-inspired goal of the second. NASA had said that the next person to walk on the lunar surface would be a woman—someone like Jessica Meir, a scientist who went by the non-military rank of doctor and would not only be the first woman on the moon, but the first mother. In 2023, she announced the upcoming birth of her daughter on social media by posting photos of herself pregnant. "This is what an astronaut looks like," she wrote.

They also looked like Wang Yaping, a former colonel in the Chinese Air Force and a veteran of two spaceflight missions, including a six-month stay on Tiangong, the Chinese space station. Like Meir, she was a mother, who in an interview after her second flight in 2022 had a message for her young daughter: "I am back with all the stars collected for you."

Here were the next generation of astronauts, lifting off in modern rockets and spacecraft in a new race to distant worlds. The first woman would carry a new flag, built to endure for generations in the low-angled light of the South Pole of the moon, and perhaps even eventually on Mars. The once blurry path was coming into focus. The only questions were: Which flag would she plant, and what would be her name?

NOTES

INTRODUCTION

2 **China has paid SpaceX:** Andrew Jones, "China Unveils Fully Reusable Starship-Like Rocket Concept," *SpaceNews,* November 14, 2024, https://spacenews.com/china-unveils-fully-reusable-starship-like-rocket-concept/.

3 **Assembling the International Space Station:** NASA, "Station Facts," https://www.nasa.gov/international-space-station/space-station-facts-and-figures/.

3 **"You need to be looking":** Elon Musk, "Making Life Multiplanetary," September 28, 2017, https://www.spacex.com/media/making_life_multiplanetary_transcript_2017.pdf.

6 **"we boldly go":** Neil deGrasse Tyson, X, https://x.com/neiltyson/status/538433684169179137.

6 **could be worth $1 trillion:** "Space: The $1.8 Trillion Opportunity for Global Economic Growth," World Economic Forum and McKinsey & Company, April 8, 2024, https://www.weforum.org/publications/space-the-1-8-trillion-opportunity-for-global-economic-growth/; see also "Creating Space," Morgan Stanley, https://www.morganstanley.com/Themes/global-space-economy#:~:text=Morgan%20Stanley%20believes%20that%20everyone,more%20accessible%20for%20female%20astronauts.

7 **"build the road to space":** Anmar Frangoul, "Jeff Bezos Calls for a Dynamic, Entrepreneurial Boom in Space," CNBC, September 6, 2017, https://www.cnbc.com/2017/09/06/jeff-bezos-blue-origin-calls-for-a-dynamic-entrepreneurial-space.html.

7 **China intends to land:** Andrew Jones, "How China Plans to Put Astronauts on the Moon by 2030," Space.com, November 27, 2024, https://www.space.com/the-universe/moon/how-china-plans-to-put-astronauts-on-the-moon-by-2030-video.

7 **the second country:** Katerina Ang, "China Becomes Second Country to Drive a

Rover on Mars," *The Washington Post,* May 22, 2021, https://www.washingtonpost
.com/world/2021/05/22/china-successfully-deploys-rover-explore-mars/.

CHAPTER 1

11 **"It's a huge honor":** "Oscar Photos: Executives (and Their Dates) on the Red Carpet," *Hollywood Reporter,* February 26, 2017, https://www.hollywoodreporter.com
/gallery/oscar-photos-executives-dates-red-carpet-980543/.

13 **outbursts that could quickly turn cruel:** Brad Stone, *The Everything Store:
Jeff Bezos and the Age of Amazon* (Little, Brown and Company, 2013), chap. 6,
Kindle.

15 **"must recognize that space":** Robert S. Walker and Peter Navarro, "Trump's Space
Policy Reaches for Mars and the Stars," *SpaceNews,* October 19, 2016, https://
spacenews.com/trumps-space-policy-reaches-for-mars-and-the-stars/.

16 **This time, a growing number:** European Space Policy Institute, "Emerging Spacefaring Nations," June 2021, https://www.espi.or.at/wp-content/uploads/2022/06
/ESPI-Report-79-Emerging-Spacefaring-Nations-Executive-Summary.pdf.

16 **there were three essential competitions:** Todd Harrison, "It's Not a Space Race, It's
the Space Olympics," American Enterprise Institute, July 31, 2024, https://www.aei
.org/research-products/working-paper/its-not-a-space-race-its-the-space
-olympics/.

17 **It later emerged:** Andrew Jacobs, "In Leaked Lecture, Details of China's News
Cleanups," *The New York Times,* June 3, 2010, https://www.nytimes.com/2010/06
/04/world/asia/04china.html.

17 **China's space agency later published:** Information Office of the State Council of
the People's Republic of China, "China's Space Activities in 2011," December 29,
2011, http://www.china.org.cn/government/whitepaper/node_7145648.htm.

18 **more than three thousand pieces of debris:** Secure World Foundation, "2007 Chinese Anti-Satellite Test Fact Sheet," November 23, 2010, https://swfound.org
/media/9550/chinese_asat_fact_sheet_updated_2012.pdf.

18 **"We have considered space a sanctuary":** Christian Davenport, "The Fight to Protect the Most Valuable Real Estate in Space," *The Washington Post,* May 9, 2016,
https://www.washingtonpost.com/business/economy/a-fight-to-protect-the-most
-valuable-real-estate-in-space/2016/05/09/df590af2-1144-11e6-8967
-7ac733c56f12_story.html.

18 **"The dream of the Chinese people":** Chris Buckley, "As Rover Lands, China Joins
Moon Club," *The New York Times,* December 14, 2013, https://www.nytimes.com
/2013/12/15/world/asia/china-lands-probe-on-the-moon-report-says.html.

19 **One official in the department:** David Shukman, "Why China Is Fixated on the
Moon," BBC, November 29, 2013, https://www.bbc.com/news/25141597.

19 **A 2011 study:** Harrison Schmitt, "Lunar Helium-3 Fusion Resource Distribution,"
University of Wisconsin–Madison, May 2, 2011, https://archive.org/details
/LunarHelium-3FusionResourceDistribution.

19 **Just 100 kilograms:** "What Is a Megawatt," National Resources Council, February 24, 2012, https://www.nrc.gov/docs/ML1209/ML120960701.pdf.

19 **In 2015, Congress passed:** Christian Davenport, "A Dollar Can't Buy You a Cup of
Coffee, but That's What NASA Intends to Pay for Some Moon Rocks," *The Washington Post,* https://www.washingtonpost.com/technology/2020/12/03/moon
-mining-contracts-named/.

21 **On December 14, 2016:** David Streitfeld, "'I'm Here to Help,' Trump Tells Tech

Executives at Meeting," *The New York Times*, December 14, 2016, https://www
.nytimes.com/2016/12/14/technology/trump-tech-summit.html.

21 **That morning, he had met:** Walter Isaacson, *Elon Musk* (Simon & Schuster, 2023),
chap. 44, Kindle.

22 **"erodes our democracy":** "Jeff Bezos: Donald Trump Comments Erode Our
Democracy," CNBC, October 20, 2016, https://www.youtube.com/watch
?v=mXx4SVvISqk.

22 **"I found today's meeting":** Anita Balakrishnan, "Trump Meets with Tech Titans
as Bezos Lauds 'Productive' Session," CNBC, December 14, 2016, https://www
.cnbc.com/2016/12/14/trump-to-tech-leaders-no-formal-chain-of-command-here
.html.

22 **"to get NASA going again":** Isaacson, *Elon Musk,* chap. 44, Kindle.

23 **"totally out of control":** Christian Davenport, "Citing High Cost, Trump Says
Boeing's Contract to Build Air Force One Should Be Canceled," *The Washington
Post,* December 6, 2016, https://www.washingtonpost.com/news/checkpoint/wp
/2016/12/06/citing-cost-overruns-trump-says-boeings-contract-to-build-air-force
-one-should-be-canceled/.

23 **In 2006, when it was scrambling:** Christian Davenport, "With a Spacecraft in
Trouble and the White House Watching, SpaceX Had to Deliver," *The Washington
Post,* March 16, 2018, https://www.washingtonpost.com/business/economy/with-a
-spacecraft-in-trouble-and-the-white-house-watching-spacex-had-to-deliver/2018
/03/15/553d89cc-2701-11e8-874b-d517e912f125_story.html.

24 **"If their companies' technology":** Mike Pence, *So Help Me God* (Simon & Schuster, 2022), chap. 41, Kindle.

25 **"I believe the dreamers":** Christian Davenport, "Jeff Bezos on Nuclear Reactors in
Space, the Lack of Bacon on Mars, and Humanity's Destiny in the Solar System,"
The Washington Post, September 15, 2016, https://www.washingtonpost.com/news
/the-switch/wp/2016/09/15/jeff-bezos-on-nuclear-reactors-in-space-the-lack-of
-bacon-on-mars-and-humanitys-destiny-in-the-solar-system/.

CHAPTER 2

26 **"Frankly, I think we":** Dan Leone, "Musk Calls Out Blue Origin, ULA for 'Phony
Blocking Tactic' on Shuttle Pad Lease," *SpaceNews,* September 25, 2013, https://
spacenews.com/37389musk-calls-out-blue-origin-ula-for-phony-blocking-tactic
-on-shuttle-pad/.

26 **Musk called it "ridiculous":** Christian Davenport, *The Space Barons: Elon Musk, Jeff
Bezos, and the Quest to Colonize the Cosmos* (PublicAffairs, 2018), chap. 11, Kindle.

27 **"The number one thing":** Interview with Amazon CEO Jeff Bezos, *The David
Rubenstein Show: Peer to Peer Conversations,* September 13, 2018, https://www
.youtube.com/watch?v=f3NBQcAqyu4.

28 **SpaceX had recently won:** Mike Gruss, "Orbital ATK, SpaceX Win Air Force Propulsion Contracts," *SpaceNews,* January 13, 2016, https://spacenews.com/orbital
-atk-spacex-win-air-force-propulsion-contracts/.

29 **They knew that Bezos:** Davenport, *The Space Barons,* chap. 11, Kindle.

31 **"We are going to continue":** Heather Kelly, "Jeff Bezos: Amazon Will Keep Working with the DoD," *CNN Business,* October 15, 2018, https://www.cnn.com/2018
/10/15/tech/jeff-bezos-wired/index.html.

32 **"During the transition":** Mike Pence, *So Help Me God* (Simon & Schuster, 2022),
chap. 41, Kindle.

36 **"Just a couple of hours ago":** "Jeff Bezos on the Importance of Expanding into the Solar System," *Aviation Week,* March 3, 2017, https://www.youtube.com/watch?v=tjz2vP3zPhE.

37 **But no photos allowed:** Zach Johnson, "No Photos Allowed! The Wildest Stories from Madonna's Legendary (and Super Exclusive) Oscar Parties," *E! News,* February 22, 2017, https://www.eonline.com/news/831140/no-photos-allowed -the-wildest-stories-from-madonna-s-legendary-and-super-exclusive-oscar -parties.

37 **The Material Girl set the tone:** Emily Smith, "Johnny Depp Parties at Madonna's Exclusive Post-Oscars Bash," *New York Post,* February 27, 2017, https://pagesix .com/2017/02/27/johnny-depp-parties-at-madonnas-exclusive-after-party/.

CHAPTER 3

38 **"My friends who want to move":** Jeff Foust, "The Cosmic Vision of Jeff Bezos," *SpaceNews,* March 5, 2019, https://spacenews.com/the-cosmic-vision-of-jeff-bezos/.

38 **"Sometimes my friends say":** Christian Davenport, "Jeff Bezos on Nuclear Reactors in Space, the Lack of Bacon on Mars, and Humanity's Destiny in the Solar System," *The Washington Post,* September 15, 2016, https://www.washingtonpost .com/news/the-switch/wp/2016/09/15/jeff-bezos-on-nuclear-reactors-in-space -the-lack-of-bacon-on-mars-and-humanitys-destiny-in-the-solar-system/.

39 **"Beauty in the office":** Gwynne Shotwell interview at the McCain Institute Sedona Forum, "Technology, Innovation and Our National Interests," May 6, 2023, https://www.mccaininstitute.org/the-sedona-forum/past-forums/the-sedona -forum-2023/.

39 **As a child growing up:** Peter Holley, "Elon Musk Has Been Missing Deadlines Since He Was a Kid," *The Washington Post,* June 6, 2018, https://www.washington post.com/news/innovations/wp/2018/06/06/elon-musk-has-been-missing-dead lines-since-he-was-a-kid/.

39 **"I do have, like, an issue":** Holley, "Elon Musk Has Been Missing Deadlines Since He Was a Kid."

40 **"At SpaceX, we specialize":** Interview with Tim Dodd, "The Everyday Astronaut," X, May 23, 2022, https://x.com/Erdayastronaut/status/1528810070645125120.

41 **At the time, the plan:** Christian Davenport, "Elon Musk on Mariachi Bands, *Zero-G* Games, and Why His Mars Plan Is Like 'Battlestar Galactica,'" *The Washington Post,* September 28, 2106, https://www.washingtonpost.com/news/the -switch/wp/2016/09/28/elon-musk-on-mariachi-bands-how-he-plans-to-make -space-travel-fun-and-why-his-mars-plan-is-like-battlestar-galactica/.

41 **But as Musk had laid out his design:** Christian Davenport, "Elon Musk Offers Glimpse of Plans to Deliver Humans to Mars," *The Washington Post,* September 27, 2016, https://www.washingtonpost.com/news/the-switch/wp/2016/09/27 /elon-musk-to-discuss-his-vision-for-how-he-plans-to-colonize-mars/.

43 **In 2005, SpaceX:** Davenport, *The Space Barons,* chap. 8, Kindle.

43 **"It only makes sense":** National Press Club Luncheon with Elon Musk, September 29, 2011, https://www.youtube.com/watch?v=xrVD3tcVWTY.

44 **A report from 1961:** "Joint Report on Facilities and Resources Required at Launch Site to Support NASA Manned Lunar Landing Program: Phase I," NASA, DOD, July 31, 1961.

47 **Disney also set up dummy corporations:** James Rufus Koren, "How Disney Used Shell Companies to Start Its Magic Kingdom," *Los Angeles Times,* April 9, 2016,

https://www.latimes.com/business/la-fi-disney-shell-companies-20160408-story
.html.

47 **less than $100 an acre:** Mark Andrews, "Disney Assembled Cast of Buyers to
Amass Land Stage for Kingdom," *Orlando Sentinel,* May 30, 1993, https://www
.orlandosentinel.com/1993/05/30/disney-assembled-cast-of-buyers-to-amass-land
-stage-for-kingdom/?clearUserState=true.

47 **Jeff Bezos did something similar:** Mylene Mangalindan, "Buzz in West Texas Is
About Jeff Bezos and His Launch Site," *The Wall Street Journal,* November 10,
2006, https://www.wsj.com/articles/SB116312683235519444.

47 **These were linked:** Davenport, *The Space Barons,* chap. 1, Kindle.

47 **He had impressed Musk:** Davenport, *The Space Barons,* chap. 10, Kindle.

47 **Later, he'd be dispatched:** Ken Thomas, Brian Schwartz, and Becky Peterson, "The
Musk Deputy Running DOGE's Cost-Cutting Drive," *The Wall Street Journal,* February 10, 2025, https://www.wsj.com/politics/policy/steve-davis-elon-musk-cost
-cutting-cc1dc7c9?mod=Searchresults_pos1&page=1.

48 **"But I would say Texas":** Eric Berger, "Musk Tells Legislators What a Spaceport
Would Do for Texas," *Houston Chronicle,* March 8, 2013, https://www.houston
chronicle.com/news/houston-texas/houston/article/musk-tells-legislators-what-a
-spaceport-would-do-4340776.php.

49 **"This is money from the heavens":** Laura B. Martinez, "City Backs SpaceX Proposal," *The Brownsville Herald,* May 4, 2012.

49 **"pump $85 million in capital":** Rick Perry, press release, "Gov. Perry Announces
State Incentives Bringing SpaceX Commercial Launch Facility, 300 Jobs to the
Brownsville Area," August 4, 2014, https://spacenews.com/gov-perry-announces
-state-incentives-bringing-spacex-commercial-launch-facility-300-jobs-to-the
-brownsville-area/.

50 **"pioneer heritage":** Rick Perry, press release, "Gov. Perry Helps Break Ground on
SpaceX Commercial Launch Facility, September 22, 2014, https://spacenews.com
/gov-perry-helps-break-ground-on-spacex-commercial-launch-facility/.

50 **"is to create the technology":** Bobby Blanchard, "SpaceX CEO Says Brownsville
Facility Might Land First Human on Mars," *The Texas Tribune,* September 22,
2014, https://www.texastribune.org/2014/09/22/brownsville-spacex-facility-might
-land-first-mars/.

51 **310,000 cubic yards:** Tiffany Huertas, "SpaceX Working to Stabilize Land at Rocket
Launch Site," ValleyCentral.com, February 11, 2016, https://www.valleycentral.com
/news/local-news/spacex-working-to-stabilize-land-at-rocket-launch-site/.

51 **"Essentially, we want":** Elon Musk, "Making Life Multiplanetary," September 28,
2017, https://www.spacex.com/media/making_life_multiplanetary_transcript
_2017.pdf.

CHAPTER 4

54 **Jeff Bezos owns a compound:** Jordan Pandy and Madeline Berg, "Tom Brady Is
Open to Selling His New Miami Mansion. It's the Latest Sign of Jeff Bezos' Effect on
His Neighbors," *Business Insider,* January 24, 2025, https://www.businessinsider.com
/tom-brady-open-selling-indian-creek-mansion-near-jeff-bezos-2025-1.

54 **He owns an $80 million penthouse:** Chris DiLella, "Exclusive: See Inside Amazon
CEO Jeff Bezos' New $80 Million NYC Mega-Home," CNBC, June 5, 2019, https://
www.cnbc.com/2019/06/05/photos-amazon-ceo-jeff-bezos-new-multimillion
-dollar-nyc-penthouse.html.

54 **His mansion in Beverly Hills:** Katy McLaughlin and Katherine Clarke, "Jeff Bezos Buys David Geffen's Los Angeles Mansion for a Record $165 Million," *The Wall Street Journal,* February 12, 2020, https://www.wsj.com/articles/jeff-bezos-buys -david-geffens-los-angeles-mansion-for-a-record-165-million-11581542020.

54 **his $78 million waterfront estate:** Mary K. Jacob, "As Jeff Bezos Buys Up Maui, Hawaiian Locals Hope for the Best," *New York Post,* November 4, 2021, https:// nypost.com/2021/11/04/as-jeff-bezos-buys-up-maui-hawaiian-locals-hope-for -best/.

54 **His home in Washington:** Emily Heil and Kathy Orton, "Jeff Bezos Is the Anony- mous Buyer of the Biggest House in Washington," *The Washington Post,* Janu- ary 12, 2017, https://www.washingtonpost.com/news/reliable-source/wp/2017/01 /12/jeff-bezos-is-the-anonymous-buyer-of-the-biggest-house-in-washington/.

54 **$500 million yacht:** Kevin Koenig, "Jeff Bezos' New Yacht Is Finally Ready to Set Sail," *The New York Times,* May 19, 2023, https://www.nytimes.com/2023/05/19 /style/bezos-yacht-koru.html.

55 **"I had tears in my eyes":** Brad Stone, *Amazon Unbound: Jeff Bezos and the Inven- tion of a Global Empire* (Simon & Schuster, 2021), chap. 11, Kindle.

55 **"any astronauts on board":** Jeff Bezos, "First Developmental Test Flight of New Shepard," April 29, 2015, https://web.archive.org/web/20150430154042/https: /www.blueorigin.com/news/press_release/blue-origin-completes-acceptance -testing-of-be-3-engine-for-new-shepard-sub.

56 **"If I'm eighty years old":** Christian Davenport, "Jeff Bezos on Nuclear Reactors in Space, the Lack of Bacon on Mars, and Humanity's Destiny in the Solar System," *The Washington Post,* September 15, 2016, https://www.washingtonpost.com/news /the-switch/wp/2016/09/15/jeff-bezos-on-nuclear-reactors-in-space-the-lack-of -bacon-on-mars-and-humanitys-destiny-in-the-solar-system/.

57 **"The thing I'm most excited about":** Christian Davenport, "Jeff Bezos' Blue Origin Launch Company to Launch from Historic Pad at Space Coast," *The Washington Post,* September 15, 2015, https://www.washingtonpost.com/news/the-switch/wp /2015/09/15/jeff-bezoss-blue-origin-space-company-to-launch-from-historic-pad -at-space-coast/.

62 **"I suggest to the USA":** Jeffrey Kluger, "More Cosmic Sabre Rattling from Russia's Space Boss," *Time,* May 6, 2022, https://time.com/6174331/rogozin-threatens -space-station/.

62 **"neighborhood watch program":** Cheryl Pellerin, "Hyten: Deterrence in Space Means No War Will Be Fought There," Department of Defense News, January 26, 2017, https://www.defense.gov/News/News-Stories/Article/Article/1061833/hyten -deterrence-in-space-means-no-war-will-be-fought-there/.

62 **"send a message to the world":** Mike Gruss, "Space Surveillance Sats Pressed into Early Service," *SpaceNews,* September 18, 2015, https://spacenews.com/space -surveillance-sats-pressed-into-early-service/.

65 **"Our contribution to national security":** Alan Boyle, "Interview: Jeff Bezos Lays Out Blue Origin's Space Vision, from Tourism to Off-Planet Heavy Industry," *GeekWire,* April, 13, 2016, https://www.geekwire.com/2016/interview-jeff-bezos/.

67 **"First of all, when Elon":** Catherine Clifford, "President of SpaceX: This Is What It's Like Working for Elon Musk," CNBC, May 16, 2018, https://www.cnbc.com /2018/05/16/spacex-president-gwynne-shotwell-what-its-like-to-work-for-elon -musk.html.

CHAPTER 5

71 "By choosing this location": Vice President Pence Hosts National Space Council, Trump White House Archive, https://www.youtube.com/watch?v=4_izFqcZ67U.

72 "We will turn our attention": Christian Davenport, "Pence Vows America Will Return to the Moon. The History of Such Promises Suggests Otherwise," *The Washington Post,* October 11, 2017, https://www.washingtonpost.com/news /retropolis/wp/2017/10/10/presidents-love-evoking-jfks-iconic-moon-speech -now-its-the-trump-administrations-turn/.

72 "In the debate that followed": National Space Council Meeting, October 5, 2017, https://www.c-span.org/program/public-affairs-event/national-space-council -meeting/488466.

74 A truck driver in 2013: Chris Matyszczyk, "Truck Driver Has GPS Jammer, Accidentally Jams Newark Airport," CNET, August 11, 2013, https://www.cnet.com /culture/truck-driver-has-gps-jammer-accidentally-jams-newark-airport/.

75 the Cuban Missile Crisis: Sandra Erwin, "House Intelligence Chair Blasts White House Over Russia's Space Nuke Threat," *SpaceNews,* June 20, 2024, https:// spacenews.com/house-intelligence-chair-blasts-white-house-over-russias-space -nuke-threat/.

75 "our allies can openly discuss": Christian Davenport, Ellen Nakashima, Abigail Hauslohner, and Shane Harris, *The Washington Post,* "With a Dire Warning, Concerns Rise About Conflict In Space with Russia," February 15, 2024, https://www .washingtonpost.com/technology/2024/02/15/space-weapons-russia-china-starlink/.

75 In 1962, the United States: William Neff, Frank Hulley-Jones, and Joel Achenbach, "What Would Happen If Russia Detonated a Nuclear Bomb In Space?" *The Washington Post,* July 6, 2024, https://www.washingtonpost.com/technology /interactive/2024/nukes-space-explosion-nuclear-weapon-russia/.

76 "secret sibling": Aaron Bateman, *Weapons In Space: Technology, Politics, and the Rise and Fall of the Strategic Defense Initiative* (MIT Press, 2024).

76 By the early 1980s: Philip Taubman, "Secrecy of U.S. Reconnaissance Office Is Challenged," *The New York Times,* March 1, 1981, https://www.nytimes.com/1981 /03/01/us/secrecy-of-us-reconnaissance-office-is-challenged.html.

76 "the money had been hidden": Tim Weiner, "A Secret Agency's Secret Budgets Yield Lost Billions, Officials Say," *The New York Times,* January 30, 1996, https:// www.nytimes.com/1996/01/30/us/a-secret-agency-s-secret-budgets-yield-lost -billions-officials-say.html.

76 In 1987, U.S. Rep. George Brown Jr.: Barton Gellman, "Remember You Didn't Hear It Here," *The Washington Post,* September 18, 1992, https://www .washingtonpost.com/archive/politics/1992/09/19/remember-you-didnt-read-it -here/586ba0c1-e896-4adb-a4d0-f719a00dad0c/.

76 The satellites performed: Craig Whitlock and Barton Gellman, "To Hunt Osama bin Laden, Satellites Watched Over Abbottabad, Pakistan and Navy SEALS," *The Washington Post,* August 19, 2013, https://www.washingtonpost.com/world /national-security/to-hunt-osama-bin-laden-satellites-watched-over-abbottabad -pakistan-and-navy-seals/2013/08/29/8d32c1d6-10d5-11e3-b4cb-fd7ce041d814 _story.html.

77 "I'm not NASA": David Martin, "The Battle Above," *60 Minutes,* CBS News, April 26, 2015, https://www.cbsnews.com/news/rare-look-at-space-command -satellite-defense-60-minutes/.

77 "industrial age warfare": Christian Davenport, "The Fight to Protect the 'Most

Valuable Real Estate in Space,' " *The Washington Post,* May 9, 2016, https://www .washingtonpost.com/business/economy/a-fight-to-protect-the-most-valuable -real-estate-in-space/2016/05/09/df590af2-1144-11e6-8967-7ac733c56f12_story .html.

78 **"It is disturbing"**: Christian Davenport, "Some in Congress Are Pushing for a 'Space Corps,' Dedicated to Fighting Wars in the Cosmos," *The Washington Post,* September 15, 2017, https://www.washingtonpost.com/news/checkpoint/wp/2017 /09/15/some-in-congress-are-pushing-for-a-space-corps-dedicated-to-fighting -wars-in-the-cosmos/.

78 **"about showing leadership"**: Lee Billings, "Q&A: Plotting U.S. Space Policy with White House Adviser Scott Pace," *Scientific American,* November 6, 2017, https:// www.scientificamerican.com/article/q-a-plotting-u-s-space-policy-with-white -house-adviser-scott-pace/.

79 **"It's a one and done stunt"**: Loren Grush, "Everyone Hates NASA's Asteroid Capture Program," *Popular Science,* August 7, 2014, https://www.popsci.com/article /technology/everyone-hates-nasas-asteroid-capture-program/.

79 **"NASA's prioritization"**: Lee Billings, "NASA's Plan to Visit an Asteroid Faces a Rocky Start," *Scientific American,* November 10, 2014, https://www.scientific american.com/article/nasa-s-plan-to-visit-an-asteroid-faces-a-rocky-start/.

CHAPTER 6

83 **The tornado touched down**: Holly Bailey, *The Mercy of the Sky: The Story of a Tornado* (Viking, 2015).

83 **Later, meteorologists would**: Seth Borenstein, "Power of Moore Tornado Dwarfs Hiroshima Bomb," Associated Press, May 21, 2013, https://apnews.com/general -news-779e55edec9f466581043606bfd5f8ea.

85 **GPS Radio Occultation**: PlanetiQ, "What Is GPS-RO?," https://planetiq.com/gps -ro-101/#:~:text=GPS%20Radio%20Occultation%20(GPS%2DRO,modeling%20 and%20space%20weather%20prediction.

86 **American Space Renaissance Act**: Christian Davenport, "Why Congress's Newest Space Advocate Says the U.S. Faces a 'Sputnik Moment,' " *The Washington Post,* April, 12, 2016, https://www.washingtonpost.com/news/the-switch/wp/2016/04 /12/why-congress-newest-space-advocate-says-the-u-s-faces-a-sputnik-moment/.

87 **that had been Bill Nelson**: Emre Kelly, "Sen. Nelson at KSC: Space Coast 'Coming Alive,' Could See Two Rocket Launches a Day," *Florida Today,* August 9, 2017, https://www.floridatoday.com/story/tech/science/space/2017/08/09/bill-nelson -blue-origin-spacex-oneweb-nasa-making-space-coast-complex-america /553942001/.

87 **"kick the tires"**: Malcolm W. Browne, "Senator Who Aims High: Edwin Jacob Garn," *The New York Times,* April 13, 1985, https://www.nytimes.com/1985/04/13 /us/man-in-the-news-senator-who-aims-high-edwin-jacob-garn.html.

87 **The astronaut corps scoffed**: Christian Davenport, "Trump Campaign Pulls Ad About SpaceX Launch After Former Astronaut Calls It Political Propaganda," *The Washington Post,* June 5, 2020, https://www.washingtonpost.com/technology/2020 /06/05/trump-campaign-nasa-ad-pulled/.

87 **"Anybody who knows me"**: Robert Doherty, "Congressman Picked for Shuttle Ride: He Won't Have Far to Go to Launch Site, *Los Angeles Times,* November 17, 1985, https://www.latimes.com/archives/la-xpm-1985-11-17-mn-6979-story.html.

88 **the Challenger disaster**: John Uri, "35 Years Ago: Remembering Challenger and

Her Crew," NASA, Johnson Space Center January 28, 2021, https://www.nasa.gov
/history/35-years-ago-remembering-challenger-and-her-crew/.

88 **"This deep respect for NASA"**: U.S. Senate Committee on Commerce, Science and
Transportation, nomination hearing, November 1, 2017, https://www.commerce
.senate.gov/2017/11/nomination-hearing-11-01-2017.

89 **"The NASA administrator"**: Christian Davenport, "Trump's Nominee for NASA
Administrator Comes Under Fire at Senate Hearing," *The Washington Post,*
November 1, 2017, https://www.washingtonpost.com/news/the-switch/wp/2017
/11/01/trumps-nominee-for-nasa-administrator-comes-under-fire-at-senate
-hearing/.

90 **"devastating for the space program"**: Jeff Foust, "Bridenstine Faces Obstacles to
Senate Confirmation," *SpaceNews,* September 3, 2017, https://spacenews.com
/bridenstine-faces-obstacles-to-senate-confirmation/.

91 **Flake was placated**: Connor O'Brien, "Senate Advances Bridenstine to Lead
NASA," *Politico,* April 18, 2018, https://www.politico.com/story/2018/04/18
/jim-bridenstine-nasa-senate-492565.

91 **Bridenstine was confirmed**: Seung Min Kim and Christian Davenport, "Senate
Confirms Trump Pick for NASA Administrator Over Democratic Objections," *The
Washington Post,* April 19, 2018, https://www.washingtonpost.com/politics/senate
-confirms-trump-pick-as-nasa-administrator-over-democratic-objections/2018
/04/19/58692c6a-43f2-11e8-baaf-8b3c5a3da888_story.html.

92 **"Administrator Bridenstine"**: "Remarks by President Trump at a Meeting with the
National Space Council and Signing of Space Policy Directive-3," June 18, 2018,
https://trumpwhitehouse.archives.gov/briefings-statements/remarks-president
-trump-meeting-national-space-council-signing-space-policy-directive-3/.

92 **"very soon we're going to Mars"**: "Trump: 'We May Even Have a Space Force,'"
The Washington Post, March 13, 2018, https://www.washingtonpost.com/video
/politics/trump-we-may-even-have-a-space-force/2018/03/13/9dcd9f9c-270b
-11e8-a227-fd2b009466bc_video.html.

93 **"Space Pearl Harbor"**: Report of the Commission to Assess United States National
Security Space Management and Organization, January 11, 2001, https://aerospace
.csis.org/wp-content/uploads/2018/09/RumsfeldCommission.pdf.

93 **"create enormous upheaval"**: Christian Davenport, "Some in Congress Are Pushing for a 'Space Corps,' Dedicated to Fighting Wars in the Cosmos," *The Washington Post,* September 15, 2017, https://www.washingtonpost.com/news/checkpoint
/wp/2017/09/15/some-in-congress-are-pushing-for-a-space-corps-dedicated-to
-fighting-wars-in-the-cosmos/.

93 **In reality, the White House**: Christian Davenport, "The White House Was Seriously Considering a 'Space Force' Long Before Trump Talked About It Publicly,"
The Washington Post, August 21, 2018, https://www.washingtonpost.com
/technology/2018/08/21/white-house-was-seriously-considering-space-force-well
-before-trump-talked-about-it-publicly/.

95 **"Everyone who was alive"**: Mary Louise Kelly, "50 Years After Apollo 11 Moon
Landing, NASA Sets Its Sights on Mars," NPR, July 15, 2019, https://www.npr.org
/2019/07/15/741281881/50-years-after-apollo-11-moon-landing-nasa-sets-its
-sights-on-mars.

96 **recalculated the odds**: Jeff Foust, "Recalculating Risk," *The Space Review,* February 13, 2017, https://www.thespacereview.com/article/3171/1.

CHAPTER 7

101 **"This is a rocket of truly huge scale":** Seth Borenstein, "Giant Rocket Being Planned," Associated Press, April 6, 2011, https://www.heraldtribune.com/story /news/2011/04/06/giant-rocket-being-planned/29008129007/.

101 **"At first it sounds real easy":** William Harwood, "SpaceX Shows Off Powerful Falcon Heavy Rocket Ahead of First Launch," CBS News, December 22, 2017, https:// www.cbsnews.com/news/spacex-shows-off-powerful-falcon-heavy-rocket/.

102 **"We were pretty naïve":** Christian Davenport, "Elon Musk Is Set to Launch His Falcon Heavy Rocket, a Flamethrower of a Different Sort," *The Washington Post,* January 30, 2018, https://www.washingtonpost.com/news/the-switch/wp/2018/01 /30/elon-musk-is-set-to-launch-his-falcon-heavy-rocket-a-flamethrower-of -another-sort/.

102 **Even Musk said the odds:** William Harwood, "Elon Musk Giddy Ahead of SpaceX's Historic Falcon Heavy Launch," CBS News, February 6, 2018, https:// www.cbsnews.com/news/elon-musk-giddy-ahead-of-spacexs-historic-falcon -heavy-launch/.

102 **"more than a fully loaded Boeing":** SpaceX Next: Falcon Heavy Press Conference, April 5, 2011, https://www.youtube.com/watch?v=DtoADdSry6g.

103 **put together a video:** SpaceX, "How Not to Land an Orbital Rocket Booster," September 14, 2017, https://www.youtube.com/watch?v=bvim4rsNHkQ.

103 **For the first flight:** Tariq Malik, "Wheel of Cheese Launched into Space on Private Spacecraft," Space.com, December 9, 2010, https://www.space.com/10459-wheel -cheese-launched-space-private-spacecraft.html.

103 **"We were talking":** Bryan Bishop, "Elon Musk and the Creator of Westworld Made an Inspirational Trailer for the Falcon Heavy Launch," *The Verge,* March 10, 2018, https://www.theverge.com/2018/3/10/17105322/elon-musk-spacex-falcon -heavy-westworld-jonathan-nolan-trailer-sxsw; South by Southwest, "Elon Musk Answers Your Questions," March 11, 2018, https://www.youtube.com /watch?v=kzlUyrccbos.

103 **The Tesla would separate:** William Harwood, "'Starman' and Tesla Heading for Deep Space," CBS News, February 7, 2018, https://www.cbsnews.com/news /starman-tesla-roadster-orbit-spacex-falcon-heavy/.

106 **Local officials were expecting:** J. D. Gallop, "Heavy Traffic, Delays Reported After SpaceX Falcon Heavy Launch," *Florida Today,* February 6, 2018, https://www .floridatoday.com/story/news/crime/2018/02/06/launch-traffic-crowded-but-no -reported-problems/310694002/.

107 **"We're very excited":** Falcon Heavy Test Flight, SpaceX, February 6, 2018, https:// www.youtube.com/watch?v=wbSwFU6tY1c.

108 **The landings of the side boosters:** "SpaceX Launch Falcon Heavy: Watch Live," *The Guardian,* February 6, 2018, https://www.youtube.com/watch?v=99llRhH71vA.

109 **2.3 million viewers on YouTube:** Micah Singleton, "SpaceX's Falcon Heavy Launch Was YouTube's Second Biggest Live Stream Ever," *The Verge,* February 6, 2018, https://www.theverge.com/2018/2/6/16981730/spacex-falcon-heavy-launch -youtube-live-stream-record/.

111 **"Europe's move":** Jan Wörner, "Europe's Move," European Space Agency, February 11, 2018, https://blogs.esa.int/janwoerner/2018/02/11/europes-move/.

111 **"totally crushed":** Charlotte Gao, "China Has Mixed Feelings About Elon Musk's Falcon Heavy Success," *The Diplomat,* February 8, 2018, https://thediplomat.com /2018/02/china-has-mixed-feelings-about-elon-musks-falcon-heavy-success/.

111 **Its three main launch sites:** Marina Koren, "Why It's a Bad Idea to Launch Rockets Over Land," *The Atlantic,* January 13, 2018, https://www.theatlantic.com /science/archive/2018/01/china-long-march-rocket-exploded/550439/.

111 **In 1996, an earlier version:** Anatoly Zak, "Disaster at Xichang," *Air & Space Magazine,* February 2013, https://www.smithsonianmag.com/air-space-magazine /disaster-at-xichang-2873673/.

112 **A few months after:** Andrew Jones, "China to Test Rocket Reusability with Planned Long March-8 Launcher," *SpaceNews,* April 30, 2018, https://spacenews .com/china-to-test-rocket-reusability-with-planned-long-march-8-launcher/.

CHAPTER 8

113 **well before Joe Rogan:** *Joe Rogan Experience,* No. 1169—Elon Musk, September 7, 2018, https://www.youtube.com/watch?v=ycPr5-27vSI.

115 **A CNN anchor noted:** Christine Romans, "Watch Elon Musk Smoke Marijuana on Podcast," CNN, September 7, 2018, https://www.cnn.com/videos/cnnmoney /2018/09/07/elon-musk-tesla-ceo-smokes-marijuana-on-podcast-vpx.cnn.

115 **MSNBC's correspondent said:** Velshi & Ruhle, "Tesla CEO Elon Musk Smokes Weed During Joe Rogan Podcast Interview," MSNC, September 7, 2018, https:// www.youtube.com/watch?v=1rS8fFbW57o.

115 ***The Wall Street Journal* reported:** Tim Higgins, "Tesla's Shares Slide After More Executives Leave, Musk Interview," *The Wall Street Journal,* September 7, 2018, https://www.wsj.com/articles/elon-musk-appears-to-smoke-marijuana-on-camera -in-lengthy-interview-1536318688.

115 **Musk settled:** Renea Merle, "Tesla's Elon Musk Settles with SEC, Paying $20 Million Fine and Resigning as Board Chairman," *The Washington Post,* September 29, 2018, https://www.washingtonpost.com/business/2018/09/29/teslas-elon-musk -settles-with-sec-paying-million-fine-resigning-board-chairman/.

117 **In the weeks that followed:** Christian Davenport, "NASA to Launch Safety Review of SpaceX and Boeing After Video of Elon Musk Smoking Pot Rankled Agency Leaders," *The Washington Post,* November 20, 2018, https://www.washingtonpost .com/business/2018/11/20/nasa-launch-safety-review-spacex-boeing-after-video -elon-musk-smoking-pot-rankled-agency-leaders/.

118 **"Every single one of those accidents":** Marina Koren, "NASA Administrator on Elon Musk: 'That Was Not Appropriate Behavior,'" *The Atlantic,* November 29, 2018, https://www.theatlantic.com/science/archive/2018/11/elon-musk-nasa -spacex-commercial-crew-safety-review/576997/.

118 **To get more power:** Christian Davenport, "Elon Musk's SpaceX Is Using a Powerful Rocket Technology. NASA Advisers Say It Could Put Lives at Risk," *The Washington Post,* May 5, 2018, https://www.washingtonpost.com/business/economy/elon-musks -space-x-is-using-a-powerful-rocket-technology-nasa-advisers-say-it-could-put -lives-at-risk/2018/05/05/f810b182-3cec-11e8-a7d1-e4efec6389f0_story.html.

119 **"there is a unanimous":** Jeff Foust, "Letter Raises Questions About SpaceX Fueling Plans and Committee Roles," *SpaceNews,* November 6, 2016, https://spacenews .com/letter-raises-questions-about-spacex-fueling-plans-and-committee-roles/.

119 **"If we're throwing a bunch":** Ashlee Vance, *Elon Musk: Tesla, SpaceX, and the Quest for a Fantastic Future* (Ecco, 2015), chap. 9, Kindle.

CHAPTER 9

122 **"not an easy choice"**: Jeff Foust, "NASA Selects Boeing and SpaceX for Commercial Crew Contracts," *SpaceNews,* September 16, 2014, https://spacenews.com /41891nasa-selects-boeing-and-spacex-for-commercial-crew-contracts/.

129 **"boil things down"**: Elon Musk, "The Mind Behind Tesla, SpaceX, Solar City," TED Talk, February 2013, https://www.ted.com/talks/elon_musk_the_mind _behind_tesla_spacex_solarcity.

129 **"Question everything"**: Gwynne Shotwell interview at the McCain Institute Sedona Forum, "Technology, Innovation and Our National Interests," May 6, 2023, https://www.mccaininstitute.org/the-sedona-forum/past-forums/the-sedona -forum-2023/.

130 **the company's "algorithm"**: Kiko Dontchev, "Running the Algorithm: SpaceX's Approach to Exponential Growth with VP of Launch Kiko Dontchev," Summit, May 15, 2023, https://www.youtube.com/watch?app=desktop&v=ZOWakxXjotg.

134 **Flying with the Russians**: Christian Davenport, "The Unsung Astronauts," *The Washington Post,* June 15, 2018, https://www.washingtonpost.com/news/business /wp/2018/06/15/feature/what-does-it-mean-to-be-a-nasa-astronaut-in-the -celebrity-space-age-of-elon-musk-and-richard-branson/.

135 **"train wreck followed"**: Clara Moskowitz, "Space Station Crew Prepares for Landing: Can Feel Like a Car Crash, Astronaut Says," NBC News, September 23, 2010, https://www.nbcnews.com/id/wbna39335840.

135 **"When the parachute opened"**: Scott Kelly, *Endurance: A Year in Space, A Lifetime of Discovery* (Alfred A. Knopf, 2017), chap. 7, Kindle.

135 **Take Jonny Kim**: Alvin Powell, "Harvard Medical School Grad to Depart Residency to Start Astronaut Training," *Harvard Gazette,* July 21, 2017, https://news .harvard.edu/gazette/story/2017/07/med-school-grad-to-trade-scrubs-for-space -suit/.

136 **"He can kill you"**: Brittany Shammas, "First He Was a Navy SEAL. Then He Went to Harvard Medical School. The Moon Could Be Next," *The Washington Post,* January 14, 2020, https://www.washingtonpost.com/lifestyle/2020/01/14/nasa astronautjonnykim/.

136 **"I'd like to see kids growing up"**: Christian Davenport, "Why NASA's Next Rockets May Say 'Budweiser' on the Side," *The Washington Post,* September 10, 2018, https://www.washingtonpost.com/technology/2018/09/10/why-nasas-next-rockers -might-say-budweiser-side/.

136 **Even the astronauts groused**: Christian Davenport, "As the Possibility of Going to Space Grows, Astronauts Still Don't Know How They Get Picked to Fly," *The Washington Post,* September 15, 2020, https://www.washingtonpost.com /technology/2020/09/15/with-more-chances-fly-space-than-ever-before-us -astronauts-are-still-unsure-how-they-get-picked/.

136 **When it came time**: Graham Southorn, "The Apollo 11 Mission Patch: How It Came to Be," *BBC Sky at Night Magazine,* July 10, 2019, https://www.skyatnight magazine.com/space-missions/the-apollo-11-mission-patch-how-it-came-to-be.

137 **"For some of you"**: "Commercial Crew Program Crew Assignment Announcement," NASA, August 3, 2108, https://images.nasa.gov/details/Commercial_Crew _Program_Crew_Assignment_Announcement_August_3_2018_686042.

139 **"This year, American"**: "Apollo Astronaut Buzz Aldrin at the 2019 State of the Union," NASA, February 6, 2019, https://www.nasa.gov/image-article/apollo -astronaut-buzz-aldrin-2019-state-of-union/.

140 **By March 2019:** Jaret Matthews, Caitlin Driscoll, Edward Fouad, Andrew Welter, Marc Jarmulowicz, and Jessica Ipnar, "Design, Development, Testing, and Flight of the Crew Dragon Docking System," IEEM Paper, 2020, https://esmats.eu /amspapers/pastpapers/pdfs/2020/matthews.pdf.

141 **"I don't think it'll be a problem":** Loren Grush, "NASA Gives SpaceX the OK to Launch New Passenger Spacecraft on Uncrewed Spaceflight," *The Verge,* February 22, 2019, https://www.theverge.com/2019/2/22/18236771/nasa-spacex-dragon -commercial-crew-dm-1-test-flight.

141 **"If something goes wrong":** *NASA & SpaceX: Journey to the Future,* Discovery and Science Channels, May 2020, https://www.amazon.com/NASA-SpaceX-Journey -Future-Special/dp/B0896Q52L5.

142 ·**"I can't believe how well":** Marie Lews, "Demo-1 Post-Splashdown Comments from Benjo Reed," NASA Commercial Crew Program Blog, March 8, 2019, https:// blogs.nasa.gov/commercialcrew/2019/03/.

143 **"I don't think we really":** Michael Sheetz, "SpaceX Completes Historic Crew Dragon Test Flight for NASA with Splashdown in the Atlantic," CNBC, March 8, 2019, https://www.cnbc.com/2019/03/08/spacex-crew-dragon-splashdown-in-the -atlantic-ocean-for-nasa.html.

CHAPTER 10

144 **Leanne Caret:** Christian Davenport and Joel Achenbach, "NASA Rocket Becomes Boeing's Latest Headache as Trump Demands Moon Mission," *The Washington Post,* March 22, 2019, https://www.washingtonpost.com/technology/2019 /03/22/nasa-rocket-becomes-boeings-latest-headache-trump-demands-moon -mission/.

144 **to $20 billion and counting:** Casey Drier, "Why We Have the SLS," The Planetary Society, August 3, 2022, https://www.planetary.org/articles/why-we-have-the-sls.

145 **Space Policy Directive-1:** "New Space Policy Directive Calls for Human Expansion Across Solar System," NASA, December 11, 2017, https://www.nasa.gov/news -release/new-space-policy-directive-calls-for-human-expansion-across-solar -system/.

145 **In 2018, China:** Joan Johnson-Freese, "China Launched More Rockets into Orbit in 2018 Than Any Other Country," *MIT Technology Review,* December 19, 2018, https://www.technologyreview.com/2018/12/19/66274/china-launched-more -rockets-into-orbit-in-2018-than-any-other-country/.

146 **"marked a new chapter":** Siobhan O'Grady, "China Launches Moon Mission, Seeking to Be First Country in Decades to Collect Lunar Rocks," *The Washington Post,* November 23, 2020, https://www.washingtonpost.com/world/asia_pacific /china-moon-mission/2020/11/23/aa479520-2d2f-11eb-9dd6-2d0179981719 _story.html.

146 **"like a dancer circling":** "Tidal Locking," NASA, https://science.nasa.gov/moon /tidal-locking/.

146 **In China, the *Global Times*:** Sarah Kaplan, Gerry Shih, and Rick Noack, "China Lands Spacecraft on the Far Side of the Moon, a Historic First," *The Washington Post,* January 3, 2019, https://www.washingtonpost.com/science/2019/01/03/china -lands-spacecraft-far-side-moon-historic-first/?noredirect=on.

146 **"Unlike mankind's mania":** Marina Koren, "Why the Far Side of the Moon Matters So Much," *The Atlantic,* January 2019, https://www.theatlantic.com/science /archive/2019/01/far-side-moon-china/579349/.

146 "The universe is an ocean": Jeffrey Kluger, "Inside the New Race to the Moon," *Time,* July 18, 2019, https://time.com/longform/race-to-the-moon/.

147 "If we don't go there now": Gordon G. Chang, "We're in a New Cold War in Space: China Plans to Take Over Moon," *Newsweek,* January 9, 2023, https://www.newsweek.com/were-new-cold-war-space-china-plans-take-over-moon-opinion-1772306.

147 Wicker opened the hearing: "The New Space Race: Ensuring U.S. Global Leadership on the Final Frontier," U.S. Senate Committee on Commerce, Science, and Transportation, May 13, 2019, https://www.commerce.senate.gov/2019/3/the-new-space-race-ensuring-u-s-global-leadership-on-the-final-frontier.

148 Among those options: Jackie Wattles, "NASA Considers Sidelining Its Boeing-Built Rocket for an Upcoming Moon Mission," CNN, March 25, 2019, https://www.cnn.com/2019/03/25/tech/nasa-boeing-sls-delays/index.html.

148 It provided thirteen thousand jobs: Davenport and Achenbach, "NASA Rocket Becomes Boeing's Latest Headache."

149 "an unsustainable trajectory": Marcia Smith, "Augustine Committee: Current NASA Human Spaceflight Program on 'Unsustainable Trajectory,' " SpacePolicy Online.com, September 8, 2009, https://spacepolicyonline.com/news/augustine-committee-current-nasa-human-space-flight-program-on-quot-unsustainable-trajectory-quot/.

149 Congress had other ideas: Christian Davenport, "After Years of Setbacks, NASA's SLS Moon Rocket Is Finally Ready to Fly," *The Washington Post,* August 27, 2022, https://www.washingtonpost.com/technology/2022/08/27/nasa-sls-moon-artemis-human-space/.

149 Dissenting in a speech: *SpaceNews* Staff, "House Gives Final Approval to NASA Authorization Act," *SpaceNews,* September 30, 2010, https://spacenews.com/house-gives-final-approval-nasa-authorization-act/.

150 "While I agree": Jeff Foust, "Industry and Lawmakers Go to Defense of SLS," *SpaceNews,* March 14, 2019, https://spacenews.com/industry-and-lawmakers-go-to-defense-of-sls/.

151 headline from *Axios*: Andrew Freedman, "In a Major Shift, NASA May Use Commercial Rockets for Next Moon Mission," *Axios,* March 13, 2019, https://www.axios.com/2019/03/13/nasa-moon-mission-commercial-rocket.

151 "Surprise NASA announcement": Jonathan O'Callaghan, "Surprise NASA Announcement Puts Future of New Mega-Rocket in Doubt," *Forbes,* March 17, 2019, https://www.forbes.com/sites/jonathanocallaghan/2019/03/17/surprise-nasa-announcement-puts-future-of-new-mega-rocket-in-doubt/.

152 "It was really good": Mike Pence, *So Help Me God* (Simon & Schuster, 2022), chap. 5, Kindle.

154 "NASA's exploration efforts": "Remarks by Vice President Mike Pence at the Fifth Meeting of the National Space Council," The White House, March 26, 2019, https://trumpwhitehouse.archives.gov/briefings-statements/remarks-vice-president-pence-fifth-meeting-national-space-council-huntsville-al/.

157 "This was NASA's architecture": Jeff Foust, "Mixed Reactions to Accelerated Moon Plan," *SpaceNews,* March 26, 2019, https://spacenews.com/mixed-reactions-to-accelerated-moon-plan/.

158 "I know you have": "Vice President Pence and Administrator Bridenstine Talk with Astronauts on Space Station," NASA, March 6, 2019, https://www.youtube.com/watch?v=gPYgwrnr644.

158 "This decision was based": Jena McGregor, "In Spacesuit Saga, Women See Their Own Stories," *The Washington Post*, March 28, 2019, https://www.washingtonpost.com/business/2019/03/28/nasas-spacesuit-saga-women-see-their-own-stories/.

CHAPTER 11

160 Two months before Pence's speech: Jeff Foust, "Blue Origin Breaks Ground for BE-4 Factory," *SpaceNews*, January 25, 2019, https://spacenews.com/blue-origin-breaks-ground-for-be-4-factory/.

160 expanding its headquarters: Alan Boyle, "Construction Is Well Under Way for Blue Origin Space Venture's Expanded Headquarters," *GeekWire*, April 29, 2019, https://www.geekwire.com/2019/construction-well-way-blue-origin-space-ventures-expanded-hq/.

162 "erodes our democracy": Anita Balakrishnan, "Amazon's Jeff Bezos Said Trump's Behavior 'Erodes Our Democracy Around the Edges,'" CNBC, October 20, 2016, https://www.cnbc.com/2016/10/20/amazons-jeff-bezos-said-trumps-behavior-is-eroding-democracy-around-the-edges.html.

163 "While I hope no one": Jeff Bezos, "Jeff Bezos on Post Purchase," *The Washington Post*, August 5, 2013, https://www.washingtonpost.com/national/jeff-bezos-on-post-purchase/2013/08/05/e5b293de-fe0d-11e2-9711-3708310f6f4d_story.html.

163 "I for one give him": Alex Weprin, "Bezos Pledges 'Open Mind' to Trump Administration," *Politico*, November 10, 2016, https://www.politico.com/blogs/on-media/2016/11/bezos-pledges-open-mind-to-trump-administration-231178.

163 Over cheese soufflé: Martin Baron, *Collision of Power: Trump, Bezos, and The Washington Post* (Flatiron Books, 2023), Prologue.

163 Trump continued to rant: Sean Burch, "Jeff Bezos Calls Out President Trump's 'Dangerous' Media Criticism," *The Wrap*, September 14, 2018, https://www.yahoo.com/entertainment/jeff-bezos-calls-president-trump-144226081.html.

164 "Rich guys, they: Christian Davenport, "Why the Trump Administration Wants to Make It Easier for Elon Musk's SpaceX and Others to Get to Space," *The Washington Post*, May 24, 2018, https://www.washingtonpost.com/news/the-switch/wp/2018/05/24/why-the-trump-administration-wants-to-make-it-easier-for-elon-musks-spacex-and-others-to-get-to-space/.

164 "I don't know if you saw": Michael Sheetz, "Trump Praises Elon Musk's SpaceX for 'Beautifully' Landing Rocket Boosters," CNBC, March, 8, 2018, https://www.cnbc.com/2018/03/08/trump-praises-elon-musk-spacex-for-beautifully-landing-rocket-boosters.html.

164 In 2014, SpaceX spent: Client Profile: SpaceX, OpenSecrets.org, https://www.opensecrets.org/federal-lobbying/clients/summary?cycle=2019&id=D000029147.

164 Blue Origin, by contrast: Client Profile: Blue Origin, OpenSecrets.org, https://www.opensecrets.org/federal-lobbying/clients/summary?cycle=2019&id=D000069501.

164 Barnes & Thornburg: "Barnes & Thornburg Latest Big Firm to Join Trump's Transition Orbit," *The American Lawyer*, November 17, 2016, https://www.law.com/americanlawyer/almID/1202772720397/.

165 Under the headline: Editorial Board, "Mike Pence, Boldly Sending America Back to Where Man Has Gone Before," *The Washington Post*, March 28, 2019, https://www.washingtonpost.com/opinions/mike-pence-boldly-sending-america-back-to-where-man-has-gone-before/2019/03/28/9e0a007a-5169-11e9-88a1-ed346f0ec94f_story.html.

166 **curry favor with Trump:** Martin Baron, "Where Jeff Bezos Went Wrong with *The Washington Post*," *The Atlantic*, March 3, 2025, https://www.theatlantic.com/ideas /archive/2025/03/bezos-appease-trump-administration/681899/.

166 **"Why am I feeling":** Alan Boyle, "Gundam? Alien? It's Jeff Bezos Piloting a Giant Robot at Amazon's MARS Conference," *GeekWire*, March 19, 2017, https://www .geekwire.com/2017/jeff-bezos-pilots-giant-robot/.

167 **"The Earth is finite":** Alina Utrata, "Lost In Space," *Boston Review*, July 14, 2021, https://www.bostonreview.net/articles/lost-in-space/.

167 **"I still believe that":** Todd Bishop and Alan Boyle, "Blue Moon and Beyond: How Jeff Bezos Plans to Take Civilization to Space, Starting with Lunar Colony," *GeekWire*, May 12, 2019, https://www.geekwire.com/2019/blue-moon-beyond-jeff -bezos-plans-take-civilization-space-starting-lunar-colony/#:~:text=We%20know %20what%20we%20want,would%20be%20an%20incredible%20civilization.

168 **"These are very large structures":** Jeff Bezos, "Blue Origin 2019: For the Benefit of Earth," May 10, 2019, https://www.youtube.com/watch?v=GQ98hGUe6FM.

169 **eavesdropping Alexa:** Geoffrey A. Fowler, "Alexa Has Been Eavesdropping on You This Whole Time," *The Washington Post*, May 6, 2019, https://www.washington post.com/technology/2019/05/06/alexa-has-been-eavesdropping-you-this-whole -time/.

169 **enormous windfalls:** Michael Sainato, "'I'm Not a Robot: Amazon Workers Protest Unsafe, Grueling Working Conditions," *The Guardian*, February 45, 2020, https:// www.theguardian.com/technology/2020/feb/05/amazon-workers-protest-unsafe -grueling-conditions-warehouse.

169 **Notoriously, Amazon:** David Streitfeld, "Inside Amazon's Very Hot Warehouse," *The New York Times*, September 19, 2011, https://archive.nytimes.com/bits.blogs .nytimes.com/2011/09/19/inside-amazons-very-hot-warehouse/.

169 **"control population on Earth":** Christian Davenport, *The Space Barons: Elon Musk, Jeff Bezos, and the Quest to Colonize the Cosmos* (PublicAffairs, 2018), chap. 15, Kindle.

171 **"[Bezos] gathered":** Lori Garver, *Escaping Gravity: My Quest to Transform NASA and Launch a New Space Age* (Diversion Books, 2022), chap. 8, Kindle.

173 **China knew the power:** Marina Koren, "Why It's a Bad Idea to Launch Rockets Over Land," *The Atlantic*, January 13, 2018, https://www.theatlantic.com/science /archive/2018/01/china-long-march-rocket-exploded/550439/.

174 **"The first time humanity":** Lee Roop, "NASA Seeks $1.6 Billion to Jump Start New Moon Shot Program Called Artemis," Alabama.com, May 13, 2019, https:// www.al.com/news/2019/05/nasa-seeks-16b-to-jump-start-new-moon-shot -program-called-artemis.html.

175 **"But what if I gave you":** Olivia Nuzzi, "How Trump Offered NASA Unlimited Funding to Go to Mars in His First Term," *New York Magazine*, January 22, 2019, https://nymag.com/intelligencer/2019/01/trump-offered-nasa-unlimited-funding -to-go-to-mars-by-2020.html.

175 **On average, it's 140 million:** Daisy Dobrijevic, "Distance to Mars: How Far Away Is the Red Planet?" Space.com, February 4, 2022, https://www.space.com/16875 -how-far-away-is-mars.html#:~:text=Mars%20close%20approach,-Close%20 approach%20provides&text=Mars%20makes%20a%20close%20approach,is %20the%20temperature%20on%20Mars?&text=%E2%80%93%20How%20was %20Mars%20made?,the%20same%20window%20of%20opportunity.

176 **"refocusing on the moon":** Elise Oggioni, "NASA CFO Says the Agency Intends to Get Back to the Moon by 2024," *Fox Business,* June 7, 2019, https://www.foxbusiness.com/fbntv/nasa-cfo-says-the-agency-intends-to-get-back-to-the-moon-by-2024.

176 **DeWit was a political operative:** Yvonne Wingett Sanchez, "Trump Taps Treasurer Jeff DeWit as State Campaign Chair," January 20, 2016, *The Republic,* https://www.azcentral.com/story/news/politics/politicalinsider/2016/01/20/treasurer-jeff-dewit-donald-trump-arizona-campaign-chair/78887388/.

176 **NASA officials scrambling:** Jeff Foust, "Trump Tweet Throws Space Policy into Chaos," *SpaceNews,* June 7, 2019, https://spacenews.com/trump-tweet-throws-space-policy-into-chaos/.

CHAPTER 12

177 **Kelly said he'd look into it:** Emre Kelly, "Smoke Seen for Miles as SpaceX's Crew Dragon Suffers 'Anomaly' at Cape Canaveral During Engine Test Fire," *Florida Today,* April 20, 2019, https://www.floridatoday.com/story/tech/science/space/2019/04/20/smoke-seen-miles-spacex-crew-dragon-suffers-anomaly-cape-canaveral/3531086002/.

180 **"The test was not satisfactory":** Jeff Foust, "Crew Dragon Parachutes Failed in Recent Test," *SpaceNews,* May 9, 2019, https://spacenews.com/crew-dragon-parachutes-failed-in-recent-test/; Eric Berger, "SpaceX Had a Problem with a Parachute Test in April," *Ars Technica,* May 8, 2019, https://arstechnica.com/science/2019/05/spacex-had-a-problem-during-a-parachute-test-in-april/.

182 **Steel could handle both:** Ryan D'Agostino, "Elon Musk: Why I'm Building the Starship Out of Stainless Steel," *Popular Mechanics,* January 22, 2019, https://www.popularmechanics.com/space/rockets/a25953663/elon-musk-spacex-bfr-stainless-steel/.

182 **Starlink satellites:** Caleb Henry, "Musk: We're Not Spinning Off Starlink," *SpaceNews,* March 9, 2020, https://spacenews.com/musk-were-not-spinning-off-starlink/.

186 **"Push the button":** Jackie Wattles, "NASA Administrator Tells Elon Musk's SpaceX 'It's Time to Deliver,'" CNN, September 28, 2019, https://www.cnn.com/2019/09/28/business/elon-musk-spacex-nasa-bridenstine-crew-dragon/index.html#:~:text=%E2%80%9CI%20am%20looking%20forward%20to,to%20deliver%2C%E2%80%9D%20he%20said.

187 **"Did he say Commercial Crew":** "Elon Musk: Starship Could Take People to Orbit Within a Year," CNN, September 29, 2019, https://www.cnn.com/videos/business/2019/09/29/elon-musk-starship-interview-orig.cnn.

188 **"drogue parachutes":** Anna C. Heiney, "Top 10 Things to Know For NASA's SpaceX Demo-2 Return," NASA, July 24, 2020, https://www.nasa.gov/humans-in-space/top-10-things-to-know-for-nasas-spacex-demo-2-return/#:~:text=Dragon%20Endeavour%20has%20two%20sets,approximately%20119%20miles%20per%20hour.

189 **"basically destroyed the check valve":** Stephen Clark, "SpaceX Points to Leaky Valve as Culprit in Crew Dragon Test Accident," *Spaceflight Now,* July 15, 2018, https://spaceflightnow.com/2019/07/15/spacex-points-to-leaky-valve-as-culprit-in-crew-dragon-test-accident/.

189 **"we've done everything possible":** Christian Davenport, "NASA Administrator

Visits SpaceX in Bid to Ease Tension in Their Relationship," October 10, 2019, https://www.washingtonpost.com/technology/2019/10/11/nasa-administrator -visits-spacex-bid-ease-tension-their-relationship/.

190 **"Parachutes are way harder"**: Christian Davenport, "NASA Astronauts Aboard SpaceX Capsule Heading to a Splashdown in the Gulf of Mexico," *The Washington Post,* August 1, 2020, https://www.washingtonpost.com/technology/2020/08/01 /nasa-spacex-return-splashdown/.

192 **Now, in January 2020:** Christian Davenport, "SpaceX Completes Key Test of its Dragon Capsule. Its First Human Spaceflight Might Come in Spring," *The Washington Post,* January 19, 2020, https://www.washingtonpost.com/technology/2020 /01/19/spacexemergencyabortttest/.

193 **"We expect there to be"**: Stephen Clark, "SpaceX Will Trigger an Intentional Rocket Failure to Prove Crew Capsule's Safety," *Spaceflight Now,* January 18, 2020, https://spaceflightnow.com/2020/01/18/spacex-will-trigger-an-intentional-rocket -failure-to-prove-crew-capsules-safety/.

193 **"Holy shit"**: *NASA & SpaceX: Journey to the Future,* Discovery and Science Channels, May 2020, https://www.amazon.com/NASA-SpaceX-Journey-Future-Special /dp/B0896Q52L5.

194 **"Obviously, they are keenly"**: Davenport, "SpaceX Completes Key Test of Its Dragon Capsule."

CHAPTER 13

195 **NASA now had a date:** Anna Heiney, "Launch Day Arrives for NASA SpaceX's Demo-2," NASA Blogs, May 27, 2020, https://blogs.nasa.gov/commercialcrew /2020/05/27/launch-day-arrives-for-nasas-spacex-demo-2/.

195 **"it probably has a lot"**: Christian Davenport, "The Company Astronaut," *The Washington Post,* July 24, 2018, https://www.washingtonpost.com/news/business /wp/2018/07/24/feature/nasa-trained-boeing-employed-chris-ferguson-hopes-to -make-history-as-a-company-astronaut/.

196 **Boeing almost lost Starliner:** Christian Davenport, "NASA Finds 'Fundamental' Software Problems in Boeing's Starliner Spacecraft, *The Washington Post,* February 7, 2020, https://www.washingtonpost.com/technology/2020/02/07/boeing -starliner-software-problems/.

196 **Boeing had recently suffered:** "Boeing Accused of Building 'Flying Coffins,'" BBC, October 30, 2019, https://www.bbc.com/news/business-50225025.

196 **"we would have caught it"**: Christian Davenport, "Boeing Admits It Failed to Test the Starliner Spacecraft Adequately Before Its Maiden Flight," *The Washington Post,* February 28, 2020, https://www.washingtonpost.com/technology/2020/02/28 /boeing-admits-starliner-testing-flaws/.

197 **The space agency had largely:** Christian Davenport, "Boeing Faced Only 'Limited' Safety Review While SpaceX Got a Full Examination," *The Washington Post,* November 18, 2019, https://www.washingtonpost.com/technology/2019/11 /18/boeing-faced-only-limited-safety-review-nasa-while-spacex-got-full -examination/.

197 **"We'll fly when we're ready"**: Jeff Foust, "Commercial Crew Companies Emphasize Safety Over Schedule," *SpaceNews,* September 14, 2016, https://spacenews .com/commercial-crew-companies-emphasize-safety-over-schedule/.

198 **"we want people to feel free"**: Jeff Foust, "Crew Dragon Ready for Historic Launch as NASA Looks Ahead to Next Mission," *SpaceNews,* May 26, 2020,

https://spacenews.com/crew-dragon-ready-for-historic-launch-as-nasa-looks
-ahead-to-next-mission/.

200 **"So I'd just like to say":** Mark Strassmann, "Elon Musk Praises NASA Astronauts
for Their 'Nerves of Steel' Ahead of Historic SpaceX Launch," CBS News, May 27,
2020, https://www.cbsnews.com/news/spacex-elon-musk-nasa-astronauts-historic
-launch/.

200 **"I think it's going to stay":** Marcia Dunn, "NASA, SpaceX Bringing Astronaut
Launches Back to Home Turf," Associated Press, May 21, 2020, https://apnews
.com/article/us-news-ap-top-news-ca-state-wire-tx-state-wire-virus-outbreak-536
cdbad77169d956e66836260347294.

201 **"a sacred honor":** Christian Davenport, "Inside SpaceX, the Willy Wonka–Like
Rocket Factory That Plans to Send Private Citizens to Space," *The Washington Post,*
https://www.washingtonpost.com/business/2018/08/14/inside-space-x-willy
-wonka-like-rocket-factory-that-eventually-could-send-private-citizens-space/.

201 **"paranoia reviews":** Christian Davenport, "SpaceX Is About to Fly Astronauts for
a Third Time. But There's Nothing Routine About It," *The Washington Post,*
April 20, 2021, https://www.washingtonpost.com/technology/2021/04/20/elon
-musk-spacex-nasa-crew-2/.

201 **"That really drives it home":** Christian Davenport, "SpaceX Is Using a Powerful
New Rocket Technology. NASA Advisers Say It Could Put Lives at Risk," *The
Washington Post,* May 5, 2017, https://www.washingtonpost.com/business
/economy/elon-musks-space-x-is-using-a-powerful-rocket-technology-nasa
-advisers-say-it-could-put-lives-at-risk/2018/05/05/f810b182-3cec-11e8-a7d1
-e4efec6389f0_story.html.

201 **The chances of losing:** Jeff Foust, "Commercial Crew Astronauts Accept Risks of
Test Flight," *SpaceNews,* May 27, 2020, https://spacenews.com/commercial-crew
-astronauts-accept-risks-of-test-flight/#:~:text=NASA's%20commercial%20
crew%20program%20set,mission%20%E2%80%94%20of%201%20in%20270.

201 **"Risk cannot be boiled":** Jeff Foust, "Recalculating Risk," *The Space Review,* February 13, 2017, https://www.thespacereview.com/article/3171/1.

202 **The Bugs:** Brendan Byrne, "Shuttle Veteran Leads First Human Space Mission
from U.S. Since 2011," NPR, May 27, 2020, https://www.npr.org/2020/05/27
/862025245/shuttle-veteran-leads-first-human-space-mission-launched-from-u-s
-since-2011.

204 **NASA and SpaceX decided:** Michael Sheetz, "SpaceX and NASA Postpone Historic Astronaut Launch Due to Bad Weather," CNBC, May 27, 2020, https://www
.cnbc.com/2020/05/27/spacex-and-nasa-postpone-historic-astronaut-launch-due
-to-bad-weather.html.

205 **"100 percent":** Christian Davenport, "Elon Musk and SpaceX Pull Off Another
Feat Few Thought Possible," *The Washington Post,* May 30, 2020, https://www
.washingtonpost.com/technology/2020/05/30/elon-musk-spacex-pull-off-another
-feat-few-thought-possible/.

206 **It was a smooth, comfortable ride:** Christian Davenport, "SpaceX Launch of
NASA Astronauts Provides a Chance to Compare the New and Old," *The Washington Post,* June 1, 2020, https://www.washingtonpost.com/technology/2020/06/01
/spacex-launch-comparisons-space-shuttle/.

207 **"I'm expecting a little":** Christian Davenport, "The Next Americans in Space," *The
Washington Post,* May 25, 2020, https://www.washingtonpost.com/technology
/2020/05/25/who-spacex-launch-astronauts/.

208 "Dragon, SpaceX, we show": Crew Demo-2 Splashdown, August 2, 2020, https://www.youtube.com/watch?v=tSJIQftoxeU.

208 "had come alive": Loren Grush, "NASA Astronaut on SpaceX Crew Dragon Return: 'Sounded Like an Animal,'" *The Verge,* August 5, 2020, https://www.theverge.com/21354742/nasa-spacex-crew-dragon-bob-behnken-doug-hurley-return.

209 "like getting hit in the back": Stephen Clark, "Dragon Astronauts Describe Sounds and Sensations of Returning to Earth," *Spaceflight Now,* August 4, 2020, https://spaceflightnow.com/2020/08/04/dragon-astronauts-describe-sounds-and-sensations-of-returning-to-earth/.

209 ceremony welcoming them home: Meghan Bartels, "'I Prayed for This One,' SpaceX's Elon Musk Says After NASA Astronauts' Splashdown Success," Space.com, August 3, 2020, https://www.space.com/elon-musk-spacex-nasa-crew-splashdown-emotions.html; "Demo-2 Astronauts Behnken and Hurley Return to Houston at Ellington Field," NASA video, August 3, 2020, https://www.youtube.com/watch?v=Rr6RxbQmcac.

CHAPTER 14

215 Jeff Bezos had vowed: Stephen Clark, "Blue Origin's Staying Power Bankrolled by Jeff Bezos' Multibillion-Dollar Investment," *Spaceflight Now,* April 6, 2017, https://spaceflightnow.com/2017/04/06/blue-origins-staying-power-bankrolled-by-jeff-bezoss-multibillion-dollar-investment/.

215 $579 million: Christian Davenport, "Jeff Bezos and Elon Musk Win Contracts for Spacecraft to Land Astronauts on the Moon," *The Washington Post,* April 30, 2020, https://www.washingtonpost.com/technology/2020/04/30/jeff-bezos-elon-musk-win-contracts-spacecraft-land-nasa-astronauts-moon/.

216 "national team": Christian Davenport, "Jeff Bezos' Blue Origin to Team Up with Aerospace Giants to Help Meet Trump's Moon Mandate," *The Washington Post,* October 22, 2019, https://www.washingtonpost.com/technology/2019/10/22/jeff-bezos-blue-origin-team-up-with-aerospace-giants-help-meet-trumps-moon-mandate/.

216 January 25, 2020: Jeff Stein and Abha Bhattarai, "White House Adviser Accuses Amazon's Jeff Bezos of Backing Out of Meeting on Fake Products," *The Washington Post,* February 5, 2020, https://www.washingtonpost.com/business/2020/02/05/white-house-adviser-accuses-amazons-jeff-bezos-backing-out-meeting-fake-products/.

217 "There was significant": Jay Greene, "Amazon Cloud Boss Chides Pentagon for Awarding Microsoft Lucrative Contract," *The Washington Post,* December 4, 2019, https://www.washingtonpost.com/technology/2019/12/04/amazon-cloud-boss-chides-pentagon-awarding-microsoft-lucrative-contract/.

217 On January 31, 2020: Editorial Board, "NASA Keeps Falling Victim to Presidential Whims," *The Washington Post,* January 31, 2020, https://www.washingtonpost.com/opinions/nasa-keeps-falling-victim-to-presidential-whims/2020/01/31/a85641e4-43ab-11ea-aa6a-083d01b3ed18_story.html.

218 "had never ordered up": Martin Baron, *Collision of Power: Trump, Bezos, and* The Washington Post (Flatiron Books, 2023), chap. 8, Kindle.

218 "This is cowardice": Marty Baron, post on X, October 25, 2024, https://x.com/PostBaron/status/1849847940761657353.

220 "It's obviously a very different": Stephen Clark, "NASA Identifies Risks in SpaceX's

Starship Lunar Lander Proposal," *Spaceflight Now,* May 1, 2020, https://spaceflight now.com/2020/05/01/nasa-identifies-risks-in-spacexs-starship-lunar-lander -proposal/.

221 **It burst:** Jeff Foust, "SpaceX Starship Suffers Testing Setback," *SpaceNews,* November 20, 2019, https://spacenews.com/spacex-starship-suffers-testing-setback/.

221 **"It will definitely get fancier":** Dave Mosher, "Inside the 'Awkward,' 'Tense,' and 'Heated' Private Meeting Between Elon Musk and Texans Whom SpaceX Is Trying to Buy Out to Fully Realize Its Vision to Reach Mars," *Business Insider,* October 15, 2019, https://www.businessinsider.com/elon-musk-spacex-boca-chica-village -private-meeting-buyouts-2019-10.

222 **"My new thing is management":** Christian Davenport and Faiz Siddiqui, "How Elon Musk Went from Sleeping on the Factory Floor to Being on the Cusp of Flying Astronauts to Space," *The Washington Post,* February 21, 2020, https://www .washingtonpost.com/technology/2020/02/21/how-elon-musk-went-sleeping -factory-being-cusp-launching-crew-into-space/.

222 **February 2020:** Christian von Preysing, "Exclusive: Interview with SpaceX Founder at Boca Chica Job Fair," KRGV, February 7, 2020, https://www.krgv.com /news/exclusive-interview-with-spacex-founder-at-boca-chica-job-fair/.

222 **"You know the term":** Christian Davenport, "Elon Musk's Improbable Mars Quest Runs Through a Border Town Concerned with More Than Getting to Space," *The Washington Post,* https://www.washingtonpost.com/technology/2019/09/30/elon -musks-improbable-mars-quest-runs-through-border-town-concerned-with -more-than-getting-space/.

223 **The first, SN-1:** Tariq Malik, "SpaceX's Starship SN1 Prototype Appears to Burst During Pressure Test," Space.com, March 1, 2020, https://www.space.com/spacex -starship-sn1-prototype-bursts-videos.html.

223 **SN-3 collapsed:** Jeff Foust, "Third Starship Prototype Destroyed in Tanking Test," *SpaceNews,* April 3, 2020, https://spacenews.com/third-starship-prototype -destroyed-in-tanking-test/.

223 **"We need to accelerate":** Michael Sheetz, "Elon Musk Tells SpaceX Employees That Its Starship Rocket Is the Top Priority Now," CNBC, June 7, 2020, https:// www.cnbc.com/2020/06/07/elon-musk-email-to-spacex-employees-starship-is-the -top-priority.html#:~:text=%22We%20need%20to%20accelerate%20 Starship,SpaceX%20unveiled%20the%20latest%20prototype.

225 **more than $30 billion:** Michael Sheetz, "SpaceX Valuation Rises to $33.3 Billion as Investors Look to Satellite Opportunity," CNBC, May 31, 2019, https://www.cnbc .com/2019/05/31/spacex-valuation-33point3-billion-after-starlink-satellites -fundraising.html.

225 **"like rebuilding the Internet":** Cecilia Kang and Christian Davenport, "SpaceX Founder Files with Government to Provide Internet Service from Space," *The Washington Post,* June 9, 2015, https://www.washingtonpost.com/business /economy/spacex-founder-files-with-government-to-provide-internet -service-from-space/2015/06/09/db8d8d02-0eb7-11e5-a0dc-2b6f404ff5cf _story.html.

225 **"successfully gone into operation":** Stephen Clark, "SpaceX's First 60 Starlink Broadband Satellites Deployed in Orbit," *Spaceflight Now,* May 24, 2019, https:// spaceflightnow.com/2019/05/24/spacexs-first-60-starlink-broadband-satellites -deployed-in-orbit/.

225 **"Total Internet connectivity revenue":** Sheetz, "SpaceX Valuation Rises to $33.3 Billion."

226 **"This was one of the hardest":** Alan Boyle, "SpaceX's Elon Musk Says 'Goodness' Will Come from Twice-Delayed Starlink Launch," *GeekWire*, May 16, 2019, https://www.geekwire.com/2019/spacex-elon-musk-starlink/.

226 **"I do believe":** Christian Davenport, "Elon Musk's SpaceX Is Striving to Win the Race to Build the Internet in Space," *The Washington Post*, May 15, 2019, https://www.washingtonpost.com/technology/2019/05/15/can-we-get-wifi-outer-space-elon-musk-others-are-trying/.

226 **Amazon planned:** Amazon Staff, "Amazon Receives FCC Approval for Project Kuiper Satellite Constellation," July 30, 2020, https://www.aboutamazon.com/news/company-news/amazon-receives-fcc-approval-for-project-kuiper-satellite-constellation.

226 **One of the first things:** Michael Sheetz, "Bezos Hired a SpaceX Vice President to Run Amazon's Satellite Internet Project After Musk Fired Him," CNBC, April 7, 2019, https://www.cnbc.com/2019/04/07/amazon-hired-former-spacex-management-for-bezos-satellite-internet.html.

227 **That year, investors:** "Start-up Space: Update on Investment in Commercial Space Ventures," BryceTech.com, https://brycetech.com/reports/report-documents/Bryce_Start_Up_Space_2020.pdf.

227 **"We wanted to be":** Irina Liu, Evan Linck, Bhavya Lal, Keith W. Crane, Xueying Han, and Thomas J. Colvin, "Evaluation of China's Commercial Space Sector," Science and Technology Policy Institute, September 2019, https://www.ida.org/research-and-publications/publications/all/e/ev/evaluation-of-chinas-commercial-space-sector.

227 **"enjoys significant support":** Makena Young and Akhil Thadani, "Low Orbit High Stakes: All-In on the LEO Broadband Competition," CSIS, December 2022, https://csis-website-prod.s3.amazonaws.com/s3fs-public/publication/221214_Young_LowOrbit_HighStakes.pdf?VersionId=vH1lp3dD7VcHGRcvuF9OdzV2WJc_KG42.

228 **sixteen rocket start-ups:** Liu et al., "Evaluation of China's Commercial Space Sector."

229 **"When I came into office":** "Remarks by President Trump at Kennedy Space Center," The White House, May 30, 2020, https://trumpwhitehouse.archives.gov/briefings-statements/remarks-president-trump-kennedy-space-center/.

230 **"My own time and thinking":** Jeff Bezos, "A Message from Our CEO and Founder," Amazon, March 21, 2020, https://www.aboutamazon.com/news/company-news/a-message-from-our-ceo-and-founder.

CHAPTER 15

232 **"was studying":** Buzz Aldrin and Wayne Warga, *Return to Earth* (Random House, 1973), p. 239.

232 **"Even more damaging":** Paul D. Spudis, "Faded Flags on the Moon," *Air & Space Magazine*, July 19, 2011, https://www.smithsonianmag.com/air-space-magazine/faded-flags-on-the-moon-32929921/.

232 **Others have surmised:** James Fincannon, "Six Flags on the Moon: What Is Their Current Condition," NASA: *Apollo Lunar Surface Journal*, 2012, https://www.nasa.gov/history/alsj/ApolloFlags-Condition.html.

232 **Nearly fifty years later:** Stephen Clark, "Video Replay of Chang'e-5 Landing on the

Moon," *Spaceflight Now,* December 2, 2020, https://spaceflightnow.com/2020/12/02/video-replay-of-change-5s-landing-on-the-moon/.

233 **"A new wave":** Andrew Jones, "China Outlines Architecture for Future Crewed Moon Landings," *SpaceNews,* October 30, 2020, https://spacenews.com/china-outlines-architecture-for-future-crewed-moon-landings/.

233 **"that could stand the harsh":** "Chinese Spacecraft Takes Off from Moon with Samples," CGTN, December 4, 2020, https://news.cgtn.com/news/2020-12-03/China-s-Chang-e-5-lunar-probe-starts-journey-back-to-Earth-VV8T6nrDnW/index.html.

233 **"The Chinese national flag":** Deng Xiaoci and Fan Anqi, "Chang'e-5 Probe Unfolds Chinese National Flag, Takes Off from Moon with Lunar Surface Samples," *Global Times,* December 3, 2020, https://www.globaltimes.cn/content/1208931.shtml.

234 **in the summer of 2019:** Christian Davenport, "NASA Is Trying to Land on the Moon. Its Biggest Challenge Might Be Congress," *The Washington Post,* September 24, 2019, https://www.washingtonpost.com/technology/2019/09/24/nasa-is-trying-land-moon-biggest-challenge-might-be-congress/.

234 **"Rhetoric about American":** Mike Wall, "Putting Astronauts on the Moon in 2024 Is a Tall Order, NASA Says," Space.com, September 20, 2019, https://www.space.com/nasa-2024-moon-mission-difficult.html.

235 **"This time when we go":** Mike Wall, "Next Footsteps on the Moon Will Be 'for All of America,' NASA Chief Says," Space.com, August 27, 2019, https://www.space.com/next-footsteps-moon-for-all-america-artemis.html.

236 **"promote peaceful purposes":** Christian Davenport, "NASA Unveils New Rules to Guide Behavior in Space and on the Lunar Surface," *The Washington Post,* May 15, 2020, https://www.washingtonpost.com/technology/2020/05/15/moon-rules-nasa-artemis/.

238 **"the biggest, most diverse coalition":** Christian Davenport, "Seven Nations Join the U.S. in Signing the Artemis Accords, Creating a Legal Framework for Behavior in Space," *The Washington Post,* October 13, 2020, https://www.washingtonpost.com/technology/2020/10/13/artemis-moon-mining-agreement-signed/.

238 **"taking steps to unleash":** "Remarks by Vice President Pence at the 8th Meeting of the National Space Council," The White House, December 9, 2020, https://trumpwhitehouse.archives.gov/briefings-statements/remarks-vice-president-pence-8th-meeting-national-space-council-cape-canaveral-fl/.

238 **"It really is amazing":** Christian Davenport, "As Pence Names Which Astronauts Will Go to the Moon, Some See a Political Ploy," *The Washington Post,* December 9, 2020, https://www.washingtonpost.com/technology/2020/12/09/pence-space-council-artemis/.

239 **"to carry out a successful":** "The Post-Apollo Space Program: Directions for the Future," NASA, September 1969, https://www.nasa.gov/history/the-post-apollo-space-program-directions-for-the-future/.

240 **"Reporters asked Trump":** David Montgomery, "Trump's Excellent Space Force Adventure," *The Washington Post Magazine,* December 3, 2019, https://www.washingtonpost.com/magazine/2019/12/03/trumps-proposal-space-force-was-widely-mocked-could-it-be-stroke-stable-genius-that-makes-america-safe-again/.

240 **"Hello, citizens of Earth":** Montgomery, "Trump's Excellent Space Force Adventure."

240 **"Wow. Space Force!"**: Sandra Erwin, "White House Has Nothing to Say on Space Force," *SpaceNews,* February 2, 2021, https://spacenews.com/white-house-has-nothing-to-say-on-space-force/.

241 **"Creating a Space Force is arguably"**: Montgomery, "Trump's Excellent Space Force Adventure."

241 **"It's concerning to see"**: Jacqueline Feldscher, "Top House Republican Demands Psaki Apologize Over 'Disgraceful' Space Force Quip," *Politico,* February 2, 2021, https://www.politico.com/news/2021/02/02/house-republican-psaki-space-force-465175.

241–42 **"I did send a tweet"**: "Press Briefing by Press Secretary Jen Psaki," The White House, February 3, 2021, https://bidenwhitehouse.archives.gov/briefing-room/press-briefings/2021/02/03/press-briefing-by-press-secretary-jen-psaki-february-3-2021/.

244 **For years, China had ignored**: Jane Perlez, "Tribunal Rejects Bejing's Claims in South China Sea," *The New York Times,* July 13, 2016, https://www.nytimes.com/2016/07/13/world/asia/south-china-sea-hague-ruling-philippines.html.

244 **the Pentagon accused China**: Terri Moon Cronk, "Chinese Seize U.S. Navy Underwater Drone in South China Sea," Defense Department News, December 16, 2016, https://www.defense.gov/News/News-Stories/Article/Article/1032823/chinese-seize-us-navy-underwater-drone-in-south-china-sea/.

244 **"an unsafe and unprofessional"**: Steven Lee Myers, "American and Chinese Warships Narrowly Avoid High-Seas Collision," *The New York Times,* October 2, 2018, https://www.nytimes.com/2018/10/02/world/asia/china-us-warships-south-china-sea.html.

244 **"Kristin, who is back"**: Eric Berger, "White House Says It Supports Artemis Program to Return to the Moon," *Ars Technica,* February 4, 2021, https://arstechnica.com/science/2021/02/senate-democrats-send-a-strong-signal-of-support-for-artemis-moon-program/.

CHAPTER 16

246 **"Rather than capitulate"**: Jeff Bezos, "No Thank You, Mr. Pecker," *Medium,* February 7, 2019, https://medium.com/@jeffreypbezos/no-thank-you-mr-pecker-146e3922310f.

247 **the single winner**: Christian Davenport, "Elon Musk's SpaceX Wins Contract to Land NASA Astronauts on the Moon," *The Washington Post,* April 16, 2021, https://www.washingtonpost.com/technology/2021/04/16/nasa-lunar-lander-contract-spacex/.

248 **"not have enough funding"**: "Source Selection Statement," NASA, April 16, 2021, https://www.nasa.gov/wp-content/uploads/2019/04/option-a-source-selection-statement-final.pdf.

251 **"executed a flawed acquisition"**: Christian Davenport, "Jeff Bezos Challenges NASA Moon-Contract Award to Elon Musk's SpaceX," *The Washington Post,* April 26, 2021, https://www.washingtonpost.com/technology/2021/04/26/jeff-bezos-challenges-nasa-moon-contract-award-elon-musks-spacex/.

251 **"It's really atypical"**: Kenneth Chang, "Jeff Bezos' Rocket Company Challenges NASA Over SpaceX Moon Lander Deal," *The New York Times,* April 26, 2021, https://www.nytimes.com/2021/04/26/science/spacex-moon-blue-origin.html?smid=url-share.

252 **"Realizing now that it gambled"**: Joey Roulette, "Blue Origin 'Gambled' with Its

Moon Lander Pricing, NASA Says in Legal Documents," *The Verge,* September 29, 2021, https://www.theverge.com/2021/9/29/22689729/blue-origin-moon-lunar -lander-price-nasa-hls-foia.

252 **"They're two years older":** Michael Sheetz, "SpaceX President Knocks Bezos' Blue Origin: 'They Have a Billion Dollars of Free Money Every Year,'" CNBC, October 25, 2019, https://www.cnbc.com/2019/10/25/spacex-shotwell-calls-out-blue -origin-boeing-lockheed-martin-oneweb.html.

253 **"hideous":** David Goldman, "Elon Musk Agrees with Antisemitic X Post That Claims Jews 'Push Hatred' Against White People," CNN, November 17, 2023, https://www.cnn.com/2023/11/15/media/elon-musk-antisemitism-white-people /index.html.

254 **"exceed the maximum":** Jeff Foust, "SpaceX Violated Launch License in Starship SN8 Launch," *SpaceNews,* February 2, 2021, https://spacenews.com/spacex -violated-launch-license-in-starship-sn8-launch/.

255 **"You did an awesome job":** Walter Isaacson, *Elon Musk* (Simon & Schuster, 2023), chap. 57, Kindle.

255 **"Try to hold a Musk-led company":** Mariella Moon, "Amazon Complains Elon Musk's Companies Don't Play by the Rules," *Endgadget,* Sept 9, 2021, https://www .engadget.com/amazon-elon-musk-rules-are-for-other-people-fcc-filing -084518762.html.

256 **Finally, in May:** Christian Davenport, "Elon Musk's SpaceX Lands Starship Space-craft in First Full Successful Test Flight," *The Washington Post,* May 5, 2021, https://www.washingtonpost.com/technology/2021/05/05/elon-musks-spacex -lands-starship-spacecraft-first-time/.

257 **"I think we're close":** Christian Davenport, "2021 Was a Huge Year for Space Ex-ploration. 2022 Could Be Even Bigger," *The Washington Post,* December 27, 2021, https://www.washingtonpost.com/technology/2021/12/27/space-events-in-2022/.

258 **"This is not a volunteer organization":** Isaacson, *Elon Musk,* chap. 59, Kindle.

259 **"I think there needs":** Jeff Foust, "Senate Bill Would Direct NASA to Select a Sec-ond HLS Company," *SpaceNews,* May 12, 2021, https://spacenews.com/senate-bill -would-direct-nasa-to-select-a-second-hls-company/.

260 **"undermines the federal":** Christian Davenport, "The Rivalry Between Jeff Bezos and Elon Musk Was Already Intense. Now It Extends to the Moon," *The Washing-ton Post,* May 21, 2021, https://www.washingtonpost.com/technology/2021/05/21 /elon-musk-jeff-bezos-moon-rivalry/.

260 **"What is Elon Musk":** Davenport, "The Rivalry Between Jeff Bezos and Elon Musk."

261 **If NASA would award:** Aaron Gregg, "Blue Origin Offers to Waive $2 Billion in NASA Payments in Bid for Moon Landing Contract," *The Washington Post,* July 26, 2021, https://www.washingtonpost.com/business/2021/07/26/blue-origin -moon-contract/.

261 **"NASA did not violate":** William Harwood, "Government Accountability Office Denies Protests of NASA–SpaceX Lunar Lander Contract," CBS News, July 30, 2021, https://www.cbsnews.com/news/gao-denies-protests-of-nasa-spacex-lunar -lander-contract/.

262 **"Blue Origin is in the position":** Ross Wilkers, "How the Judge Took Apart Blue Origin's Lunar Lander Lawsuit," *Washington Technology,* November 19, 2021, https://www.washingtontechnology.com/2021/11/how-the-judge-took-apart-blue -origins-lunar-lander-lawsuit/355502/.

262 **"We've lost nearly seven"**: Ashlet Strickland, "NASA Says Moon Landing Goal Pushed to 2025 Due to Blue Origin Litigation, Other Factors," November 9, 2021, CNN, https://www.cnn.com/2021/11/09/world/nasa-artemis-program-update-scn/index.html.

263 **"They're going to be landing"**: Jeff Foust, "Nelson Uses Chinese Mars Landing as a Warning to Congress," *SpaceNews,* May 19, 2021, https://spacenews.com/nelson-uses-chinese-mars-landing-as-a-warning-to-congress/.

263 **"to cooperate and contribute"**: Christian Davenport, "As China's Space Ambitions Grow, NASA Tells Congress It Needs More Money to Compete," *The Washington Post,* June 17, 2021, https://www.washingtonpost.com/technology/2021/06/17/china-space-race-nasa/.

263 **Eventually other countries**: Andrew Jones, "China Wants 50 Countries Involved in Its ILRS Moon Base," *SpaceNews,* July 23, 2024, https://spacenews.com/china-wants-50-countries-involved-in-its-ilrs-moon-base/.

264 **"As long as technological"**: Andrew Jones, "Chinese Crewed Moon Landing Possible by 2030, Says Senior Space Figure," *SpaceNews,* November 15, 2021, https://spacenews.com/chinese-crewed-moon-landing-possible-by-2030-says-senior-space-figure/.

CHAPTER 17

266 **"one-man publicity circus"**: Geraldine Fabrikant, "Of All That He Sells, He Sells Himself Best," *The New York Times,* June 1, 1997, https://www.nytimes.com/1997/06/01/business/of-all-that-he-sells-he-sells-himself-best.html.

266 **"was like being strapped"**: Richard Branson, *Losing My Virginity: How I Survived, Had Fun, and Made a Fortune Along the Way* (Crown Business, 1998), p. 217.

266 **a harrowing accident**: Irene Klotz, "U.S. Investigators Blame Virgin Galactic Crash on Lax Pilot Training," Reuters, July 28, 2015, https://www.reuters.com/article/lifestyle/science/us-investigators-blame-virgin-galactic-crash-on-lax-pilot-training-idUSKCN0Q21YS/.

266 **its CEO called "flawless"**: Christian Davenport, "Richard Branson's Virgin Galactic Reports Reaching Space for the Third Time," *The Washington Post,* May 22, 2021, https://www.washingtonpost.com/technology/2021/05/22/richard-branson-virgin-galactic-space-flight/.

266 **"One good thing"**: Davenport, "Richard Branson's Virgin Galactic Reports Reaching Space for the Third Time."

267 **"I've been itching to go"**: Christian Davenport, "Billionaires' Race to Space: Virgin Galactic's Richard Branson Now Set to Beat Blue Origin's Bezos to Space," *The Washington Post,* July 1, 2021, https://www.washingtonpost.com/technology/2021/07/01/branson-bezos-space-race/.

269 **"Space Tourism Rivalry"**: Joey Roulette, "Space Tourism Rivalry Gets Extremely Petty Ahead of Branson's Spaceflight," *The Verge,* July 9, 2021, https://www.theverge.com/2021/7/9/22570287/space-tourism-virgin-galactic-branson-blue-origin-bezos.

269 **interview on CNBC**: "Jeff Who?" https://x.com/CNBC/status/1410292775900401669.

269 **"We made a pot of tea"**: Nick Rufford, "Richard Branson: 'I woke to Find Elon Musk in my Kitchen at 2 a.m.," *The Sunday Times,* January 7, 2023, https://www.thetimes.com/life-style/article/richard-branson-i-woke-to-find-elon-musk-in-my-kitchen-at-2am-f0s70gq8l.

270 **"If space travel lost"**: Jeff Shesol, "Don't Cede the Space Race to China and the Billionaires," *The New York Times,* February 18, 2022, https://www.nytimes.com/2022/02/18/opinion/space-china-billionaires.html.

271 **"Just magical"**: Rachael Nail, "Richard Branson Successfully Reaches Space Onboard Virgin's SpaceShipTwo," *Florida Today,* July 11, 2021, https://www.floridatoday.com/story/tech/science/space/2021/07/11/richard-branson-successfully-reaches-space-onboard-virgins-spaceshiptwo/7921405002/.

271 **"Man, I gotta pee"**: Ben Evans, "'Man, I Gotta Pee': 55 Years Since Freedom 7 Began America's Adventure in Space," AmericaSpace.com, https://www.americaspace.com/2016/04/30/man-i-gotta-pee-55-years-since-freedom-7-began-americas-adventure-in-space-part-1/.

272 **wasn't the only requirement**: Elizabeth Howell, "Want to Bid on Blue Origin's Space Tourist Seat Auction? Be Sure to Read the Fine Print," Space.com, May 19, 2021, https://www.space.com/blue-origin-space-tourist-auction-terms.

272 **Initially, Virgin Galactic charged**: Micah Maidenberg, "Virgin Galactic Spaceflight Tickets to Start at $450,000 a Seat," *Wall Street Journal,* August 6, 2021, https://www.wsj.com/articles/virgin-galactic-space-flight-tickets-to-start-at-450-000-a-seat-11628262330.

272 **$28 million**: Christian Davenport, "A Seat to Fly with Jeff Bezos to Space Sells for $28 Million at Auction," *The Washington Post,* June 12, 2021, https://www.washingtonpost.com/technology/2021/06/12/jeff-bezos-blue-origin-auction/.

273 **Spaceflight Participant Program**: Alan Ladwig interviewed by Sandra Johnson, "NASA Headquarters Oral History Project," June 7, 2017, https://www.nasa.gov/wp-content/uploads/2024/11/ladwiga-6-7-17.pdf?emrc=52e770.

273 **"Anyone who has lived"**: Michael Collins, "Riding the Beast," *The Washington Post,* January 29, 1986, https://www.washingtonpost.com/archive/politics/1986/01/30/riding-the-beast/0d7963e9-fe01-452d-a2d1-3bbfc03a8721/.

274 **"We didn't take any shortcuts"**: Christian Davenport, "Quiet and Secretive Blue Origin Hopes to Start New Chapter with Jeff Bezos' Space Flight," *The Washington Post,* July 18, 2021, https://www.washingtonpost.com/technology/2021/07/18/jeff-bezos-blue-origin-space-future/.

274 **"gradually stepping up"**: Davenport, "Quiet and Secretive Blue Origin Hopes to Start New Chapter with Jeff Bezos' Space Flight."

275 **a last-minute alteration**: Brad Stone, *Amazon Unbound: Jeff Bezos and the Invention of a Global Empire* (Simon & Schuster, 2021), Epilogue, Kindle.

276 **"Orbiting Earth"**: Kiona N. Smith, "What Yuri Gagarin Saw from Orbit Changed Him Forever," *Forbes,* April 12, 2021, https://www.forbes.com/sites/kionasmith/2021/04/12/what-yuri-gagarin-saw-from-orbit-changed-him-forever/.

276 **"It suddenly struck me"**: "Astronaut Friday: Neil Armstrong," Space Center Houston, August 16, 2019, https://spacecenter.org/astronaut-friday-neil-armstrong/.

276 **"It just seemed so beautiful"**: Christian Davenport and Julie Vitovskaya, "50 Astronauts, in Their Own Words," *The Washington Post,* June 19, 2019, https://www.washingtonpost.com/graphics/2019/national/50-astronauts-life-in-space/.

276 **"OH. MY. GOD."**: "Replay: First Human Flight Press Conference," Blue Origin, July 20, 2021, https://www.youtube.com/watch?v=Kmpb7xJJ10I.

CHAPTER 18

277 **"Guys, if you're willing"**: Micah Maidenberg and Doug Cameron, "Jeff Bezos and Blue Origin Crew Discuss Their Experience in Space," *The Wall Street Journal,*

July 20, 2021, https://www.wsj.com/articles/jeff-bezos-and-blue-origin-crew-discuss-their-experience-in-space-11626813845.

279 **Jeff Bezos's employees:** Christian Davenport, "Inside Blue Origin: Employees Say Toxic, Dysfunctional 'Bro Culture' Led to Mistrust, Low Morale, and Delays at Jeff Bezos' Space Venture," *The Washington Post,* October 11, 2021, https://www.washingtonpost.com/technology/2021/10/11/blue-origin-jeff-bezos-delays-toxic-workplace/.

280 **Alexandra Abrams:** Christian Davenport, "Blue Origin Fired a Senior Executive, Citing Inappropriate Behavior. Current and Former Employees Say It's Part of the Company's Toxic Culture," *The Washington Post,* September 30, 2021, https://www.washingtonpost.com/business/2021/09/30/blue-origin-sexist-work-culture/.

280 **posted by the whistleblower site** *Lioness:* Alexandra Abrams, "Bezos Wants to Create a Better Future in Space. His Company Blue Origin Is Stuck in a Toxic Past," *Lioness,* September 30, 2021, https://www.lioness.co/post/bezos-wants-to-create-a-better-future-in-space-his-company-blue-origin-is-stuck-in-a-toxic-past.

281 **an exodus of talent:** Michael Sheetz, "Turmoil at Bezos' Blue Origin: Talent Exodus Came After CEO's Push For Full Return To the Office," CNBC, October 1, 2021, https://www.cnbc.com/2021/10/01/jeff-bezos-blue-origin-talent-exodus-ceo-pushed-return-to-office.html.

282 **"reasonably good":** Richard Waters, "Elon Musk: Interview with FT's Person of the Year," *Financial Times,* December 15, 2021, https://www.ft.com/content/a7f75d25-d710-4aaa-9f57-49e24d67744d.

283 **"If that went wrong":** Christian Davenport, "The Path Forward: Private Space Travel with Jared Isaacman," *The Washington Post,* October 3, 2022, https://www.washingtonpost.com/washington-post-live/2022/10/03/transcript-path-forward-private-space-travel-with-jared-isaacman/.

284 **"If it was an airplane":** *Space Titans: Musk, Bezos, Branson,* Discovery and Science Channels, November 4, 2021, https://www.imdb.com/title/tt16102738/.

288 **SpaceX, meanwhile:** Elizabeth Howell, "SpaceX in 2021: Elon Musk's Space Company Set Records for Reusability and More," Space.com, December 28, 2021, https://www.space.com/spacex-record-breaking-2021-year.

288 **"The Raptor production":** Michael Sheetz, "Elon Musk Tells SpaceX Employees That Starship Engine Crisis Is Creating a 'Risk of Bankruptcy,'" CNBC, November 30, 2021, https://www.cnbc.com/2021/11/30/elon-musk-to-spacex-starships-raptor-engine-crisis-risks-bankruptcy.html#:~:text=%22The%20Raptor%20production%20crisis%20is,every%20two%20weeks%20next%20year.%22.

289 **"Physics does not care":** Walter Isaacson, *Elon Musk* (Simon & Schuster, 2023), chap. 59, Kindle.

290 **"We think, and so":** Christian Davenport, "Jeff Bezos' Blue Origin Will Get a Second Chance to Compete in NASA's Moon Program," *The Washington Post,* March 23, 2022, https://www.washingtonpost.com/technology/2022/03/23/nasa-moon-lander-blue-origin-spacex/.

290 **"If Mr. Bezos wants":** Bryan Bender, "Sanders Looks to Shoot Down Bezos' Moon Plans," *Politico,* March 31, 2022, https://www.politico.com/news/2022/03/31/bernie-sanders-jeff-bezos-moon-plans-00022035.

290 **"has become little more than an ATM":** Bernie Sanders, "Jeff Bezos Is Worth $160bn—Yet Congress Might Bail Out His Space Company," *The Guardian,* April 22, 2022, https://www.theguardian.com/commentisfree/2022/apr/22/jeff-bezos-space-elon-musk-billionaires-bernie-sanders.

290 **Instead, it authorized:** Jeff Foust, "Senate Passes NASA Authorization Act," *Space-News,* June 9, 2021, https://spacenews.com/senate-passes-nasa-authorization-act-2/.

291 **"DARE MIGHTY THINGS":** "Mars Decoder Ring," NASA Jet Propulsion Laboratory, February 23, 2021, https://www.jpl.nasa.gov/images/pia24431-mars-decoder-ring/.

292 **This was the flight:** "Artemis I Press Kit," NASA, https://www.nasa.gov/specials/artemis-i-press-kit/.

293 **In 2023, India:** Andrew Jones, "Chandrayaan-3: India Becomes Fourth Country to Land on the Moon," *SpaceNews,* August 23, 2023, https://spacenews.com/chandrayaan-3-india-becomes-fourth-country-to-land-on-the-moon/.

294 **Japan became the fifth:** Kantaro Komiya, "Japan Praises 'Pinpoint' Moon Landing by Its SLIM Probe," Reuters, January 25, 2024, https://www.reuters.com/world/asia-pacific/japan-says-slim-spacecrafts-pinpoint-moon-landing-is-success-2024-01-25/.

294 **"a race about the race":** Christian Davenport, "Will China Beat the United States Back to the Moon? It's Possible," *The Washington Post,* November 13, 2023, https://www.washingtonpost.com/technology/2023/11/13/china-nasa-moon-landing-first/.

294 **"It's historic because":** Christian Davenport, "NASA's Artemis I Orion Spacecraft Returns to Earth," *The Washington Post,* December 11, 2022, https://www.washingtonpost.com/technology/2022/12/11/orion-nasa-spacecraft-return/.

295 **traveling at Mach 32:** "Artemis I: Mission Facts," NASA, https://www.nasa.gov/mission/artemis-i/.

295 **"Initial indications":** Jeff Foust, "Orion Splashes Down to End Artemis I," *Space-News,* December 11, 2022, https://spacenews.com/orion-splashes-down-to-end-artemis-1/.

CHAPTER 19

297 **"To make long-term presence":** "Blue Alchemist Powers Our Lunar Future," Blue Origin, February 10, 2023, https://www.blueorigin.com/news/blue-alchemist-powers-our-lunar-future.

299 **"develop technologies":** Abbey A. Donaldson, "NASA Partners with American Companies on Key Moon Exploration Tech," NASA, July 25, 2023, https://www.nasa.gov/news-release/nasa-partners-with-american-companies-on-key-moon-exploration-tech/.

299 **Bezos was funding:** Christian Davenport, "Bezos' Blue Origin Wins NASA Contract to Land Astronauts on the Moon," *The Washington Post,* May 19, 2023, https://www.washingtonpost.com/technology/2023/05/19/moon-lander-contract-bezos/.

302 **"I think it's going to be":** "The Interview: From Amazon to Space—Jeff Bezos Talks Innovation, Progress and What's Next," *New York Times* DealBook Summit, December 4, 2024, https://www.youtube.com/watch?v=s71nJQqzYRQ.

302 **"the most important single thing":** Bill Murphy Jr., "Jeff Bezos: 'The Most Important Single Thing Is to Focus Obsessively on the Customer,'" *Inc.,* February 4, 2023, https://www.inc.com/bill-murphy-jr/bezos-most-important-single-thing-focus-obsessively-on-customer.html.

303 **In 2023, it launched:** Jeff Foust, "SpaceX to Just Miss Goal of 100 Falcon Launches in 2023," *SpaceNews,* December 19, 2023, https://spacenews.com/spacex-to-just-miss-goal-of-100-falcon-launches-in-2023/.

303 **more than 2.5 million pounds:** Data provided by BryceTech.com.

303 **September 2024:** Christian Davenport, "SpaceX Polaris Astronauts Complete First Spacewalk by Private Citizens," *The Washington Post,* September 12, 2024, https://www.washingtonpost.com/technology/2024/09/12/spacex-polaris-dawn-spacewalk-astronauts/.

304 **"could turn out to be":** Mike Wall, "Why Did SpaceX's Starship Debut Launch Cause So Much Damage to the Pad?," Space.com, April 24, 2023, https://www.space.com/spacex-starship-damage-starbase-launch-pad.

305 **"the pressure that was built":** B. Dotson, P. Metzger, J. Hafner, A. Shackelford, K. Birkenfeld, D. Britt, A. Ford, R. Truscott, S. Truscott, J. Zavaleta, J. Zemke, K. Purvis, M. Scudder, C. Johnson, J. Galloway, and J. DeShetler, "A New Launch Pad Failure Mode: Analysis of Fine Particles from the Launch of the First Starship Orbital Test Flight," University of Central Florida et al., https://arxiv.org/pdf/2403.10788.

305–6 **"Navigational hazards from rocket launching activity":** Local Notice to Mariners, U.S. Department of Homeland Security, U.S. Coast Guard, October 2, 2024, https://www.navcen.uscg.gov/sites/default/files/pdf/lnms/LNM0840g2024.pdf.

306 **"thousands of distinct":** Will Robinson-Smith, "Live Updates: SpaceX to Attempt First Booster Catch During the Starship 5 Mission," October 13, 2024, https://spaceflightnow.com/2024/10/13/live-coverage-spacex-to-launch-5th-flight-test-of-starship-from-starbase-in-southern-texas/.

306 **"We're only going to attempt":** Starship's Fifth Flight Test, SpaceX, October 13, 2024, https://www.spacex.com/launches/mission/?missionId=starship-flight-5.

307 **"Vehicle is pitching downrange":** Starship: Fifth Flight Test, SpaceX, November 6, 2024, https://www.youtube.com/watch?v=hI9HQfCAw64.

310 **SpaceX won them all:** Stephen Clark, "SpaceX Prevails over ULA, Wins Military Launch Contract Worth $733 Million," *Ars Technica,* October 18, 2024, https://arstechnica.com/space/2024/10/spacex-sweeps-latest-round-of-military-launch-contracts/.

310 **"prepared rigorously":** Christian Davenport, "Blue Origin's New Glenn Rocket Prepared for Pivotal Test Launch," *The Washington Post,* January 10, 2025, https://www.washingtonpost.com/business/2025/01/10/blue-origin-new-glenn-bezos-musk/.

312 **Country Cookin' Diner:** Patricia Tolley, " 'Very Nice and Gracious': Jeff Bezos Visits Space Coast Restaurant After Blue Origin Launch," ClickOrlando.com, https://www.clickorlando.com/news/local/2025/01/19/very-nice-and-gracious-jeff-bezos-visits-space-coast-restaurant-after-blue-origin-launch/.

313 **When Bezos went:** Maggie Haberman, "Billionaire Rivals Bezos and Musk Are Said to Have Dined with Trump at Mar-A-Lago," *The New York Times,* December 19, 2024, https://www.nytimes.com/2024/12/19/us/politics/trump-elon-musk-bezos-mar-a-lago.html.

313 **"pursue our manifest destiny":** Jonathan Lambert, "It's America's 'Manifest Destiny' to Plant a Flag on Mars, Trump Says," NPR, January 20, 2025, https://www.npr.org/2025/01/20/g-s1-43861/mars-nasa-manifest-destiny-trump-inauguration.

313 **$38 billion of taxpayer money:** Desmond Butler, Trisha Thadani, Emmanuel Martinez, Aaron Gregg, Luis Melgar, Jonathan O'Connell, and Dan Keating, "Elon Musk's Business Empire Is Built on $38 Billion in Government Funding," *The Washington Post,* February 26, 2025, https://www.washingtonpost.com/technology/interactive/2025/elon-musk-business-government-contracts-funding/.

314 **"a real job":** Hannah Panreck, "Trump Applauds Jeff Bezos' Changes at Washing-

ton Post in Rare Media Praise," Fox News, March 17, 2025, https://www.foxnews
.com/media/trump-applauds-jeff-bezos-changes-washington-post-rare-media
-praise.

314 **$40 million for the rights:** Rebecca Balhaus, Dana Mattioli, and Annie Linskey,
"How the Trumps Turned an Election Victory into a Cash Bonanza," *Wall Street
Journal*, February 13, 2025, https://www.wsj.com/politics/elections/trump-family
-election-cash-bonanza-2f5f8714.

314 **"calmer than he was":** Theodore Schleifer and Katie Robertson, "Jeff Bezos, a Past
Trump Foe, Is Optimistic About a Second Term," *The New York Times,* December 4, 2024, https://www.nytimes.com/2024/12/04/business/dealbook/jeff-bezos
-trump-amazon-washington-post.html.

314 **"There is no doubt":** Margaret Sullivan, "Jeff Bezos Is Muzzling *The Washington
Post*'s Opinion Section. That's a Death Knell," *The Guardian*, February 25, 2025,
https://www.theguardian.com/commentisfree/2025/feb/26/jeff-bezos-washington
-post-opinion.

314 **In a blistering essay:** Martin Baron, "Where Jeff Bezos Went Wrong with *The
Washington Post*," *The Atlantic,* March 3, 2025, https://www.theatlantic.com/ideas
/archive/2025/03/bezos-appease-trump-administration/681899/.

315–16 **shrapnel raining down:** Jackie Wattles, "The Most Powerful Rocket Ever Built
Exploded Over a Populated Island. Residents Are Still Dealing with the Fallout,"
CNN, February 1, 2025, https://www.cnn.com/2025/01/30/science/spacex-starship
-explosion-debris-turks-caicos/index.html.

EPILOGUE

319 **In late 2024:** Jeff Foust, "NASA Further Delays Next Artemis Missions," *Space-
News,* December 5, 2024, https://spacenews.com/nasa-further-delays-next-artemis
-missions/.

319–20 **"Should the same issue":** Christian Davenport, "NASA Moon Capsule Suffered
Extensive Damage During 2022 Test Flight," *The Washington Post,* May 2, 2024,
https://www.washingtonpost.com/technology/2024/05/02/artemis-orion-capsule
-damage/.

320 **"will be well ahead":** Joey Roulette, "NASA Announces Further Delays in Artemis
Moon Missions," Reuters, December 5, 2024, https://www.reuters.com/technology
/space/nasa-announces-further-delays-artemis-moon-missions-2024-12-05/.

320 **Odysseus:** Christian Davenport, "Intuitive Machine's Moon Mission to End Early
After Spacecraft Tumbled," *The Washington Post,* February 27, 2024, https://www
.washingtonpost.com/technology/2024/02/27/lunar-lander-odysseus-mission
-ends/.

321 **The spacecraft carried:** Katie Hunt, "Findings from the First Lunar Far Side Samples Raise New Questions About Moon's History," CNN, November 15, 2024,
https://www.cnn.com/2024/11/15/science/far-side-moon-samples-chang-e-6
/index.html.

321 **China's scientists worked:** Nectar Gan and Simone McCarthy, "China's Chang'e-6
Probe Lifts Off with Samples from Moon's Far Side in Historic First," CNN, June 4,
2024, https://www.cnn.com/2024/06/04/china/china-change6-moon-lift-off-intl
-hnk/index.html.

321 **a small Chinese flag:** Leonard David, "China's Chang'e 6 Mission Carried a Stone
Flag to the Moon's Far Side," Space.com, June 11, 2024, https://www.space.com
/china-change-6-moon-flag-basalt.

322   **"We're going to win"**: Mike Waltz, X, January 18, 2025, https://x.com /michaelgwaltz/status/1880752606437589104.

322   **"never settle for second place"**: Jared Isaacman, X.com, December 2, 2024, https://x.com/rookisaacman/status/1864346915183157636?lang=en.

323   **"I am back"**: "Exclusive: Taikonaut Wang Yaping to Her Daughter, 'Mommy's Back,'" CGTN, April 16, 2022, https://www.facebook.com/ChinaGlobalTV Network/videos/301379752144602/.

ACKNOWLEDGMENTS

This book relies on interviews with more than a hundred sources, many of whom spoke with me multiple times over many hours and shared emails, text messages, photographs, documents, transcripts of meetings, and even personal journals. In many cases, I conducted interviews on a deep-background basis, meaning I could use the information but not attribute it to anyone.

I'm grateful to all my sources, but would like to acknowledge a few in particular. Former NASA administrator Jim Bridenstine was candid and accessible and put up with my incessant questions. I owe a debt as well to Gabe Sherman, the former NASA chief of staff; and Matthew Rydin, the former press secretary, both of whom were extremely helpful. Scott Pace, the executive secretary of the National Space Council during President Trump's first term, shared his insights as well. I'm also grateful to former senator Bill Nelson, who led the space agency under President Biden, as well as his deputy, Pam Melroy. Doug Hurley spent many hours discussing his illustrious career and his daring spaceflight mission with Bob Behnken.

As he did for my last book, *The Space Barons*, Jeff Bezos sat for an interview with me despite the immense demands on his time. He also allowed me to interview several of Blue Origin's top executives, including CEO Dave Limp. Neal Karlinsky, Bill Kircos, and Rosemarie Esposito were gracious and accommodating in arranging interviews and helping me fact-check. Bezos owns my employer, *The Washington Post,* and we cover him without fear or favor, a standard that applies to this book as well.

While Elon Musk agreed to an interview for this book, by the printing deadline, I got a brief period and only was able to ask a handful of questions. Fortunately, I had interviewed him at several key moments covered in the narrative for the *Post* and was able to rely on those conversations to fill in gaps. SpaceX's James Gleeson was accommodating and a great resource for helping ensure accuracy. I also thank him for allowing the use of several of SpaceX's photos.

I've been fortunate to work at the *Post* since 2000. The newsroom has been an instrument of continuous learning for me, and I was lucky to have worked under Mark Seibel, a thoughtful, patient professor of an editor. He was kind enough to read multiple drafts of each chapter, helping ensure the manuscript was clear and focused. I'm grateful to Lori Montgomery for allowing me the time to work on the project despite unrelenting waves of news. Thanks also to Sandhya Somashekhar, Christopher Rowland, Sergio Non, Robbie DiMesio, Helen Fessenden, and Aaron Gregg. Finally, I was lucky to travel to many space assignments with the "WaPo Space Cadre," video journalists Whitney Shefte and Whitney Leaming and photojournalist Jonathan Newton, who made road trips fun.

During the course of writing the first parts of this book, I was fortunate enough to be offered a fellowship once again at the Woodrow Wilson International Center for Scholars. Thank you to Mark Green and Robert Litwak for welcoming me and providing such amazing resources. Librarians Janet Spikes and Michelle Kamalich are researchers nonpareil who helped me unearth some obscure books and documents. While there, I also reconnected with James Reston Jr., who before his unfortunate passing kindly included me in a weekly discussion among historians, academics, and journalists that was always entertaining and broadened my worldview.

Pat Bedingfield always checked on my progress to ensure I was hitting along on a straight and true path. Thomas Clark has also been supportive, and I'm fortunate to get to see him and the rest of his band of early-morning tennis players on Fridays.

While we compete against one another on a daily basis, the space press corps is a wonderfully collegial bunch, and I've been honored to cover a series of momentous events (often at odd hours) alongside such smart colleagues, including Eric Berger, Alan Boyle, Morgan Brennan, Brendan Byrne, Kenneth Chang, Stephen Clark, Miles Doran, Nadia Drake, Tim Fernholz, Kristin Fisher, Jeff Foust, Loren Grush, William Hardwood, David Kerley, Irene Klotz, Marina Koren, Micah Maidenberg, Tariq Malik, Robert Pearlman, Will Robinson-Smith, Joey Roulette, Michael Sheetz, Gina Sunseri, and Jackie Wattles. While reporting this book, I ran into sources who had also cooperated with Berger for his excellent work about SpaceX, *Reentry: SpaceX, Elon Musk, and the Reusable Rockets that Launched a Second Space Age,* and shared similar information with me. I also relied on my own previous reporting on space for the *Post.* For the scenes depicting the Moore tornado, I consulted Holly Bailey's thorough and compelling account of the storm, *The Mercy of the Sky: The Story of a Tornado.*

This is my third book with Rafe Sagalyn, my agent, who is there at every step, from conception to execution to publication, while smoothing out so many bumps along the way. At Penguin Random House I'm so grateful for Leah Trouwborst, my excellent editor, who provided the first edits, broadening the scope of the book and making it immensely more ambitious. Nick Summers parachuted in with an exacting eye and unfailing red pen that vastly improved the manuscript. While juggling multiple projects, Paul Whitlatch still found time to guide the book through its final touches with rigor and patience. I'm also grateful to associate editor Amy Li for keeping me on track and Lawrence Krauser for such a careful copyedit. Penny Simon and Tara Gilbride were essential in bringing this book to a wider audience

Finally, and most of all, I am grateful for my family. My children, Annie, Harrison, and Piper, inspire me with their kindness, enthusiasm, and perseverance. I am so proud of them, and it's the honor of my life to be their father. My wife, Heather, is an unwavering force whom I've known since we were kids at summer camp. It is my great privilege—and luck—to have found a love that so endures. Love you and love you.

INDEX

ABOUT THE AUTHOR

Christian Davenport has been a staff writer at *The Washington Post* for more than twenty-five years, most recently covering NASA and the space industry. He is a recipient of an Emmy Award for his work on the Discovery and Science Channels covering SpaceX's first human spaceflight mission, and a Peabody Award for his work on veterans with traumatic brain injuries. He has been on reporting teams that were finalists for the Pulitzer Prize three times. As a frequent radio and television commentator, he has appeared on MSNBC, CNN, CBS, *PBS NewsHour*, and several NPR shows.